The Geometry Toolbox
for Graphics and Modeling

The Geometry Toolbox
for Graphics and Modeling

Gerald E. Farin
Dianne Hansford

A K Peters
Natick, Massachusetts

Editorial, Sales, and Customer Service Office

A K Peters, Ltd.
63 South Avenue
Natick, MA 01760

http://www.akpeters.com

Library of Congress Cataloging-in-Publication Data

Farin, Gerald E.
The geometry toolbox for graphics and modeling / Gerald Farin, Dianne Hansford.
p. cm.
Includes bibliographical references and index.
ISBN 1-56881-074-1
1. Geometry – Study and teaching. 2. Computer graphics.
I. Hansford, Dianne. II. Title
QA462.F26 1997
516.3'5 – dc21 97-41259
CIP

Printed in the United States of America
02 01 00 99 98 10 9 8 7 6 5 4 3 2 1

To our advisors
R.E. Barnhill and W. Boehm

Contents

Chapter 16

Curves 243

Appendix A

PostScript Tutorial 257

Appendix B

Selected Problem Solutions 269

Bibliography 281

Index 283

Preface

We live in a visual age. Television is a part of everyday life, and computers play an increasingly important role. The World Wide Web is becoming ever more useful for information gathering, and much of the information is represented as images.

Several branches of science and engineering spur the increasing importance of visual information: computer graphics; communications technology; and that oldest branch of mathematics, geometry, combined with its more computer-oriented companion, linear algebra.

Geometry and linear algebra are essential for many fields, including engineering, scientific computation, computer graphics. These geometric concepts have been taught for decades using dry linear algebra, typically presented from a theoretical mathematical viewpoint; rarely is the subject's geometric nature explored in great detail.

The Geometry Toolbox approaches linear algebra from a geometric viewpoint and geometry from an algorithmic viewpoint. Every matrix or vector equation has an underlying geometric meaning, and we focus on this geometric meaning rather than on plug-and-chug exercises in matrix arithmetic. *The Geometry Toolbox* is intended for anyone who needs to learn the basic concepts of geometry and linear algebra, be it as a first-time student or as a practitioner whose skills are a bit rusty. Its theoretical level is kept to a minimum — we often substitute examples and images for exact proofs.

Preview of Contents

With the aid of many illustrations, Chapters 1 through 9 focus on two-dimensional concepts, and then these concepts (when they apply) are reinforced by covering them in three dimensions in Chapters 10 through 13. Since the three-dimensional world lends itself to concepts that do not exist in two dimensions, these are explored as well. Higher dimensions are visited only to introduce a systematic method for solving a general linear system in Chapter 14. Triangles (Chapter 8) and polygons (Chapter 15) are discussed as fundamental geometric entities and for their importance in creating computer images. The basics of generating curves are presented in Chapter 16; it is a nice example of how geometric concepts may be applied.

Exercises are listed at the end of each chapter, and selected solutions are given in Appendix B. Appendix A offers a tutorial on the Postscript language. This brief tutorial provides enough information to enable the reader to create his/her own Postscript images, similar to the ones in the book.

The illustrations in the book are figures and sketches. The figures are typically more complex and illustrate an application of a concept, while the sketches are a tool to illustrate the details of a concept. The sketches are hand-drawn, illustrating the power of a quickly drawn sketch to aid in understanding and retaining concepts.

Classroom Use

The Geometry Toolbox is intended for use at the freshman/sophomore undergraduate level. It serves as an introduction to linear algebra for engineers or computer scientists as well as a general introduction to geometry. It is also an ideal preparation for Computer Graphics and Geometric Modeling.

As a one-semester introduction to linear algebra, the following Chapters should be covered: 1 through 7, 9, 12, and 14.

As a one-semester introduction to geometry, Chapters 1 through 6, 8, and 10 through 13 are useful.

As an introduction to Computer Graphics and Geometric Modeling, we suggest using Chapters 1 through 6, 8, 15, and 16.

WWW Site

The Geometry Toolbox has a web site:[1]

http://eros.cagd.eas.asu.edu/ farin/gbook/gbook.html.

This web site contains general information and updates; it also contains most of the PostScript files and any data files referred to in the text. See the Web page for details on downloading these files.

PostScript is a two-dimensional language; more adventurous readers may want to use our Geomview examples. To do this, visit the Geometry Center at the University of Michigan:

http://www.geom.umn.edu/software/geomview/.

[1] Links to this site may also be found on the publisher's and the authors' home pages.

(Unfortunately this package is currently only available for Unix.) The Geomview files are also available on *The Geometry Toolbox* Web site.

Acknowledgements

We would like to thank W. Boehm, H. Eaton, R. Goldman, J. Hanson, D. Holliday, R. Pidiparti, A. Razdan, B. Steinberg, M. Throl, H. Wolters, and A. Worsey. We also gratefully acknowledge the pleasant environment provided by everyone at A K Peters, which made writing this book a fun experience for us.

Gerald Farin
Dianne Hansford

April 1998
Paradise Valley, AZ

1 Descartes' Discovery

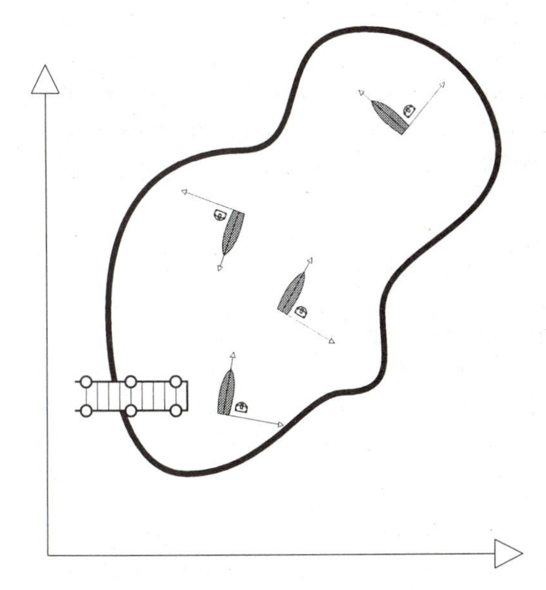

Figure 1.1.
Local and global coordinate systems: a moving boat's local coordinates change relative to a lake's global coordinates.

There is a collection of old German tales about an alleged town called Schilda, sometime in the 17th century, whose inhabitants were not known for their intelligence. Here is one story [16]:

> An army was approaching Schilda and would likely conquer it. The town council, in charge of the town treasure, had to hide it from the invaders. What better way than to sink it in the nearby town lake? So the town council members board the town boat, head for the middle of the lake, and sink the treasure. The town treasurer gets out his pocket knife and cuts a deep notch in the boat's rim, right where the treasure went down. Why would he do that, the other council members wonder? "So that we will remember

where we sunk the treasure — otherwise we'll never find it later!" replies the treasurer. Everyone is duly impressed at such planning genius!

Eventually, the war is over and the town council embarks on the town boat again, this time to reclaim the treasure from the lake. Once out on the lake, the treasurer's plan suddenly does not seem so smart anymore. No matter where they went, the notch in the boat's rim told them they had found the treasure!

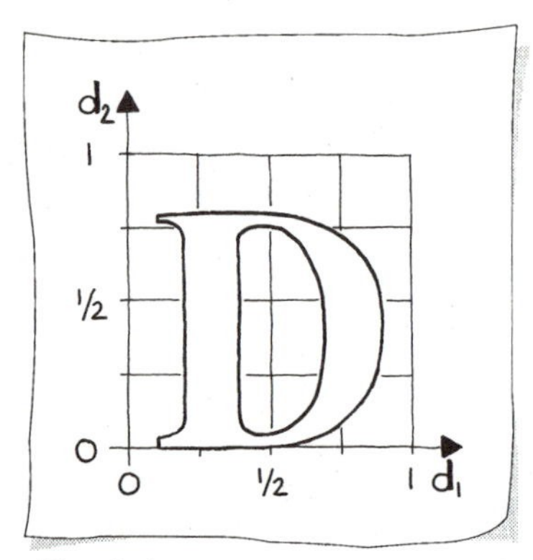

Sketch 1
A local coordinate system.

The French philosopher René Descartes (1596-1650)[1] would have known better: he invented the theory of *coordinate systems*.The treasurer marked the sinking of the treasure accurately, by marking it on the boat, a *local* coordinate system. But by neglecting the boat's position relative to the lake, the *global* coordinate system, he lost it all! (See Figure 1.1.)

The remainder of this chapter is about the interplay of local and global coordinate systems.

1.1 Local and Global Coordinates: 2D

This book is written using the LaTeX typesetting system (see [9] or [17]) which converts the page to be output to a page description language called PostScript (see [14]). It tells a laser printer where to position all the characters and symbols that go on a particular page. For the first page of this chapter, there is a PostScript command that positions the letter **D** in the chapter heading.

In order to do this, one needs a two-dimensional, or 2D, *coordinate system.* Its origin is simply the lower left corner of the page, and the $x-$ and $y-$axes are formed by the horizontal and vertical paper edges meeting there. Once we are given this coordinate system, we can position objects in it, such as our letter **D**.

The **D**, on the other hand, was designed by font designers who obviously did not know about its position on this page or of its actual size. They used their own coordinate system, and in it, the letter **D** is described by a set of points, each having coordinates relative to **D**'s coordinate system, as shown in Sketch 1.

We call this system a *local coordinate system*, as opposed to the *global coordinate system* which is used for the whole page. Positioning letters on a page thus requires mastering the interplay of the global and local systems.

Sketch 2
Global and local systems.

[1] See also http://laf.cioe.com/~jheinze/descartes.html.

Following Sketch 2, let's make things more formal: Let (x_1, x_2) be coordinates[2] in a global coordinate system, called the $[\mathbf{e}_1, \mathbf{e}_2]$-system.[3] Let (u_1, u_2) be coordinates in a local system, called the $[\mathbf{d}_1, \mathbf{d}_2]$-system. Let an object in the local system be enclosed by a box with lower left corner $(0, 0)$ and upper right corner $(1, 1)$. This means that the object "lives" in the *unit square* of the local system, i.e., a square of edge length one, and with its lower left corner at the origin.[4]

We wish to position our object into the global system so that it fits into a box with lower left corner $(\min_1, \min_2)$ and upper right corner $(\max_1, \max_2)$, called the *target box* (drawn with heavy lines in Sketch 2.This is accomplished by assigning to any coordinate (u_1, u_2) in the local system corresponding target coordinates (x_1, x_2) in the global system. The corresponding formula is quite simple:

$$x_1 = (1 - u_1)\min_1 + u_1\max_1, \tag{1.1}$$

$$x_2 = (1 - u_2)\min_2 + u_2\max_2. \tag{1.2}$$

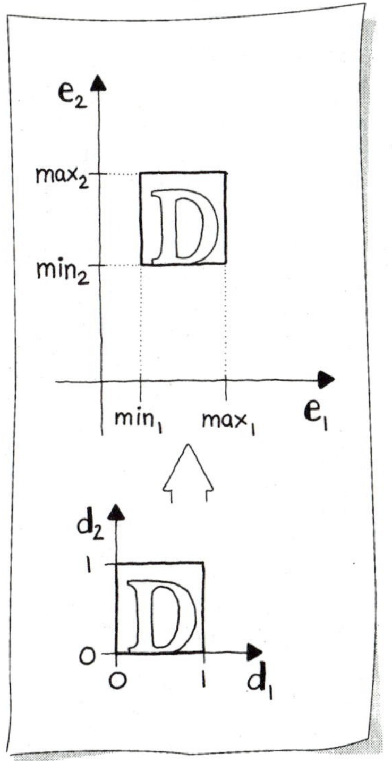

Sketch 3
Local and global D.

We say that the coordinates (u_1, u_2) are *mapped* to thecoordinates (x_1, x_2). Sketch 3 illustrates how the letter **D** is mapped.

Let's check that this actually works: the coordinates $(u_1, u_2) = (0, 0)$ in the local system must go to the coordinates $(x_1, x_2) = (\min_1, \min_2)$ in the global system. We obtain

$$x_1 = (1 - 0) \cdot \min_1 + 0 \cdot \max_1 = \min_1,$$
$$x_2 = (1 - 0) \cdot \min_2 + 0 \cdot \max_2 = \min_2.$$

Similarly, the coordinates $(u_1, u_2) = (1, 0)$ in the local system must go to the coordinates $(x_1, x_2) = (\max_1, \min_2)$ in the global system. We obtain

$$x_1 = (1 - 1) \cdot \min_1 + 1 \cdot \max_1 = \max_1,$$
$$x_2 = (1 - 0) \cdot \min_2 + 0 \cdot \max_2 = \min_2.$$

Example 1.1 Let the target box (see Sketch 4) be given by

$$(\min_1, \min_2) = (1, 3) \quad \text{and} \quad (\max_1, \max_2) = (3, 5).$$

[2]You may be used to calling coordinates (x, y) — the (x_1, x_2) notation will streamline the material in this book; it also makes writing programs easier.

[3]The boldface notation will be explained in the next chapter.

[4]Restricting ourselves to the unit square for the local system makes this first chapter easy — we will later relax this restriction.

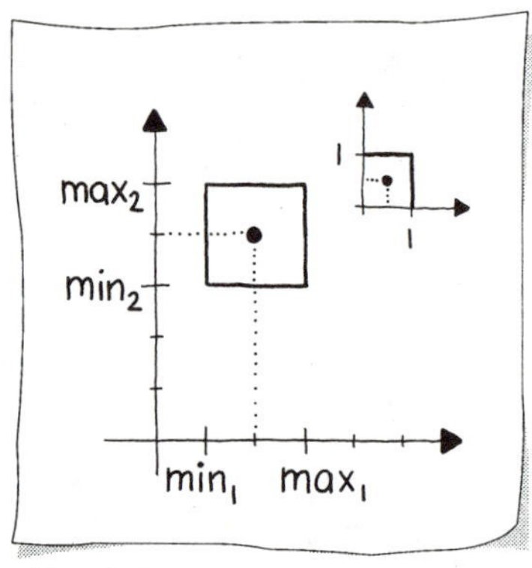

Sketch 4
Map local unit square to a target box.

The coordinates $(1/2, 1/2)$ can be thought of as the "midpoint" of the local unit square. Let's look at the result of the mapping:

$$x_1 = \left(1 - \frac{1}{2}\right) \cdot 1 + \frac{1}{2} \cdot 3 = 2,$$

$$x_2 = \left(1 - \frac{1}{2}\right) \cdot 3 + \frac{1}{2} \cdot 5 = 4.$$

This is the "midpoint" of the target box. You see here how the geometry in the unit square is replicated in the target box.

A different way of writing (1.1) and (1.2) is as follows:
Define $\Delta_1 = \max_1 - \min_1$ and $\Delta_2 = \max_2 - \min_2$. Now we have

$$x_1 = \min_1 + u_1 \Delta_1, \tag{1.3}$$

$$x_2 = \min_2 + u_2 \Delta_2. \tag{1.4}$$

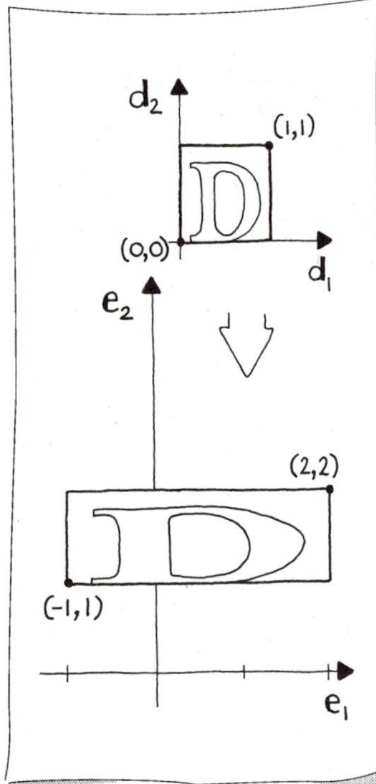

Sketch 5
A distortion.

A note of caution: if the target box is not a square, then the object from the local system will be distorted. We see this in the following example, illustrated by Sketch 5. The target box is given by

$$(\min_1, \min_2) = (-1, 1) \quad \text{and} \quad (\max_1, \max_2) = (2, 2).$$

You can see how the local object is stretched in the $\mathbf{e}_1$−direction by being put into the global system. Check for yourself that the corners of the unit square (local) still get mapped to the corners of the target box (global)!

In general, if $\Delta_1 > 1$, then the object will be stretched in the $\mathbf{e}_1$−direction, and it will be shrunk if $0 < \Delta_1 < 1$. The case of $\max_1$ smaller than $\min_1$ is not often encountered: it would result in a reversal of the object in the $\mathbf{e}_1$−direction. The same applies, of course, to the $\mathbf{e}_2$−direction. An example of several boxes containing the letter **D** is shown in Figure 1.2. Just for fun, we have included one target box with $\max_1$ smaller than $\min_1$!

This method, by the way, acts strictly on a "don't need to know" basis: we do not need to know the relationship between the local and global systems. In many cases (as in the typesetting example), there actually isn't any. Of course, one must be aware where the actual object is located in the local unit square. If it is not nicely centered, as in Figure 1.2, we might have the situation shown in Figure 1.3.

A good example for our "unit square to target box" mapping is found in computer graphics. When you open a window on your computer screen, you might want to view a particular image in it. The image is stored in a

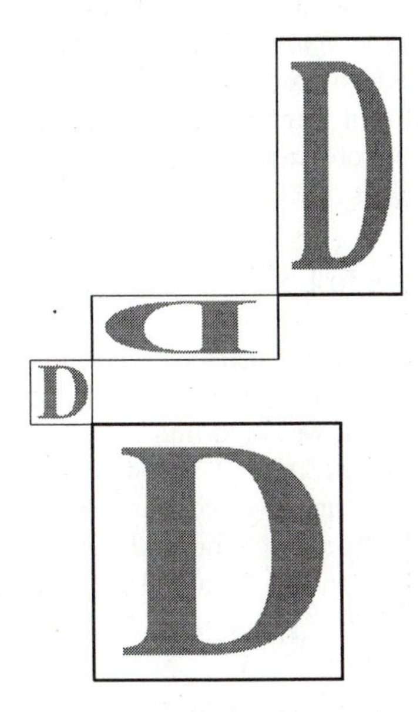

Figure 1.2.
Target boxes: the letter **D** is mapped several times.

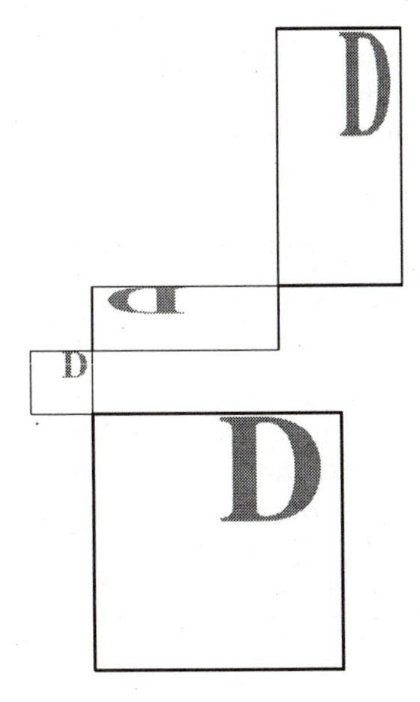

Figure 1.3.
Target boxes: the letter **D** is not centered in the unit square.

local coordinate system; suppose it is stored with *extents* $(0, 0)$ and $(1, 1)$, then it utilizes *normalized coordinates*. The target box is now given by the extents of your window, which are given in terms of *screen coordinates* and the image is mapped to it using (1.1) and (1.2). Screen coordinates are typically given in terms of *pixels*;[5] a typical computer screen would have about 700×1000 pixels.

1.2 Going from Global to Local

When discussing global and local systems in 2D, we used a target box to position (and possibly distort) the unit square in a local $[\mathbf{d}_1, \mathbf{d}_2]$-system. For given coordinates (u_1, u_2), we could find coordinates (x_1, x_2) in the global system using (1.1) and (1.2), or (1.3) and (1.4).

How about the inverse problem: given coordinates (x_1, x_2) in the global system, what are its local (u_1, u_2) coordinates? The answer is relatively easy: compute u_1 from (1.3), and u_2 from (1.4), resulting in

$$u_1 = \frac{x_1 - \min_1}{\Delta_1}, \tag{1.5}$$

$$u_2 = \frac{x_2 - \min_2}{\Delta_2}. \tag{1.6}$$

Applications for this process arise any time you use a mouse to communicate with a computer. Suppose several icons are displayed in a window. When you click on one of them, how does your computer actually know which one? The answer: it uses equations (1.5) and (1.6) to determine its position.

Example 1.2 An example is best suited to illustrate this: Let a window on a computer screen have screen coordinates

$$(\min_1, \min_2) = (120, 300) \quad \text{and} \quad (\max_1, \max_2) = (600, 820).$$

The window is filled with 21 icons, arranged in a 7×3 pattern (see Figure 1.4). A mouse click returns screen coordinates (200, 709). Which icon was clicked? The computations that take place are as follows:

$$u_1 = \frac{200 - 120}{480} \approx 0.17,$$

$$u_2 = \frac{709 - 300}{520} \approx 0.79,$$

according to (1.5) and (1.6).

[5]The term is short for "picture element."

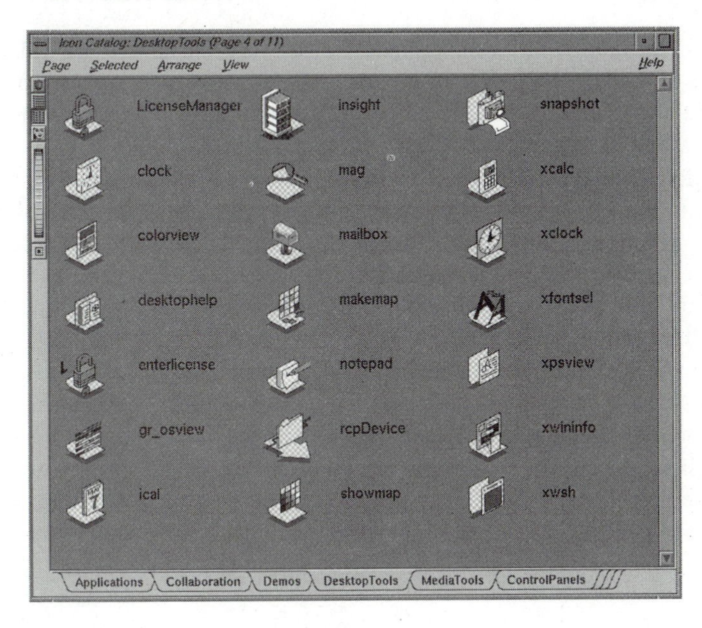

Figure 1.4.
Selecting an icon: global to local coordinates.

The u_1-partition of normalized coordinates is

$$0, 0.33, 0.67, 1.$$

The value 0.17 for u_1 is between 0.0 and 0.33, so an icon in the first column was picked. The u_2-partition of normalized coordinates is

$$0, 0.14, 0.29, 0.43, 0.57, 0.71, 0.86, 1.$$

The value 0.79 for u_2 is between 0.71 and 0.86, so the clock icon in the second row of the first column was picked.

1.3 Local and Global Coordinates: 3D

These days, almost all engineering objects are designed using a CAD (Computer Aided Design) system. Every object is defined in a coordinate system, and usually many individual objects need to be integrated into one large system. Take the example of designing a large commercial airplane.

It is defined in a three-dimensional (or 3D) coordinate system with its origin at the frontmost part of the plane, the $\mathbf{e}_1$−axis pointing toward the rear, the $\mathbf{e}_2$−axis pointing to the right (that is, if you're sitting in the plane), and the $\mathbf{e}_3$−axis is pointing upward. See Sketch 6.

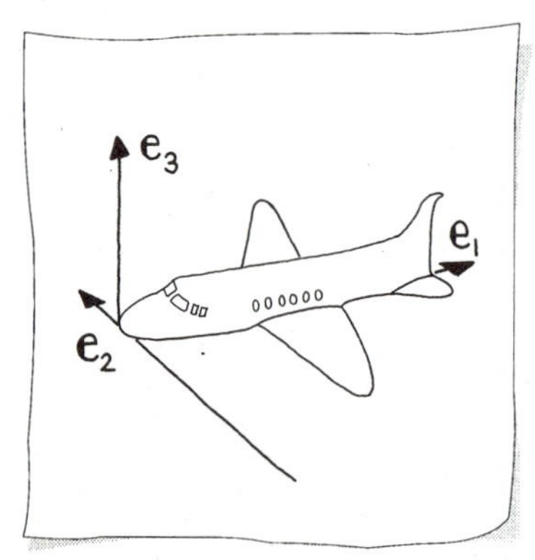

Sketch 6
Airplane coordinates.

Before the plane is built, it undergoes intense computer simulation in order to find its optimal shape. As an example, consider the engines: these may vary in size, and also their exact locations under the wings need to be specified. An engine is defined in a local coordinate system, and it is then moved to its proper location. This process will have to be repeated for all engines.

Another example would be the seats in the plane: the manufacturer would design just one — then multiple copies of it are put at the right locations in the plane's design.

Following Sketch 7, and making things more formal again: we are given a local 3D coordinate system, called the $[\mathbf{d}_1, \mathbf{d}_2, \mathbf{d}_3]$-system, with coordinates (u_1, u_2, u_3). We assume that the object under consideration is located inside the *unit cube*, i.e., all of its defining points satisfy

$$0 \le u_1, u_2, u_3 \le 1.$$

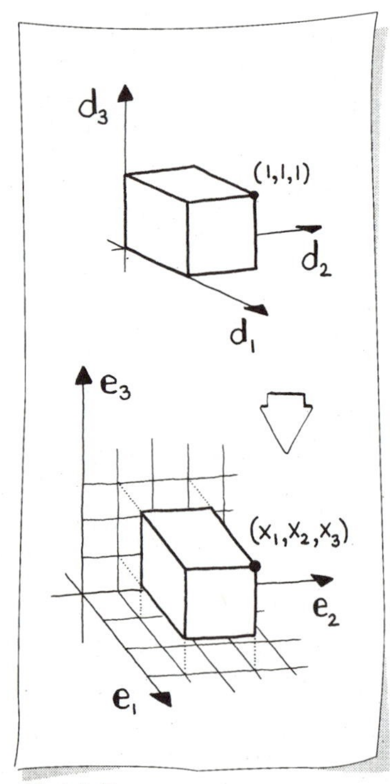

Sketch 7
Global and local 3D systems.

This cube is to be mapped onto a *3D target box* in the global $[\mathbf{e}_1, \mathbf{e}_2, \mathbf{e}_3]$-system. Let the target box be given by its lower corner $(\min_1, \min_2, \min_3)$ and its upper corner $(\max_1, \max_2, \max_3)$. How do we map coordinates (u_1, u_2, u_3) from the local unit cube into the corresponding target coordinates (x_1, x_2, x_3) in the target box? Exactly as in the 2D case, just with one more equation:

$$x_1 = (1 - u_1)\min_1 + u_1 \max_1, \tag{1.7}$$
$$x_2 = (1 - u_2)\min_2 + u_2 \max_2, \tag{1.8}$$
$$x_3 = (1 - u_3)\min_3 + u_3 \max_3. \tag{1.9}$$

As an easy exercise, check that the corners of the unit cube are mapped to the corners of the target box!

The analog to (1.3) and (1.4) is given by the rather obvious

$$x_1 = \min_1 + u_1 \Delta_1, \tag{1.10}$$
$$x_2 = \min_2 + u_2 \Delta_2, \tag{1.11}$$
$$x_3 = \min_3 + u_3 \Delta_3. \tag{1.12}$$

As in the 2D case, if the target box is not a cube, object distortions will result — this may be desired or not.

1.4 Stepping Outside the Box

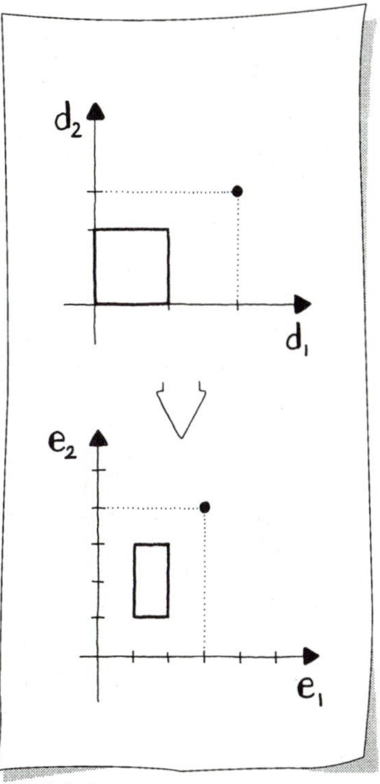

Sketch 8
A 2D coordinate outside a box.

We have restricted all objects to be within the unit square or cube; as a consequence, their images were inside the respective target boxes. This notion helps with an initial understanding, but it is not at all essential. Let's look at a 2D example, with the target box (see Sketch 8) given by

$$(\text{min}_1, \text{min}_2) = (1, 1) \quad \text{and} \quad (\text{max}_1, \text{max}_2) = (2, 3).$$

The coordinates $(u_1, u_2) = (2, 3/2)$ are not inside the $[\mathbf{d}_1, \mathbf{d}_2]$-system unit square. Yet we can map it using (1.1) and (1.2):

$$x_1 = -\text{min}_1 + 2\text{max}_1 = 3,$$
$$x_2 = -\frac{1}{2}\text{min}_2 + \frac{3}{2}\text{max}_2 = 4.$$

Since the initial coordinates (u_1, u_2) were not inside the unit square, the mapped coordinates (x_1, x_2) are not inside the target box. The notion of mapping a square to a target box is a useful concept for mentally visualizing what is happening — but it is not actually a restriction to the coordinates that we can map!

Example 1.3 Without much belaboring, it is clear the same holds for 3D. An example should suffice: the target box is given by

$$(\text{min}_1, \text{min}_2, \text{min}_3) = (1, 0, 1) \quad \text{and}$$
$$(\text{max}_1, \text{max}_2, \text{max}_3) = (2, 1, 2),$$

and we want to map the coordinates

$$(u_1, u_2, u_3) = (-1, -1, -1)$$

The result, illustrated by Sketch 9, is computed using (1.7) - (1.9): it is

$$(x_1, x_2, x_3) = (0, -1, 0).$$

1.5 Creating Coordinates

Suppose you have an interesting real object, like a model of a cat. A friend of yours in Hollywood would like to use this cat in her latest hi-tech animated movie. Such movies only use mathematical descriptions of

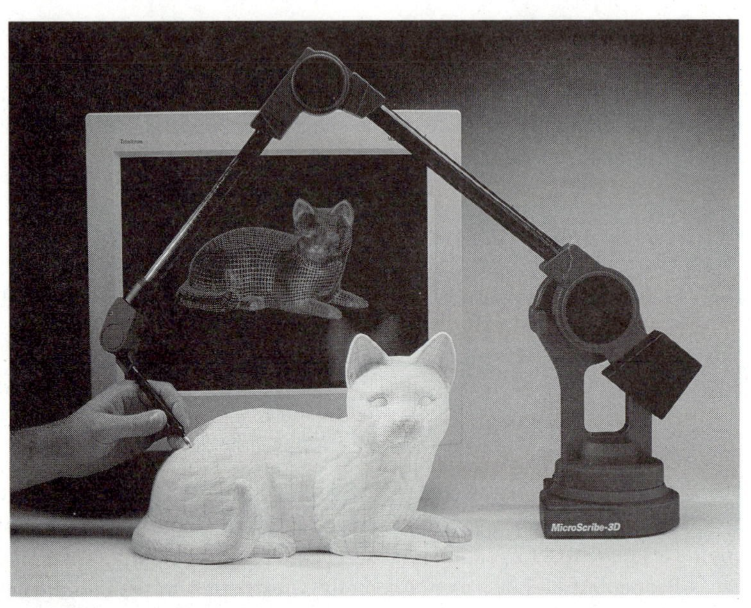

Figure 1.5.
Creating coordinates: a cat is turned into math. (Microscribe-3D from Immersion Corporation.)

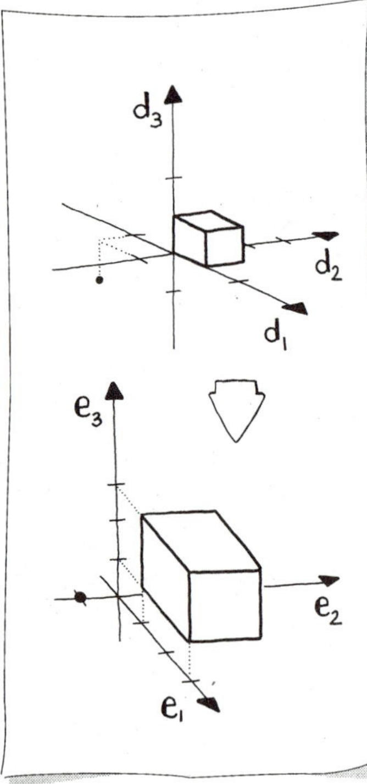

Sketch 9
A point outside a 3D box.

objects — everything must have coordinates! You might recall the movie *Toy Story*. It is a computer-animated movie, meaning that the characters and objects in every scene must have a mathematical representation.

So how do you give your cat model coordinates? This is done with a *CMM*, or *coordinate measuring machine,* see Figure 1.5. It was taken from the website www.immerse.com. The CMM is essentially an arm which is able to record the position of its tip by keeping track of the angles of its joints.

Your cat model is placed on a table and somehow fixed so it does not move during digitizing. You let the CMM's arm touch three points on the table; they will be the origin and the $\mathbf{e}_1-$ and $\mathbf{e}_2-$coordinate axes of a 3D coordinate system. From now on, those three points will be associated with coordinates $(0,0,0)$, $(1,0,0)$, and $(0,1,0)$, respectively. The $\mathbf{e}_3-$axis (vertical to the table) is then computed automatically. When you now touch your cat model with the tip of the CMM's arm, it will associate three coordinates with that position and record them. You repeat this for several hundred points, and you have your cat in the box!

This process is called *digitizing*. In the end, the cat has been "discretized," or turned into a finite number of coordinate triples. This set of points is called a *point cloud.*

Someone else will now have to build a mathematical model of your cat.[6] The mathematical model will next have to be put into scenes of the movie — but all that's needed for that are 3D coordinate transformations! (See Chapters 12 and 13.)

1.6 Exercises

1. Let coordinates of triangle vertices in the local $[\mathbf{d}_1, \mathbf{d}_2]$-system unit square be given by

$$\begin{aligned}(u_1, u_2) &= (0.1, 0.1),\\ (v_1, v_2) &= (0.9, 0.2),\\ (w_1, w_2) &= (0.4, 0.7).\end{aligned}$$

   a) If the $[\mathbf{d}_1, \mathbf{d}_2]$-system unit square is mapped to the target box with

$$(\text{min}_1, \text{min}_2) = (1, 2) \quad \text{and} \quad (\text{max}_1, \text{max}_2) = (3, 3),$$

   where are the coordinates of the triangle vertices mapped?

   b) What (u_1, u_2) coordinates correspond to $(x_1, x_2) = (2, 2)$?

2. Let the $[\mathbf{d}_1, \mathbf{d}_2, \mathbf{d}_3]$-system unit cube be mapped to the 3D target box with

$$(\text{min}_1, \text{min}_2, \text{min}_3) = (1, 1, 1) \quad \text{and} \quad (\Delta_1, \Delta_2, \Delta_3) = (1, 2, 4).$$

   Where will the coordinates $(u_1, u_2, u_3) = (0.5, 0, 0.7)$ be mapped?

3. Take the file `Boxes.ps` from the web site[7] and play with the PostScript `translate` commands to move some of the **D**-boxes around. Refer to the PostScript Tutorial (Appendix A) at the end of the book for some basic hints on how to use PostScript.

4. Given local coordinates $(2, 2)$ and $(-1, -1)$, find the global coordinates with respect to the target box with

$$(\text{min}_1, \text{min}_2) = (1, 1) \quad \text{and} \quad (\text{max}_1, \text{max}_2) = (7, 3).$$

   Make a sketch of the local and global systems. Connect the coordinates in each system with a line and compare.

[6] This type of work is called *Geometric Modeling* or *Computer Aided Geometric Design*, see [5] or [12].

[7] The web site for *The Geometry Toolbox* is http://eros.cagd.eas.asu.edu/~farin/gbook/gbook_home.html.

2 Here and There: Points and Vectors in 2D

Figure 2.1.
Hurricane Andrew approaching south Louisiana. (Image courtesy of EarthWatch Communications, Inc. http://www.earthwatch.com.)

In 1992 Hurricane Andrew hit the northwestern Bahamas, south Florida, and south-central Louisiana with such great force that it is rated as the most expensive hurricane to date, causing more than $25 billion of damage.[1] In the hurricane image (Figure 2.1) air is moving rapidly, spiraling in a counterclockwise fashion. What isn't so clear from this image is that the air moves faster as it approaches the eye of the hurricane. This air

[1]More on hurricanes can be found at the National Hurricane Center via http://www.nhc.noaa.gov.

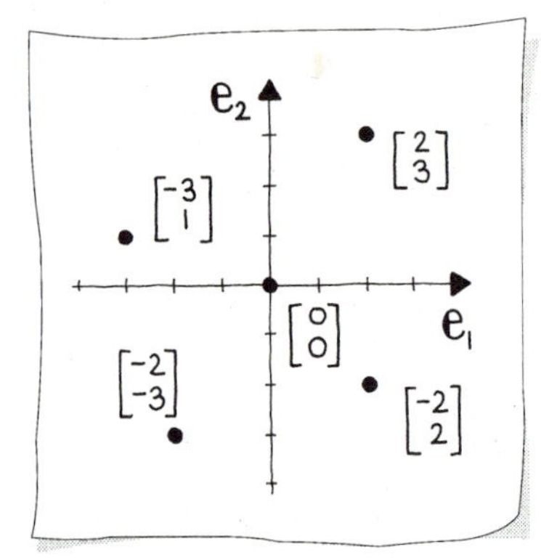

Sketch 10
Points and their coordinates.

movement is best described by points and vectors: at any location (point), air moves in a certain direction and with a certain speed (velocity vector).

This hurricane image is a good example of how helpful 2D geometry can be in a 3D world. Of course a hurricane is a 3D phenomenon; however by analyzing 2D slices, or cross sections, we can develop a very informative analysis. Many other applications call for two-dimensional geometry only. The purpose of this chapter is to define the two most fundamental tools we need to work in a 2D world: points and vectors.

2.1 Points and Vectors

The most basic geometric entity is the *point*. A point is a reference to a *location*. Sketch 10 illustrates examples of points. In the text, boldface lowercase letters represent points, e.g.,

$$\mathbf{p} = \begin{bmatrix} p_1 \\ p_2 \end{bmatrix}. \tag{2.1}$$

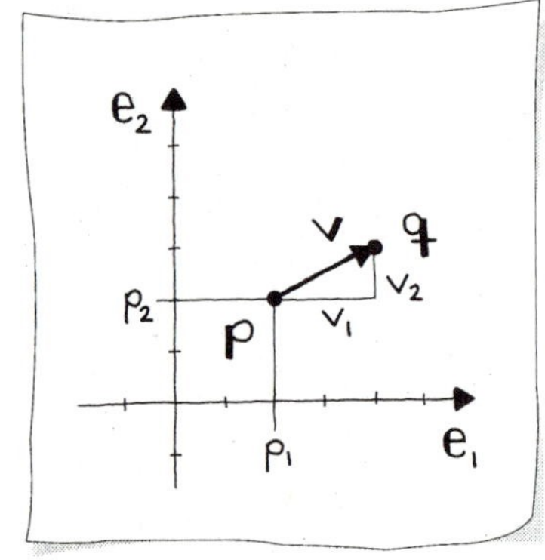

Sketch 11
Two points and a vector.

The location of $\mathbf{p}$ is p_1-units along the $\mathbf{e}_1$-axis and p_2-units along the $\mathbf{e}_2$-axis. So you see that a point's *coordinates*, p_1 and p_2, are dependent upon the location of the coordinate origin. We use the boldface notation so there is a noticeable difference between a one-dimensional (1D) number, or *scalar* p. To clearly identify $\mathbf{p}$ as a point, the notation $\mathbf{p} \in \mathbb{E}^2$ is used. This means that a 2D point "lives" in 2D Euclidean space $\mathbb{E}^2$.

Now let's move away from our reference point. Following Sketch 11, suppose the reference point is $\mathbf{p}$, and when moving along a straight path, our target point is $\mathbf{q}$. The directions from $\mathbf{p}$ would be to follow the *vector* $\mathbf{v}$. Our notation for a vector is the same as for a point: boldface lowercase letters.

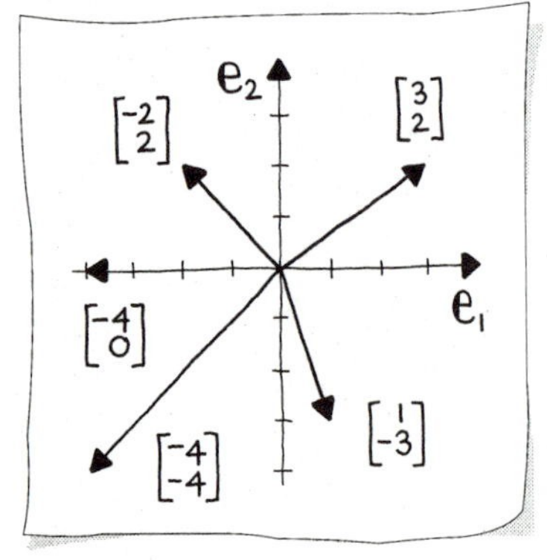

Sketch 12
Vectors and their components.

To get to $\mathbf{q}$ we say,

$$\mathbf{q} = \mathbf{p} + \mathbf{v}. \tag{2.2}$$

To calculate this, add each component separately, that is

$$\begin{bmatrix} q_1 \\ q_2 \end{bmatrix} = \begin{bmatrix} p_1 \\ p_2 \end{bmatrix} + \begin{bmatrix} v_1 \\ v_2 \end{bmatrix} = \begin{bmatrix} p_1 + v_1 \\ p_2 + v_2 \end{bmatrix}.$$

For example, in Sketch 11, we have

$$\begin{bmatrix} 4 \\ 3 \end{bmatrix} = \begin{bmatrix} 2 \\ 2 \end{bmatrix} + \begin{bmatrix} 2 \\ 1 \end{bmatrix}.$$

The *components* of $\mathbf{v}$, v_1 and v_2, indicate how many units to move along the $\mathbf{e}_1$- and $\mathbf{e}_2$-axis, respectively. This means that $\mathbf{v}$ can be defined as

$$\mathbf{v} = \mathbf{q} - \mathbf{p}. \tag{2.3}$$

This defines a vector as a difference of two points which describes a *direction and a distance*, or a *displacement*. Examples of vectors are illustrated in Sketch 12.

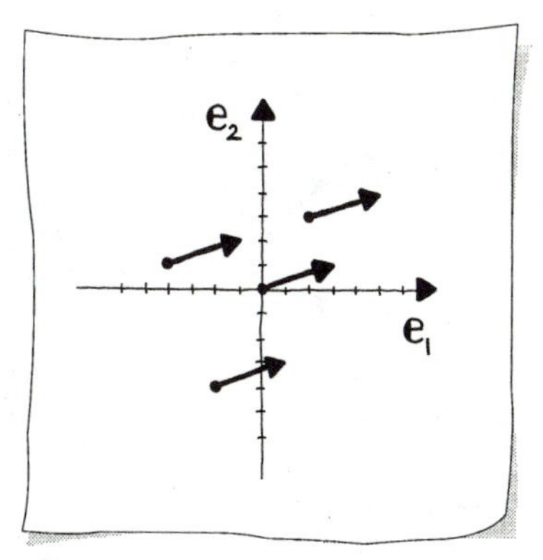

Sketch 13
Instances of one vector.

How to determine a vector's length is covered in Section 2.5. Above we described this length as a distance. Alternatively, this length can be described as speed: then we have a *velocity vector*.[2] Yet another interpretation is that the length represents acceleration: then we have a *force vector*.

A vector has a *tail* and a *head*. As in Sketch 11, the tail is typically displayed positioned at a point, or *bound to a point* in order to indicate the geometric significance of the vector. However, unlike a point, a vector does *not* define a position. Two vectors are equal if they have the same component values, just as points are equal if they have the same coordinate values. Thus, considering a vector as a difference of two points, there are any number of vectors with the same direction and length. See Sketch 13 for an illustration.

A special vector worth mentioning is the *zero vector*,

$$\mathbf{0} = \begin{bmatrix} 0 \\ 0 \end{bmatrix}.$$

This vector has no direction or length.

Other somewhat special vectors include

$$\mathbf{e}_1 = \begin{bmatrix} 1 \\ 0 \end{bmatrix} \quad \text{and} \quad \mathbf{e}_2 = \begin{bmatrix} 0 \\ 1 \end{bmatrix}.$$

In the sketches, these vectors are not always drawn true to length for clarity.

To clearly identify $\mathbf{v}$ as a vector, we write $\mathbf{v} \in \mathbb{R}^2$. This means that a 2D vector "lives" in a 2D *linear space* $\mathbb{R}^2$. (Other names for $\mathbb{R}^2$ are real or vector spaces.)

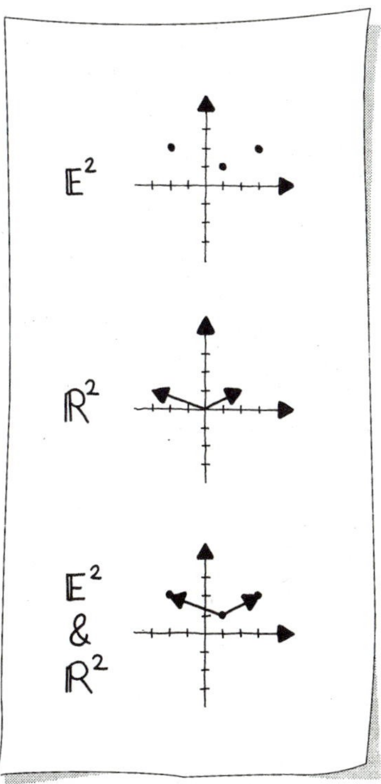

Sketch 14
Euclidean and linear spaces displayed separately and together.

2.2 What's the Difference?

When writing a point or a vector we use boldface lowercase letters; when programming we use the same data structure, e.g., arrays. This makes it

[2]This is what we'll use to continue the hurricane Andrew example.

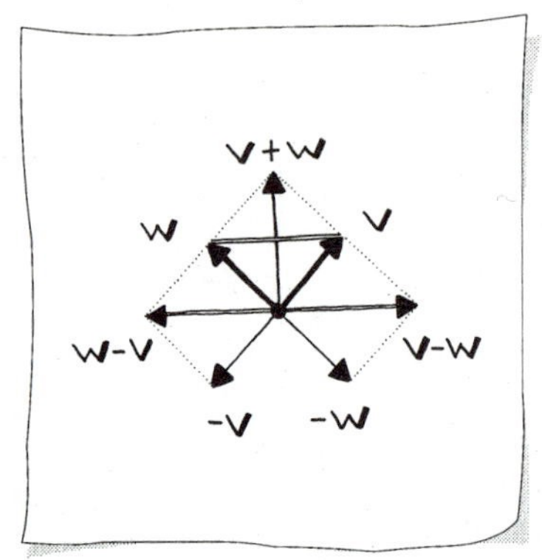

Sketch 15
Parallelogram rule.

appear that points and vectors can be treated in the same manner. Not so! Points and vectors are different geometric entities. This is reiterated by saying they live in different spaces, $\mathbb{E}^2$ and $\mathbb{R}^2$. As shown in Sketch 14, for convenience and clarity elements of Euclidean and linear spaces are typically displayed together.

The primary reason for differentiating between points and vectors is to achieve geometric constructions which are *coordinate independent*. Such constructions are manipulations applied to geometric objects that produce the same result regardless of the location of the coordinate origin. (Example: the midpoint of two points.) This idea becomes clearer by analyzing some fundamental manipulations of points and vectors. Let's use $\mathbf{p}, \mathbf{q} \in \mathbb{E}^2$ and $\mathbf{v}, \mathbf{w} \in \mathbb{R}^2$.

1. Subtracting two points yields a vector as depicted in Sketch 11 and (2.3). This is a coordinate independent operation.

2. Adding or subtracting two vectors yields another vector. See Sketch 15 which illustrates the *parallelogram rule*: the vectors $\mathbf{v} - \mathbf{w}$ and $\mathbf{v} + \mathbf{w}$ are the diagonals of the parallelogram defined by $\mathbf{v}$ and $\mathbf{w}$. This is a coordinate independent operation since vectors are defined as a difference of points.

3. Adding two points is not a coordinate independent operation. As depicted in Sketch 16, the result of such an addition would be dependent on the coordinate origin. In other words, it is not a well-defined operation – it is ambiguous and will therefore not be allowed.

4. Multiplying by a scalar s is called *scaling*. Scaling a vector is a well-defined operation. The result $s\mathbf{v}$ adjusts the length by the scaling factor. The direction is unchanged if $s > 0$ and reversed for $s < 0$. If $s = 0$ then the result is the zero vector. Sketch 17 illustrates some examples of scaling a vector.

5. Scaling a point is not a well-defined operation because it is not a coordinate independent operation (see Sketch 18.)

6. Adding a vector to a point yields another point as in Sketch 11 and (2.2). This is a coordinate independent operation.

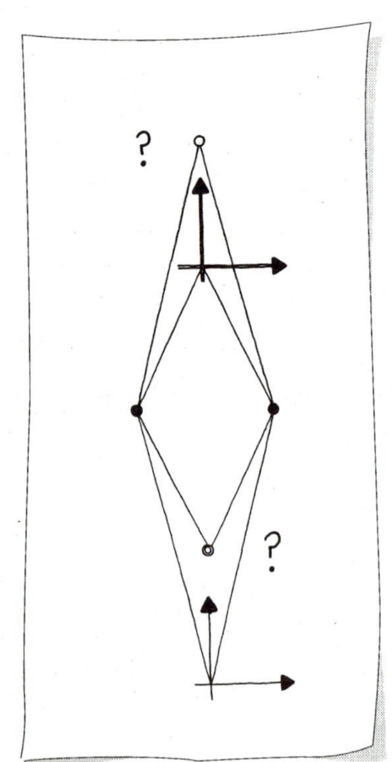

Sketch 16
Addition of points is ambiguous.

Any coordinate independent combination of two or more points and/or vectors can be grouped to fall into one or more of items 1, 2, 4, or 6. See the Exercises for examples.

2.3 Vector Fields

A good way to visualize the interplay between points and vectors is through the example of *vector fields*. In general, we speak of a vector field if every point in a given region is assigned a vector. We have already encountered an example of this in Figure 2.1: hurricane Andrew! Recall that at each location (point) we could describe the air velocity (vector). Our previous image did not actually tell us anything about the air speed, although we could presume something about the direction. This is where a vector field is helpful. Shown in Figure 2.2 is a vector field simulating hurricane Andrew. By plotting all the vectors the same length and using shading to indicate speed, the vector field can be more informative than the photograph.

Other important applications of vector fields arise in the areas of automotive and aerospace design: before a car or an airplane is built, it undergoes extensive aerodynamic simulations. In these simulations, the vectors that characterize the flow around an object are computed from complex differential equations. In Figure 2.3 we have another example of a vector field.

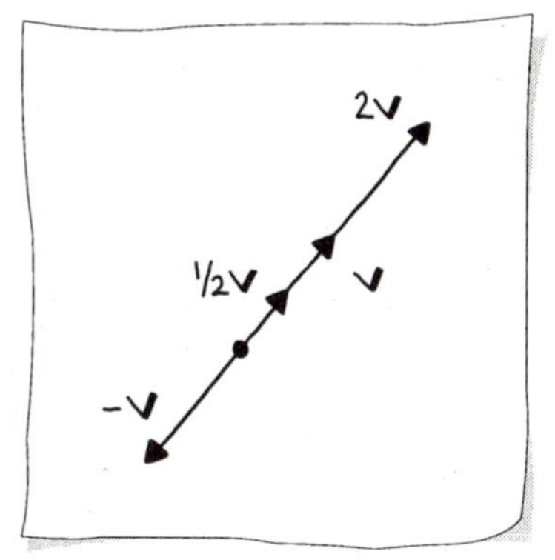

Sketch 17
Scaling a vector.

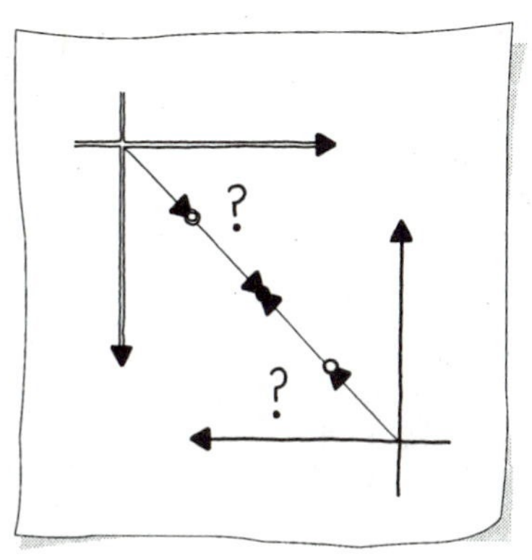

Sketch 18
Scaling of points is ambiguous.

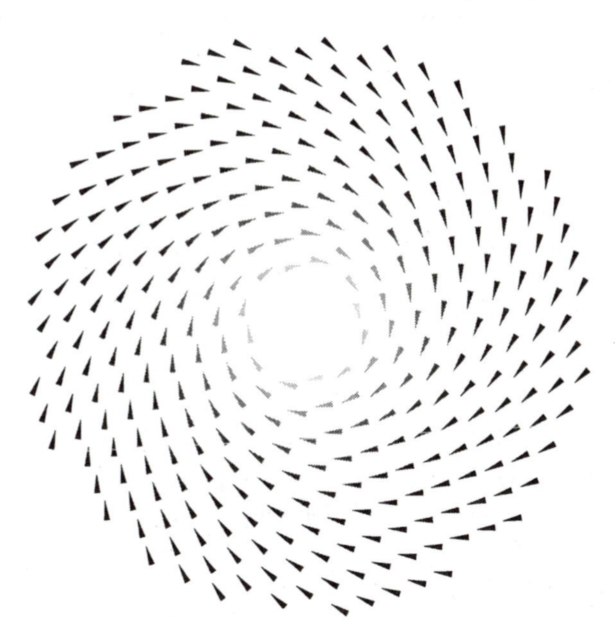

Figure 2.2.
Vector field: simulating hurricane air velocity. Lighter shading indicates greater velocity.

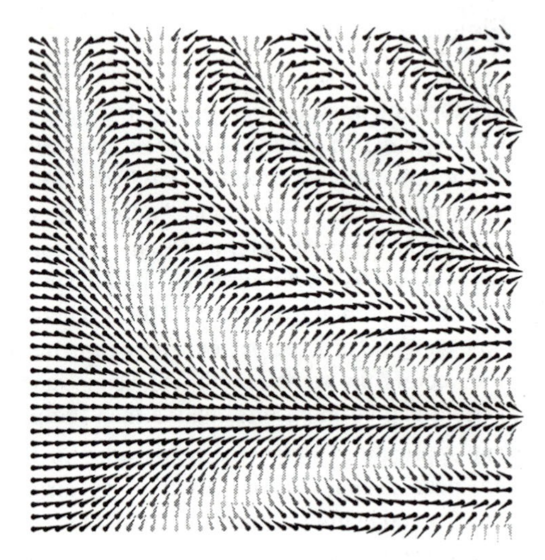

Figure 2.3.
Vector field: every point has an associated vector. Lighter shading indicates greater vector length.

2.4 Combining Points

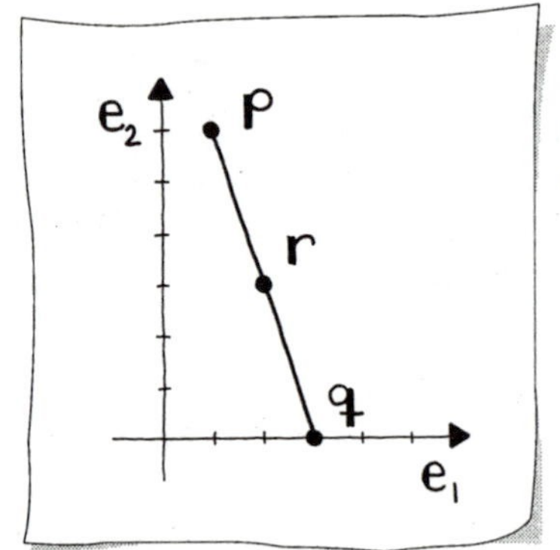

Sketch 19
The midpoint of two points.

Seemingly contrary to item 3, Section 2.2, there actually is a way to combine two points such that we get a (meaningful) third one. Take the example of the midpoint $\mathbf{r}$ of two points $\mathbf{p}$ and $\mathbf{q}$; more specifically, take

$$\mathbf{p} = \begin{bmatrix} 1 \\ 6 \end{bmatrix}, \quad \mathbf{r} = \begin{bmatrix} 2 \\ 3 \end{bmatrix}, \quad \mathbf{q} = \begin{bmatrix} 3 \\ 0 \end{bmatrix},$$

as shown in Sketch 19.

Let's start with a known coordinate independent operation as in item 6, Section 2.2. Define $\mathbf{r}$ by adding an appropriately scaled version of the vector $\mathbf{v} = \mathbf{q} - \mathbf{p}$ to the point $\mathbf{p}$:

$$\mathbf{r} = \mathbf{p} + \frac{1}{2}\mathbf{v}$$

$$\begin{bmatrix} 2 \\ 3 \end{bmatrix} = \begin{bmatrix} 1 \\ 6 \end{bmatrix} + \frac{1}{2}\begin{bmatrix} 2 \\ -6 \end{bmatrix}.$$

Expanding, this shows that $\mathbf{r}$ can also be defined as

$$\mathbf{r} = \frac{1}{2}\mathbf{p} + \frac{1}{2}\mathbf{q}$$
$$\begin{bmatrix} 2 \\ 3 \end{bmatrix} = \frac{1}{2}\begin{bmatrix} 1 \\ 6 \end{bmatrix} + \frac{1}{2}\begin{bmatrix} 3 \\ 0 \end{bmatrix}.$$

This is a legal expression for a combination of points!

There is nothing magical about the factor 1/2, however. Adding a (scaled) vector to a point is a well-defined, coordinate independent operation that yields another point. Any point of the form

$$\mathbf{r} = \mathbf{p} + t\mathbf{v} \tag{2.4}$$

is on the line through $\mathbf{p}$ and $\mathbf{q}$. Just as above, we may rewrite this as

$$\mathbf{r} = \mathbf{p} + t(\mathbf{q} - \mathbf{p})$$

and then

$$\mathbf{r} = (1 - t)\mathbf{p} + t\mathbf{q}. \tag{2.5}$$

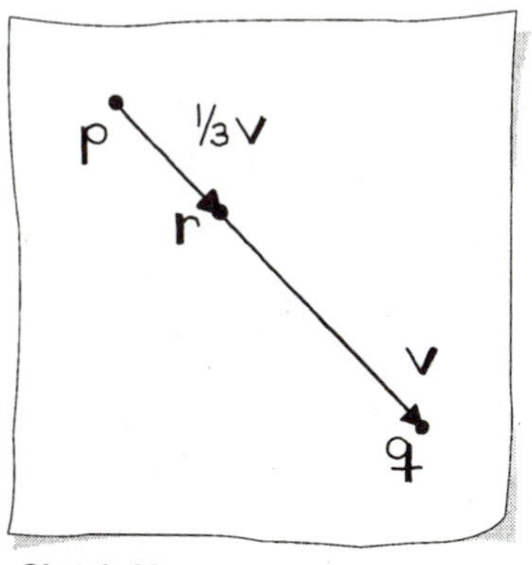

Sketch 20
Barycentric combinations: $t = 1/3$.

Sketch 20 gives an example with $t = 1/3$.

The scalar values $(1-t)$ and t are *coefficients*. A weighted sum of points where the coefficients sum to one is called a *barycentric combination*. In this special case, where one point $\mathbf{r}$ is being expressed in terms of two others, $\mathbf{p}$ and $\mathbf{q}$, $1 - t$ and t are called the *barycentric coordinates* of $\mathbf{r}$.

A barycentric combination allows us to construct $\mathbf{r}$ anywhere on the line defined by $\mathbf{p}$ and $\mathbf{q}$. If we would like to restrict $\mathbf{r}$'s position to the *line segment* between $\mathbf{p}$ and $\mathbf{q}$, then we allow only *convex combinations*: t must satisfy $0 \leq t \leq 1$. To define points outside of the line segment between $\mathbf{p}$ and $\mathbf{q}$, we need values of $t < 0$ or $t > 1$.

The position of $\mathbf{r}$ is said to be in the *ratio* of $t : (1 - t)$ or $t/(1 - t)$. Some examples are illustrated in Sketch 21.

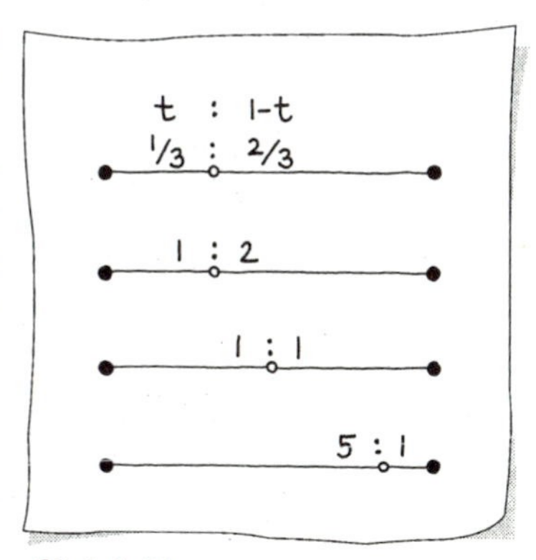

Sketch 21
Examples of ratios.

We can create barycentric combinations with *more than two points*. Let's look at three points $\mathbf{p}$, $\mathbf{q}$, and $\mathbf{r}$, which are not collinear.

Any point $\mathbf{s}$ can be formed from

$$\mathbf{s} = \mathbf{r} + t_1(\mathbf{p} - \mathbf{r}) + t_2(\mathbf{q} - \mathbf{r}).$$

This is a coordinate independent operation of point + vector + vector. Expanding and regrouping, we can also define $\mathbf{s}$ as

$$\begin{aligned} \mathbf{s} &= t_1\mathbf{p} + t_2\mathbf{q} + (1 - t_1 - t_2)\mathbf{r} \\ &= t_1\mathbf{p} + t_2\mathbf{q} + t_3\mathbf{r}. \end{aligned} \tag{2.6}$$

Thus the point $\mathbf{s}$ is defined by a barycentric combination with coefficients t_1, t_2, and $t_3 = 1 - t_1 - t_2$ with respect to $\mathbf{p}$, $\mathbf{q}$, and $\mathbf{r}$, respectively. This is another special case where the barycentric combination coefficients correspond to *barycentric coordinates.* We will encounter barycentric coordinates again in Chapter 8.

We can also *combine points* so that the result is a vector. For this, we need the coefficients to sum to zero. We encountered a simple case of this in (2.3). Suppose you have the equation

$$\mathbf{e} = \mathbf{r} - 2\mathbf{p} + \mathbf{q}, \qquad \mathbf{r}, \mathbf{p}, \mathbf{q} \in \mathbb{E}^2.$$

Does $\mathbf{e}$ have a geometric meaning? Looking at the sum of the coefficients, $1 - 2 + 1 = 0$, we would conclude by the rule above that $\mathbf{e}$ is a vector. How to see this? By rewriting the equation as

$$\mathbf{e} = (\mathbf{r} - \mathbf{p}) + (\mathbf{q} - \mathbf{p}),$$

it is clear that $\mathbf{e}$ is a vector formed from (vector + vector).

2.5 Length of a Vector

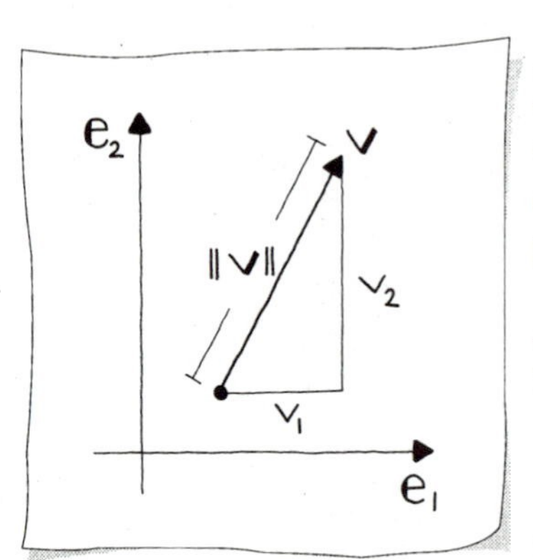

Sketch 22
Length of a vector.

As mentioned in Section 2.1, the length of a vector can represent distance, velocity, or acceleration. We need a method for finding the length of a vector, or the *magnitude.* As illustrated in Sketch 22, a vector defined the displacement necessary (with respect to the $\mathbf{e}_1$- and $\mathbf{e}_2$-axis) to get from a point at the tail of the vector to a point at the head of the vector.

In Sketch 22 we have formed a right triangle. The square of the length of the hypotenuse of a right triangle is well known from the *Pythagorean Theorem.* Denote the *length* of a vector $\mathbf{v}$ as $\|\mathbf{v}\|$. Then

$$\|\mathbf{v}\|^2 = v_1^2 + v_2^2.$$

Therefore, the magnitude of $\mathbf{v}$ is

$$\|\mathbf{v}\| = \sqrt{v_1^2 + v_2^2}. \tag{2.7}$$

Notice that if we scale the vector by an amount k then

$$\|k\mathbf{v}\| = k\|\mathbf{v}\|. \tag{2.8}$$

A *normalized vector* $\mathbf{w}$ has *unit length,* that is

$$\|\mathbf{w}\| = 1.$$

Normalized vectors are also known as *unit vectors*. To *normalize* a vector simply means to scale a vector so that it has unit length. If $\mathbf{w}$ is to be our unit length version of $\mathbf{v}$ then

$$\mathbf{w} = \frac{\mathbf{v}}{\|\mathbf{v}\|}.$$

Each component of $\mathbf{v}$ is divided by the scalar value $\|\mathbf{v}\|$. This scalar value is always *nonnegative*, which means that its value is zero or greater. It can be zero! You must check the value before dividing to be sure it is greater than your *zero divide tolerance*.[3]

We utilized unit vectors in Figures 2.2 and 2.3. The vectors in those figure are drawn as unit vectors. Gray scales are used to indicate their magnitudes: that way the vectors do not overlap, and the figure is easier to understand.

Example 2.1 Start with

$$\mathbf{v} = \begin{bmatrix} 5 \\ 0 \end{bmatrix}.$$

Applying (2.7), $\|\mathbf{v}\| = \sqrt{5^2 + 0^2} = 5$. Then the normalized version of $\mathbf{v}$ is defined as

$$\mathbf{w} = \begin{bmatrix} 5/5 \\ 0/5 \end{bmatrix} = \begin{bmatrix} 1 \\ 0 \end{bmatrix}.$$

Clearly $\|\mathbf{w}\| = 1$, so this is a normalized vector. Since we have only scaled $\mathbf{v}$ by a positive amount, the direction of $\mathbf{w}$ is the same as $\mathbf{v}$.

There are infinitely many unit vectors. Imagine drawing them all, emanating from the origin. The figure that you will get is a circle of radius one! See Figure 2.4.

To find the *distance* between two points we simply form a vector defined by the two points, e.g., $\mathbf{v} = \mathbf{q} - \mathbf{p}$, and apply (2.7).

Example 2.2 Let

$$\mathbf{q} = \begin{bmatrix} -1 \\ 2 \end{bmatrix} \text{ and } \mathbf{p} = \begin{bmatrix} 1 \\ 0 \end{bmatrix}.$$

[3]The zero divide tolerance is the absolute value of the smallest number by which you can divide confidently. When we refer to checking that a value is greater than this number, it means to check the absolute value.

Figure 2.4.
Unit vectors: they define a circle.

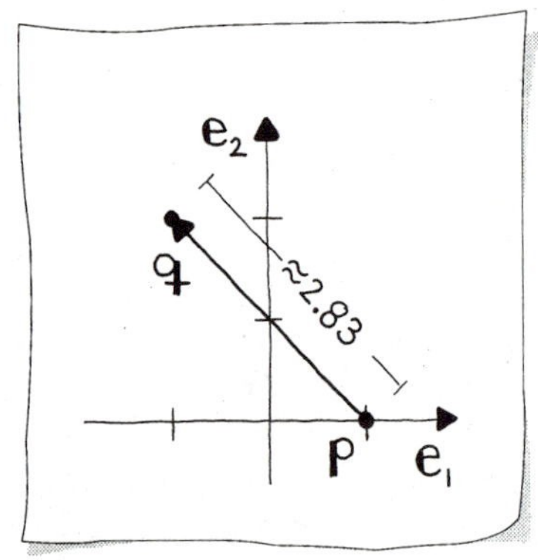

Sketch 23
Distance between two points.

Then

$$\mathbf{q} - \mathbf{p} = \begin{bmatrix} -2 \\ 2 \end{bmatrix}$$

and

$$\|\mathbf{q} - \mathbf{p}\| = \sqrt{(-2)^2 + 2^2} = \sqrt{8} \approx 2.83.$$

Sketch 23 illustrates this example.

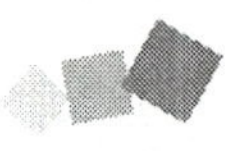

Let's revisit barycentric coordinates and ratios. Suppose we have three collinear points, **p**, **q**, and **r** as illustrated in Sketch 24. The points have the following locations.

$$\mathbf{p} = \begin{bmatrix} 2 \\ 4 \end{bmatrix}, \quad \mathbf{r} = \begin{bmatrix} 6.5 \\ 7 \end{bmatrix}, \quad \mathbf{q} = \begin{bmatrix} 8 \\ 8 \end{bmatrix}.$$

What are the barycentric coordinates of **r** with respect to **p** and **q**? To answer this, recall the relationship between the ratio and the barycentric coordinates. The barycentric coordinates t and $(1 - t)$ define **r** as

$$\begin{bmatrix} 6.5 \\ 7 \end{bmatrix} = (1-t)\begin{bmatrix} 2 \\ 4 \end{bmatrix} + t\begin{bmatrix} 8 \\ 8 \end{bmatrix}.$$

The ratio indicates the location of $\mathbf{r}$ relative to $\mathbf{p}$ and $\mathbf{q}$ in terms of relative distances. Suppose the ratio is $s_1 : s_2$. If we scale s_1 and s_2 such that they sum to one, then s_1 and s_2 are the barycentric coordinates t and $(1-t)$, respectively. By calculating the distances between points:

$$l_1 = \|\mathbf{r} - \mathbf{p}\| \approx 5.4,$$
$$l_2 = \|\mathbf{q} - \mathbf{r}\| \approx 1.8,$$
$$l_3 = l_1 + l_2 \approx 7.2,$$

we find that

$$t = l_1/l_3 = 0.75 \quad \text{and}$$
$$(1-t) = l_2/l_3 = 0.25.$$

These are the barycentric coordinates. Let's verify this:

$$\begin{bmatrix} 6.5 \\ 7 \end{bmatrix} = 0.25 \times \begin{bmatrix} 2 \\ 4 \end{bmatrix} + 0.75 \times \begin{bmatrix} 8 \\ 8 \end{bmatrix}.$$

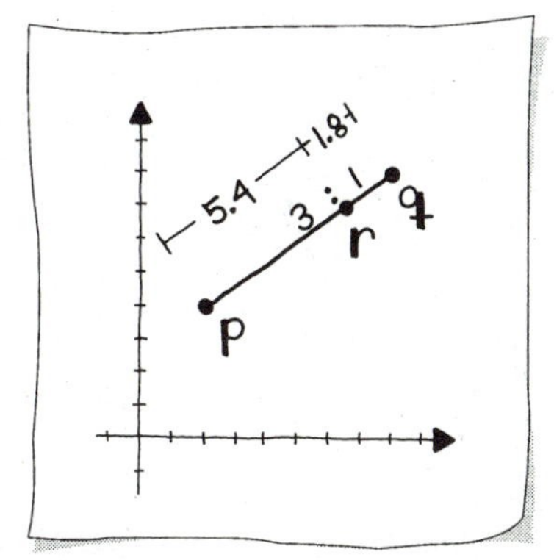

Sketch 24
Barycentric coordinates in relation to lengths.

In physics, $\mathbf{r}$ is known as the *center of gravity* of two points $\mathbf{p}$ and $\mathbf{q}$ with weights $1-t$ and t, respectively.

2.6 Independence

Two vectors $\mathbf{v}$ and $\mathbf{w}$ describe a parallelogram, as shown in Sketch 15. It may happen that this parallelogram has zero area; then the two vectors are parallel. In this case, we have a relationship of the form $\mathbf{v} = c\mathbf{w}$. If two vectors are parallel, then we call them *linearly dependent*. Otherwise, we say that they are *linearly independent.*

Two linearly independent vectors may be used to write any other vector $\mathbf{u}$ as a *linear combination*:

$$\mathbf{u} = r\mathbf{v} + s\mathbf{w}.$$

How to find r and s is described in Chapter 5. Two linearly independent vectors in 2D are also called a *basis* for $\mathbb{R}^2$.

If $\mathbf{v}$ and $\mathbf{w}$ are linearly dependent, then you cannot write all vectors as a linear combination of them, as the following example shows.

Example 2.3 Let

$$\mathbf{v} = \begin{bmatrix} 1 \\ 2 \end{bmatrix} \quad \text{and} \quad \mathbf{w} = \begin{bmatrix} 2 \\ 4 \end{bmatrix}.$$

If we tried to write the vector $\mathbf{u} = \begin{bmatrix} 1 \\ 0 \end{bmatrix}$ as $\mathbf{u} = r\mathbf{v} + s\mathbf{w}$, then this would lead to

$$1 = r + 2s, \tag{2.9}$$
$$0 = 2r + 4s. \tag{2.10}$$

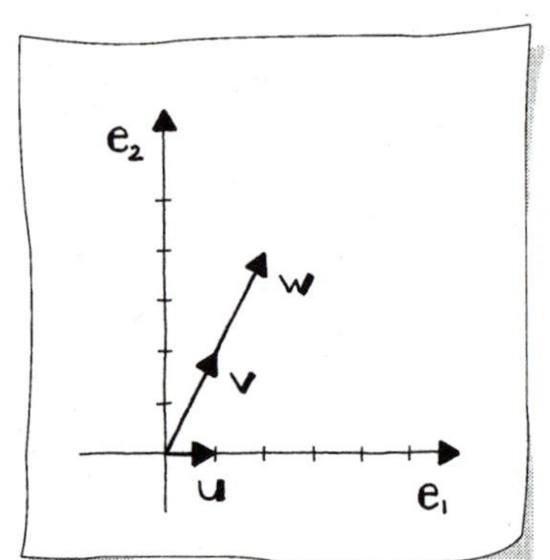

Sketch 25
Dependent vectors.

If we multiply the first equation by a factor of 2, the two right-hand sides will be equal. Equating the new left-hand sides now results in the expression $2 = 0$. This shows that $\mathbf{u}$ cannot be written as a linear combination of $\mathbf{v}$ and $\mathbf{w}$. (See Sketch 25.)

2.7 Dot Product

Given two vectors $\mathbf{v}$ and $\mathbf{w}$, we might ask:

- Are they the *same* vector?
- Are they *perpendicular* to each other?
- What *angle* do they form?

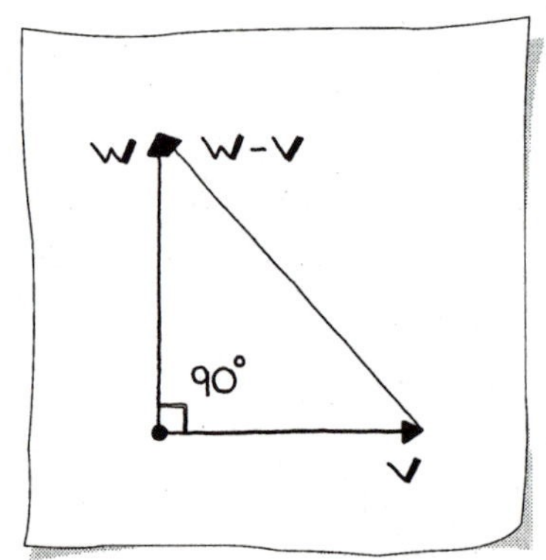

Sketch 26
Perpendicular vectors.

The *dot product* is the tool to resolve these questions.

To motivate the dot product, let's start with the Pythagorean Theorem and Sketch 26. We know that two vectors $\mathbf{v}$ and $\mathbf{w}$ are perpendicular if and only if

$$\|\mathbf{v} - \mathbf{w}\|^2 = \|\mathbf{v}\|^2 + \|\mathbf{w}\|^2. \tag{2.11}$$

Writing the components in (2.11) explicitly

$$((v_1 - w_1)^2 + (v_2 - w_2)^2) = (v_1^2 + v_2^2) + (w_1^2 + w_2^2),$$

and then expanding, bringing all terms to the left-hand side of the equation yields

$$(v_1^2 - 2v_1w_1 + w_1^2) + (v_2^2 - 2v_2w_2 + w_2^2) - (v_1^2 + v_2^2) - (w_1^2 + w_2^2) = 0,$$

which reduces to

$$v_1w_1 + v_2w_2 = 0. \tag{2.12}$$

We find that perpendicular vectors have the property that the sum of the products of their components is zero. The short-hand vector notation for (2.12) is

$$\mathbf{v} \cdot \mathbf{w} = 0. \tag{2.13}$$

This result has an immediate application: a vector $\mathbf{w}$ perpendicular to a given vector $\mathbf{v}$ can be formed as

$$\mathbf{w} = \begin{bmatrix} -v_2 \\ v_1 \end{bmatrix}$$

(switching components and negating the sign of one). Then $\mathbf{v} \cdot \mathbf{w}$ becomes $v_1(-v_2) + v_2 v_1 = 0$.

If we take two arbitrary vectors $\mathbf{v}$ and $\mathbf{w}$, then $\mathbf{v} \cdot \mathbf{w}$ will in general not be zero. But we can compute it anyway, and define

$$s = \mathbf{v} \cdot \mathbf{w} = v_1 w_1 + v_2 w_2, \tag{2.14}$$

to be the *dot product* of $\mathbf{v}$ and $\mathbf{w}$. Notice that the dot product returns a scalar s, which is why it is also called a *scalar product*. (Mathematicians have yet another name for the dot product — an *inner product*.) From (2.14) it is clear that

$$\mathbf{v} \cdot \mathbf{w} = \mathbf{w} \cdot \mathbf{v}.$$

This is called the *symmetry property*. Other properties of the dot product are given in the Exercises.

In order to understand the geometric meaning of the dot product of two vectors, let's construct a triangle from two vectors $\mathbf{v}$ and $\mathbf{w}$ as illustrated in Sketch 27.

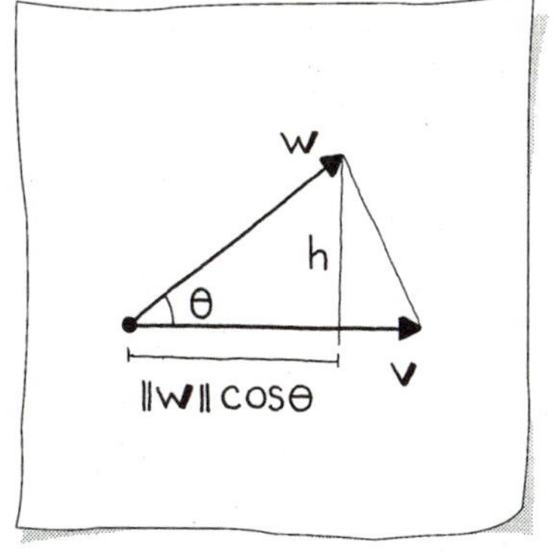

Sketch 27
Geometry of the dot product.

From trigonometry, we know that the height h of the triangle can be expressed as

$$h = \|\mathbf{w}\| \sin(\theta).$$

Squaring both sides results in

$$h^2 = \|\mathbf{w}\|^2 \sin^2(\theta).$$

Using the identity

$$\sin^2(\theta) + \cos^2(\theta) = 1,$$

we have

$$h^2 = \|\mathbf{w}\|^2 (1 - \cos^2(\theta)). \tag{2.15}$$

We can also express the height h with respect to the other right triangle in Sketch 27 and by using the Pythagorean Theorem:

$$h^2 = \|\mathbf{v} - \mathbf{w}\|^2 - (\|\mathbf{v}\| - \|\mathbf{w}\| \cos\theta)^2. \tag{2.16}$$

Equating (2.15) and (2.16) and simplifying, we have the expression,

$$\|\mathbf{v} - \mathbf{w}\|^2 = \|\mathbf{v}\|^2 + \|\mathbf{w}\|^2 - 2\|\mathbf{v}\|\|\mathbf{w}\| \cos\theta. \tag{2.17}$$

We have just proved the *Law of Cosines*.

We can formulate another expression for $\|\mathbf{v} - \mathbf{w}\|^2$ by explicitly writing out

$$\begin{aligned} \|\mathbf{v} - \mathbf{w}\|^2 &= (\mathbf{v} - \mathbf{w}) \cdot (\mathbf{v} - \mathbf{w}) \\ &= \|\mathbf{v}\|^2 - 2\mathbf{v} \cdot \mathbf{w} + \|\mathbf{w}\|^2 \end{aligned} \tag{2.18}$$

By equating (2.17) and (2.18) we find that

$$\mathbf{v} \cdot \mathbf{w} = \|\mathbf{v}\|\|\mathbf{w}\| \cos\theta. \tag{2.19}$$

Here is another expression for the *dot product* — it is a very useful one! Rearranging (2.19), the cosine of the angle between the two vectors can be determined as

$$\cos\theta = \frac{\mathbf{v} \cdot \mathbf{w}}{\|\mathbf{v}\|\|\mathbf{w}\|}. \tag{2.20}$$

Examining a plot of the cosine function in Figure 2.5, some sense can be made of (2.20).

First we consider the special case of perpendicular vectors: Recall the dot product was zero, which makes $\cos(90°) = 0$, just as it should be.

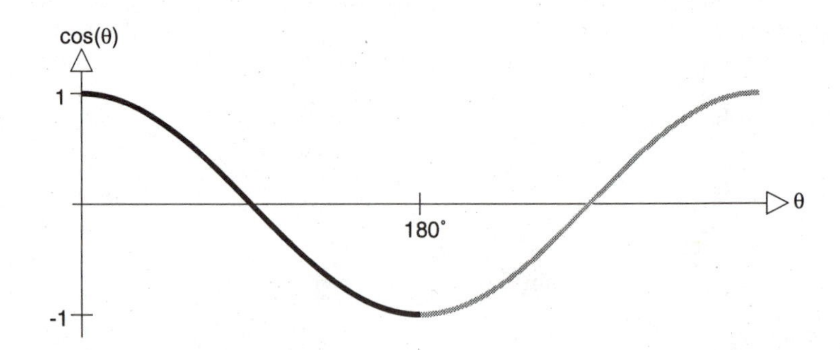

Figure 2.5.
The cosine function.

If $\mathbf{v}$ has the same (or opposite) direction as $\mathbf{w}$, that is $\mathbf{v} = k\mathbf{w}$, then (2.20) becomes

$$\cos\theta = \frac{k\mathbf{w} \cdot \mathbf{w}}{\|k\mathbf{w}\|\|\mathbf{w}\|}.$$

Using (2.8), we have

$$\cos\theta = \frac{k\|\mathbf{w}\|^2}{k\|\mathbf{w}\|\|\mathbf{w}\|} = \pm 1.$$

Again examining Figure 2.5, we see this corresponds to either $\theta = 0°$ or $\theta = 180°$, for vectors of the same or opposite direction, respectively.

The cosine values from (2.20) range between ± 1; this corresponds to angles between $0°$ and $180°$ (or $0 - \pi$ radians). Thus the smaller angle between the two vectors is measured. This is clear from the derivation: the angle θ enclosed by completing the triangle defined by the two vectors must be less than $180°$. Three types of angles can be formed: *right*, *acute*, or *obtuse*, as illustrated in counterclockwise order from twelve o'clock in Sketch 28. The $\cos(\theta)$ value of these angles is zero, positive, and negative, respectively.

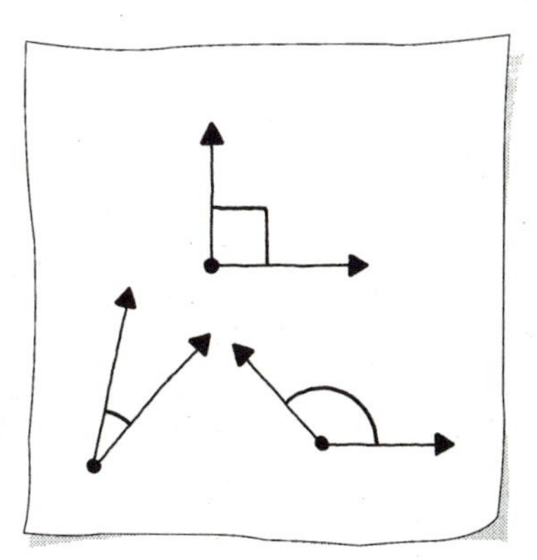

Sketch 28
Three types of angles.

If the actual angle θ needs to be calculated, then the arccosine function has to be invoked: $\theta = \operatorname{acos}(s)$. One word of warning: in some math libraries, if $s > 1$ or $s < -1$ then an error occurs and a non-usable result (NaN — Not a Number) is returned. Thus, if s is calculated, it is best to check that its value is within the appropriate range. It is not uncommon that an intended value of $s = 1.0$ actually is something like $s = 1.0000001$ due to *roundoff*. Thus the arccosine function should be used with caution. In many instances, as in comparing angles, the cosine of the angle is all you need!

Example 2.4 Let's calculate the angle between the two vectors illustrated in Sketch 29, forming an obtuse angle:

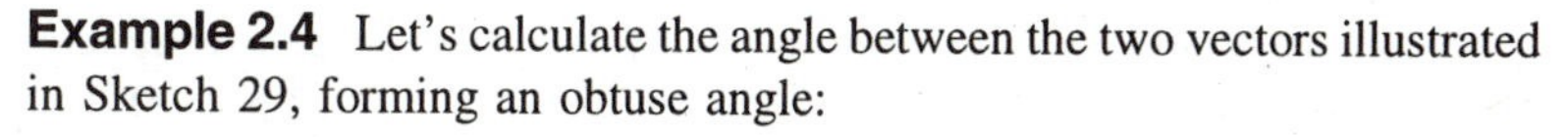

$$\mathbf{v} = \begin{bmatrix} 2 \\ 1 \end{bmatrix} \quad \text{and} \quad \mathbf{w} = \begin{bmatrix} -1 \\ 0 \end{bmatrix}.$$

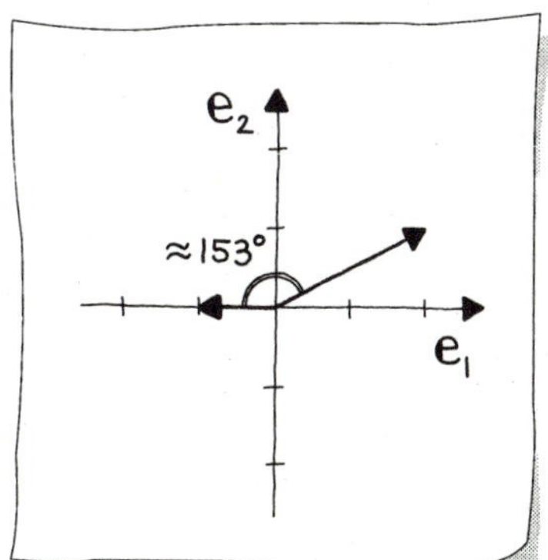

Sketch 29
The angle between two vectors.

Calculate the length of each vector,

$$\|\mathbf{v}\| = \sqrt{2^2 + 1^2} = \sqrt{5}$$
$$\|\mathbf{w}\| = \sqrt{-1^2 + 0^2} = 1.$$

The cosine of the angle between the vectors is calculated using (2.20) as

$$\cos(\theta) = \frac{(2 \times -1) + (1 \times 0)}{\sqrt{5} \times 1} = \frac{-2}{\sqrt{5}} \approx -0.8944$$

Then

$$\arccos(-0.8944) \approx 153.4^\circ$$

To convert an angle given in degrees to radians multiply by $\pi/180^\circ$. (Recall that $\pi \approx 3.14159$ radians.) This means that

$$2.677 \text{ radians} \approx 153.4^\circ \times \frac{\pi}{180^\circ}.$$

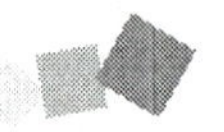

2.8 Inequalities

Here are two important inequalities when dealing with vector lengths.

Let's start with the expression from (2.19), i.e.,

$$\mathbf{v} \cdot \mathbf{w} = \|\mathbf{v}\|\|\mathbf{w}\| \cos\theta.$$

Squaring both sides gives

$$(\mathbf{v} \cdot \mathbf{w})^2 = \|\mathbf{v}\|^2\|\mathbf{w}\|^2 \cos^2\theta.$$

Noticing that $0 \leq \cos^2\theta \leq 1$, we conclude that

$$(\mathbf{v} \cdot \mathbf{w})^2 \leq \|\mathbf{v}\|^2\|\mathbf{w}\|^2. \tag{2.21}$$

This is called the *Cauchy-Schwartz inequality*. This inequality is fundamental in the study of more general vector spaces than those presented here. See a text such as [1] for a more general and thorough derivation of this inequality.

Suppose we would like to find an inequality which describes the relationship between the length of two vectors $\mathbf{v}$ and $\mathbf{w}$ and the length of their sum $\mathbf{v} + \mathbf{w}$. In other words, how does the length of the third side of a triangle relate to the lengths of the other two? Let's begin with expanding $\|\mathbf{v} + \mathbf{w}\|^2$:

$$\begin{aligned}
\|\mathbf{v} + \mathbf{w}\|^2 &= (\mathbf{v} + \mathbf{w}) \cdot (\mathbf{v} + \mathbf{w}) \\
&= \mathbf{v} \cdot \mathbf{v} + 2\mathbf{v} \cdot \mathbf{w} + \mathbf{w} \cdot \mathbf{w} \\
&\leq \mathbf{v} \cdot \mathbf{v} + 2|\mathbf{v} \cdot \mathbf{w}| + \mathbf{w} \cdot \mathbf{w} \\
&\leq \mathbf{v} \cdot \mathbf{v} + 2\|\mathbf{v}\|\|\mathbf{w}\| + \mathbf{w} \cdot \mathbf{w} \\
&= \|\mathbf{v}\|^2 + 2\|\mathbf{v}\|\|\mathbf{w}\| + \|\mathbf{w}\|^2 \\
&= (\|\mathbf{v}\| + \|\mathbf{w}\|)^2.
\end{aligned} \tag{2.22}$$

Taking square roots gives

$$\|\mathbf{v}+\mathbf{w}\| \leq \|\mathbf{v}\| + \|\mathbf{w}\|,$$

which is known as the *triangle inequality*. It states the intuitively obvious fact that the sum of any two edge lengths in a triangle is never smaller than the length of the third edge; see Sketch 30 for an illustration.

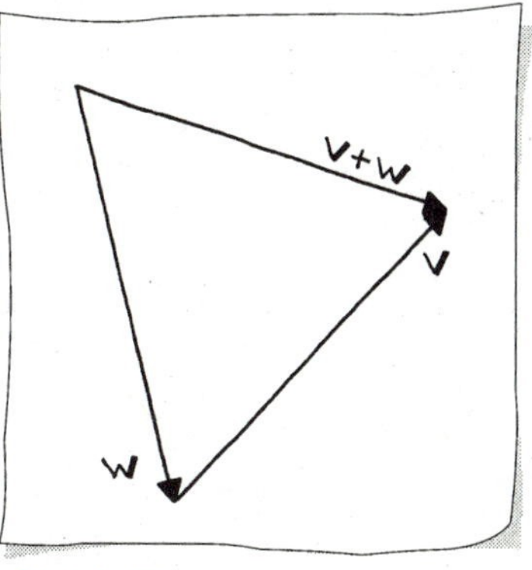

Sketch 30
The triangle inequality.

2.9 Exercises

1. Illustrate the parallelogram rule applied to the vectors
$$\mathbf{v} = \begin{bmatrix} -2 \\ 1 \end{bmatrix} \quad \text{and} \quad \mathbf{w} = \begin{bmatrix} 2 \\ 1 \end{bmatrix}.$$

2. Define your own $\mathbf{p}, \mathbf{q} \in \mathbb{E}^2$ and $\mathbf{v}, \mathbf{w} \in \mathbb{R}^2$. Determine which of the following expressions are geometrically meaningful. Illustrate those that are.

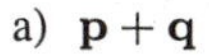

a) $\mathbf{p}+\mathbf{q}$ b) $\frac{1}{2}\mathbf{p}+\frac{1}{2}\mathbf{q}$
c) $\mathbf{p}+\mathbf{v}$ d) $3\mathbf{p}+\mathbf{v}$
e) $\mathbf{v}+\mathbf{w}$ f) $2\mathbf{v}+\frac{1}{2}\mathbf{w}$
g) $\mathbf{v}-2\mathbf{w}$ h) $\frac{3}{2}\mathbf{p}-\frac{1}{2}\mathbf{q}$

3. Illustrate a point with barycentric coordinates $(1/2, 1/4, 1/4)$ with respect to three other points.

4. If all convex combinations of three points were formed, describe the shape of the area in which this fourth point must lie.

5. What is the length of the vector $\mathbf{v} = \begin{bmatrix} -4 \\ -3 \end{bmatrix}$?

6. Find the distance between the points
$$\mathbf{p} = \begin{bmatrix} 3 \\ 3 \end{bmatrix} \quad \text{and} \quad \mathbf{q} = \begin{bmatrix} -2 \\ -3 \end{bmatrix}.$$

7. Normalize the vector $\mathbf{v} = \begin{bmatrix} -4 \\ -3 \end{bmatrix}$.

8. Show that the following properties of the dot product hold for $\mathbf{u}, \mathbf{v}, \mathbf{w} \in \mathbb{R}^2$.
$$\mathbf{v} \cdot (s\mathbf{w}) = s(\mathbf{v} \cdot \mathbf{w}) \qquad \text{homogeneous}$$
$$(\mathbf{v}+\mathbf{w}) \cdot \mathbf{u} = \mathbf{v} \cdot \mathbf{u} + \mathbf{w} \cdot \mathbf{u} \qquad \text{distributive}$$

9. Compute the angle (in degrees) formed by the vectors
$$\mathbf{v} = \begin{bmatrix} 5 \\ 5 \end{bmatrix} \quad \text{and} \quad \mathbf{w} = \begin{bmatrix} 3 \\ -3 \end{bmatrix}.$$

Lining Up — 2D Lines 3

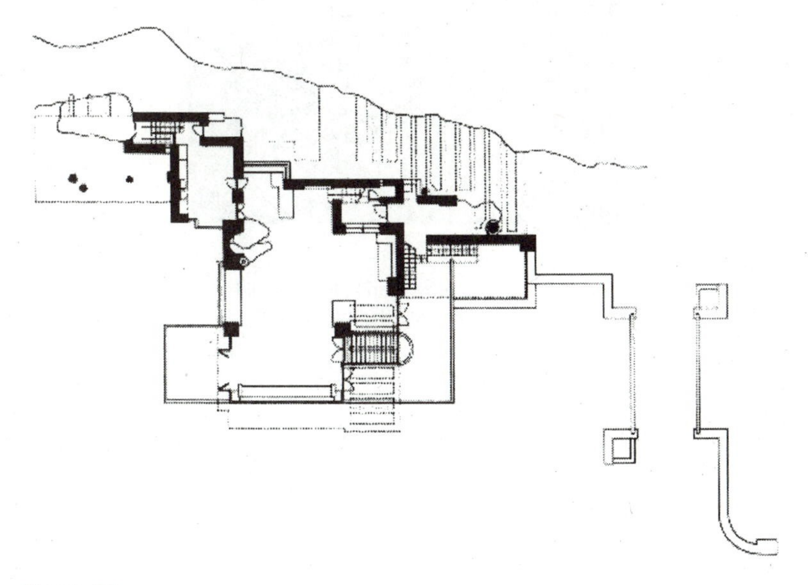

Figure 3.1.
Fallingwater: the floorplan of this building by Frank Lloyd Wright.
(Source: http://www.archinform.de/projekte/974.htm.)

"Real" objects are three-dimensional, or 3D. So why should we consider 2D objects, such as the 2D lines in this chapter? Consider Figure 3.1. It shows North America's most celebrated residential buildling: *Fallingwater*, designed by Frank Lloyd Wright. Clearly the building, illustrated in Figure 3.2, is 3D; yet in order to describe it, we need 2D floor plans outlining the building's structure. These floorplans consist almost entirely of 2D lines.

Figure 3.2.
Fallingwater: a picture of this building by Frank Lloyd Wright.
(Source: http://www.archinform.de/projekte/974.htm.)

3.1 Defining a Line

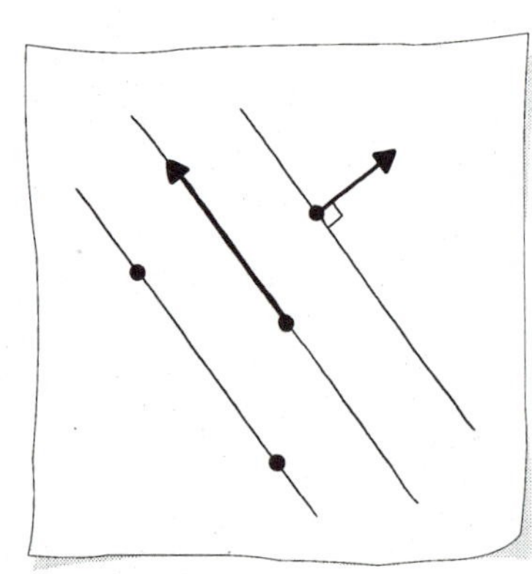

Sketch 31
Elements to define a line.

As illustrated in Sketch 31, two elements of 2D geometry define a line:

- two points,
- a point and a vector parallel to the line, or
- a point and a vector perpendicular to the line.

The unit vector that is perpendicular to a line is referred to as the *normal* to the line. Figure 3.3 shows two *families of lines*: one family of lines shares a common point and the other family of lines shares the same normal. Just as there are different ways to specify a line geometrically, there are different mathematical representations: parametric, implicit and explicit. Each representation will be examined and the advantages of each will be explained. Additionally, we will explore how to convert from one form to another.

Figure 3.3.
Families of lines: one family shares a common point and the other shares a common normal.

3.2 Parametric Equation of a Line

The *parametric equation of a line* $\mathbf{l}(t)$ has the form

$$\mathbf{l}(t) = \mathbf{p} + t\mathbf{v}, \tag{3.1}$$

where $\mathbf{p} \in \mathbb{E}^2$ and $\mathbf{v} \in \mathbb{R}^2$. The scalar value t is the *parameter*. (See Sketch 32. Evaluating (3.1) for a specific parameter $t = \hat{t}$, generates a point on the line.

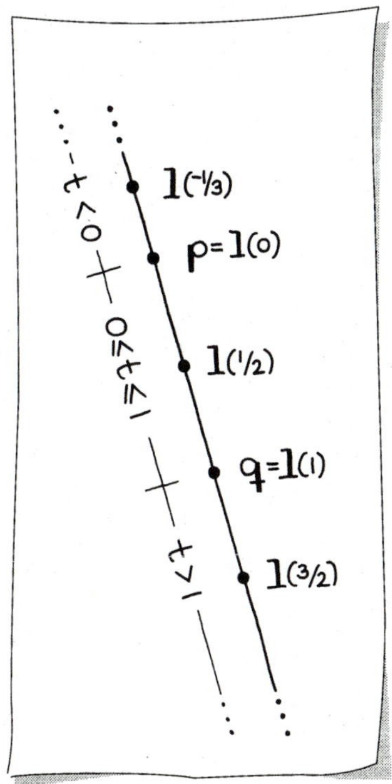

Sketch 32
Parametric form of a line.

We encountered (3.1) in Section 2.4 in the context of barycentric coordinates. Interpreting $\mathbf{v}$ as a difference of points, $\mathbf{v} = \mathbf{p} - \mathbf{q}$, this equation was reformulated as

$$\mathbf{l}(t) = (1-t)\mathbf{p} + t\mathbf{q}. \tag{3.2}$$

A parametric line can be written either in the form of (3.1) or (3.2). The latter is typically referred to as *linear interpolation.*

One way to interpret the parameter t is as *time*; at time $t = 0$ we will be at point $\mathbf{p}$ and at time $t = 1$ we will be at point $\mathbf{q}$. Sketch 32 illustrates that as t varies between zero and one, $t \in [0, 1]$, points are generated on the line between $\mathbf{p}$ and $\mathbf{q}$. Recall from Section 2.4 that these values of t constitute a *convex combination.* If the parameter is a negative number, that is $t < 0$, the direction of $\mathbf{v}$ reverses, generating points on the line "behind" $\mathbf{p}$. The case $t > 1$ is similar: this scales $\mathbf{v}$ so that it is elongated, which generates points "past" $\mathbf{q}$. In the context of linear interpolation, when $t < 0$ or $t > 1$, it is called *extrapolation.*

The parametric form is very handy for computing points on a line. For example, to compute ten equally spaced points on the line segment between $\mathbf{p}$ and $\mathbf{q}$, simply define ten values of $t \in [0, 1]$ as

$$t = i/9, \qquad i = 0 \ldots 9.$$

(Be sure this is a floating point calculation when programming!) Equally spaced parameter values correspond to equally spaced spaced points. See the Exercises for a programming problem.

Example 3.1 Compute five points on the line defined by the points

$$\mathbf{p} = \begin{bmatrix} 1 \\ 2 \end{bmatrix} \quad \text{and} \quad \mathbf{q} = \begin{bmatrix} 6 \\ 4 \end{bmatrix}.$$

Define $\mathbf{v} = \mathbf{q} - \mathbf{p}$, then the line is defined as

$$\mathbf{l}(t) = \begin{bmatrix} 1 \\ 2 \end{bmatrix} + t \begin{bmatrix} 5 \\ 2 \end{bmatrix}$$

Generate five t-values as

$$t = i/4, \qquad i = 0 \ldots 4.$$

Plug each t-value into the formulation for $\mathbf{l}(t)$:

$$i = 0, \quad t = 0, \qquad \mathbf{l}(0) = \begin{bmatrix} 1 \\ 2 \end{bmatrix};$$

$$i = 1, \quad t = 1/4, \quad \mathbf{l}(1/4) = \begin{bmatrix} 9/4 \\ 5/2 \end{bmatrix};$$

$$i = 2, \quad t = 2/4, \quad \mathbf{l}(2/4) = \begin{bmatrix} 7/2 \\ 3 \end{bmatrix};$$

$$i = 3, \quad t = 3/4, \quad \mathbf{l}(3/4) = \begin{bmatrix} 19/4 \\ 7/2 \end{bmatrix};$$

$$i = 4, \quad t = 1, \qquad \mathbf{l}(1) = \begin{bmatrix} 6 \\ 4 \end{bmatrix}.$$

Plot these values for yourself to verify them.

As you can see, the position of the point $\mathbf{p}$ and the direction and length of the vector $\mathbf{v}$ determine which points on the line are generated as we increment through $t \in [0, 1]$. This particular artifact of the parametric equation of a line is called the *parametrization*. The parametrization is related to the speed at which a point traverses the line. We may affect this speed by scaling $\mathbf{v}$: the larger the scale factor, the faster the point's motion!

3.3 Implicit Equation of a Line

Another way to represent the same line is to use the *implicit equation of a line*. For this representation, we start with a point $\mathbf{p}$, and as illustrated in Sketch 33, construct a vector $\mathbf{a}$ that is perpendicular to the line.

For any point $\mathbf{x}$ on the line, it holds that

$$\mathbf{a} \cdot (\mathbf{x} - \mathbf{p}) = 0. \tag{3.3}$$

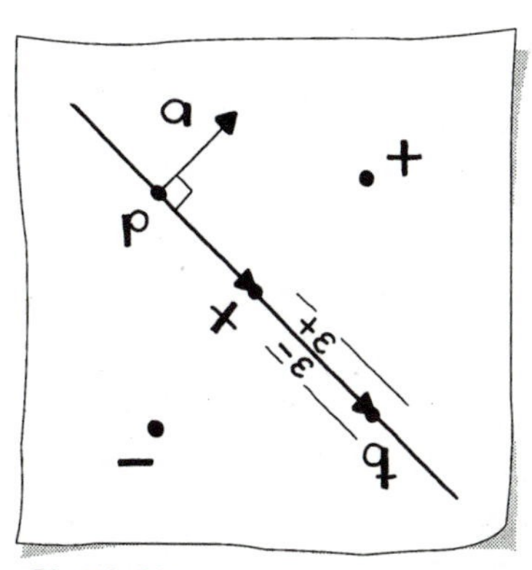

Sketch 33
Implicit form of a line.

This says that $\mathbf{a}$ and the vector $(\mathbf{x} - \mathbf{p})$ are perpendicular. If $\mathbf{a}$ has unit length, it is called the *normal* to the line, and then (3.3) is the *point normal form* of a line. Expanding this equation, we get

$$a_1 x_1 + a_2 x_2 + (-a_1 p_1 - a_2 p_2) = 0.$$

Commonly this is written as

$$a x_1 + b x_2 + c = 0, \tag{3.4}$$

where

$$a = a_1, \tag{3.5}$$

$$b = a_2, \tag{3.6}$$

$$c = -a_1 p_1 - a_2 p_2. \tag{3.7}$$

Equation (3.4) is called the *implicit equation of the line.*

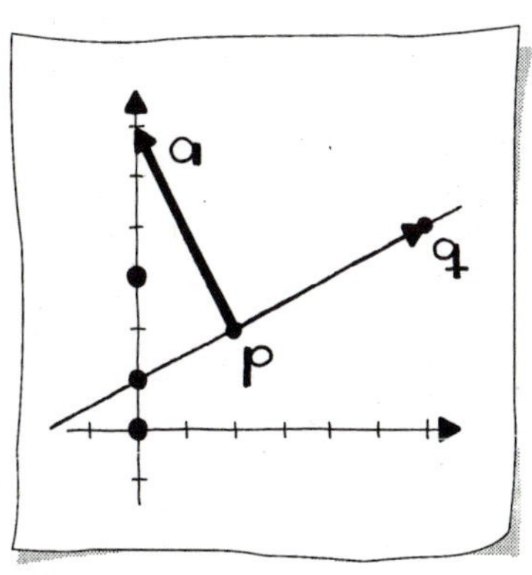

Sketch 34
Implicit construction.

Example 3.2 Following Sketch 34, suppose we know two points,

$$\mathbf{p} = \begin{bmatrix} 2 \\ 2 \end{bmatrix} \quad \text{and} \quad \mathbf{q} = \begin{bmatrix} 6 \\ 4 \end{bmatrix},$$

on the line. To construct the coefficients a,b, and c in (3.4), first form the vector

$$\mathbf{v} = \mathbf{q} - \mathbf{p} = \begin{bmatrix} 4 \\ 2 \end{bmatrix}.$$

Now construct a vector $\mathbf{a}$ that is perpendicular to $\mathbf{v}$:

$$\mathbf{a} = \begin{bmatrix} -v_2 \\ v_1 \end{bmatrix} = \begin{bmatrix} -2 \\ 4 \end{bmatrix}. \tag{3.8}$$

Note, equally as well, we could have chosen $\mathbf{a}$ to be $\begin{bmatrix} 2 \\ -4 \end{bmatrix}$. The coefficients a and b in (3.5) and (3.6) are now defined as $a = -2$ and $b = 4$. With $\mathbf{p}$ as defined above, solve for c as in (3.7). In this example,

$$c = 2 \times 2 - 4 \times 2 = -4.$$

The implicit equation of the line is complete:

$$-2x_1 + 4x_2 - 4 = 0.$$

The implicit form is very useful for deciding if an arbitrary point lies on the line. To test if a point $\mathbf{x}$ is on the line, just plug its coordinates into (3.4). If the value f of the left-hand side of this equation,

$$f = ax_1 + bx_2 + c,$$

is zero then the point is on the line.

A numerical caveat is needed here. Checking equality with floating point numbers should never be done. Instead a tolerance ϵ around zero must be used. What is a meaningful tolerance in this situation? We'll see in Section 3.7 that

$$d = \frac{f}{\|\mathbf{a}\|} \tag{3.9}$$

reflects the true distance of $\mathbf{x}$ to the line. Now the tolerance has a physical meaning, which makes it much easier to specify. Sketch 33 illustrates the physical relationship of this tolerance to the line.

The sign of d indicates on which side of the line the point lies. This sign is dependent upon the definition of $\mathbf{a}$. (Remember there were two possible orientations.) Positive d corresponds to the point on the side of the line to which $\mathbf{a}$ points.

Example 3.3 Let's continue with our example for the line

$$-2x_1 + 4x_2 - 4 = 0,$$

as illustrated in Sketch 34. We want to test if the point $\mathbf{x} = \begin{bmatrix} 0 \\ 1 \end{bmatrix}$ lies on the line. First, calculate

$$\|\mathbf{a}\| = \sqrt{-2^2 + 4^2} = \sqrt{20}.$$

The distance is

$$d = (-2 \times 0 + 4 \times 1 - 4)/\sqrt{20} = 0/\sqrt{20} = 0,$$

which indicates the point is on the line.

Test the point $\mathbf{x} = \begin{bmatrix} 0 \\ 3 \end{bmatrix}$. For this point,

$$d = (-2 \times 0 + 4 \times 3 - 4)/\sqrt{20} = 8/\sqrt{20} \approx 1.79.$$

Checking Sketch 33, this is a positive number, indicating that it is on the same side of the line as the direction of $\mathbf{a}$. Check for yourself that d does indeed reflect the actual distance of this point to the line.

Test the point $\mathbf{x} = \begin{bmatrix} 0 \\ 0 \end{bmatrix}$. Calculating the distance for this point, we get

$$d = (-2 \times 0 + 4 \times 0 - 4)/\sqrt{20} = -4/\sqrt{20} \approx -0.894.$$

Checking Sketch 33, this is a negative number, indicating it is on the opposite side of the line as the direction of $\mathbf{a}$.

Examining (3.4) you might notice that a *horizontal* line takes the form

$$bx_2 + c = 0.$$

This line intersects the $\mathbf{e}_2$-axis at $-b/c$. A *vertical* line takes the form

$$ax_1 + c = 0.$$

This line intersects the $\mathbf{e}_1$-axis at $-a/c$. Using the implicit form, these lines are in no need of special handling.

3.4 Explicit Equation of a Line

The *explicit equation of a line* is the third possible representation. The explicit form is closely related to the implicit form in (3.4). It expresses x_2 as a function of x_1: rearranging the implicit equation we have

$$x_2 = -\frac{a}{b}x_1 - \frac{c}{b}.$$

A more typical way of writing this is

$$x_2 = \hat{a}x_1 + \hat{b}.$$

where $\hat{a} = a/b$ and $\hat{b} = c/b$.

The coefficients have geometric meaning: $\hat{a}$ is the *slope* of the line and $\hat{b}$ is the $\mathbf{e}_2$-intercept. Sketch 35 illustrates the geometry of the coefficients for the line

$$x_2 = 1/3x_1 + 2.$$

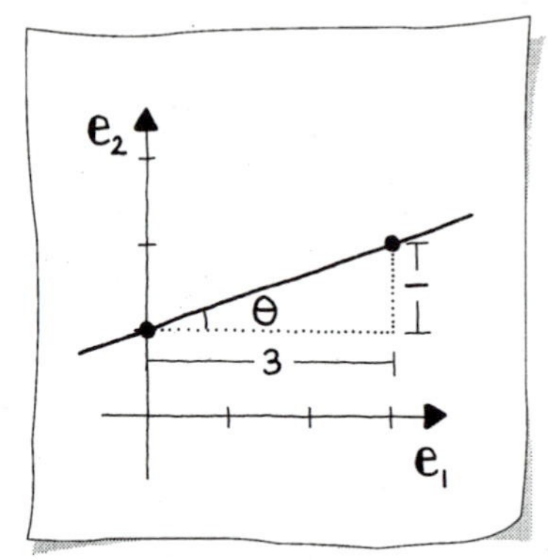

Sketch 35
A line in explicit form.

The slope measures the steepness of the line as a ratio of the change in x_2 to a change in x_1: "rise/run", or more precisely $\tan(\theta)$. The $\mathbf{e}_2$-intercept indicates that the line passes through $(0, \hat{b})$.

Immediately a drawback of the explicit form is apparent. If the "run" is zero then the (vertical) line has infinite slope. This makes life very difficult when programming! When we study transformations (e.g., changing the orientation of some geometry) in Chapter 6, infinite slope actually arises often.

The primary popularity of the explicit form comes from the study of calculus. Additionally, in computer graphics, this form is popular when *pixel* calculation is necessary. Examples are Bresenham's line drawing algorithm and scan line polygon fill algorithms (see [10] or [7]).

3.5 Converting between Parametric and Implicit Equations

As we have discussed, there are advantages to both the parametric and implicit representations of a line. Depending on the geometric algorithm, it may be convenient to use one form rather than the other. We'll ignore the explicit form, since we said it isn't very useful for general 2D geometry.

3.5.1 Parametric to Implicit.

Given: the line $\mathbf{l}$ in parametric form,

$$\mathbf{l} : \mathbf{l}(t) = \mathbf{p} + t\mathbf{v}.$$

Find: the coefficients a, b, c that define the implicit equation of the line

$$\mathbf{l} : ax_1 + bx_2 + c = 0,$$

Solution: first form a vector $\mathbf{a}$ that is perpendicular to the vector $\mathbf{v}$. Choose

$$\mathbf{a} = \begin{bmatrix} -v_2 \\ v_1 \end{bmatrix}.$$

This determines the coefficients a and b, as in (3.5) and (3.6), respectively. Simply let $a = a_1$ and $b = a_2$. Finally, solve for the coefficient c as in (3.7). Taking $\mathbf{p}$ from $\mathbf{l}(t)$ and $\mathbf{a}$, form

$$c = -(a_1 p_1 + a_2 p_2).$$

We stepped through a numerical example of this in the derivation of the implicit form in Section 3.3, and it is illustrated in Sketch 34. In this example, $\mathbf{l}(t)$ is given as

$$\mathbf{l}(t) = \begin{bmatrix} 2 \\ 2 \end{bmatrix} + t \begin{bmatrix} 4 \\ 2 \end{bmatrix}.$$

3.5.2 Implicit to Parametric.

Given: the line $\mathbf{l}$ in implicit form,

$$\mathbf{l} : ax_1 + bx_2 + c = 0.$$

Find: the line $\mathbf{l}$ in parametric form,

$$\mathbf{l} : \mathbf{l}(t) = \mathbf{p} + t\mathbf{v}.$$

Solution: recognize that we need one point on the line and a vector parallel to the line. The vector is easy: simply form a vector perpendicular to $\mathbf{a}$ of the implicit line. For example, we could set

$$\mathbf{v} = \begin{bmatrix} b \\ -a \end{bmatrix}.$$

Next, find a point on the line. Two candidate points are the intersections with the $\mathbf{e}_1$- or $\mathbf{e}_2$-axis,

$$\begin{bmatrix} -c/a \\ 0 \end{bmatrix} \quad \text{or} \quad \begin{bmatrix} 0 \\ -c/b \end{bmatrix},$$

respectively. For numerical stability, let's choose the intersection closest to the origin. Thus we choose the former if $|a| > |b|$, and the latter otherwise.

Example 3.4 Revisit the numerical example from the implicit form derivation in Section 3.3; it is illustrated in Sketch 34. The implicit equation of the line is

$$-2x_1 + 4x_2 - 4 = 0.$$

We want to find a parametric equation of this line,

$$\mathbf{l} : \mathbf{l}(t) = \mathbf{p} + t\mathbf{v}.$$

First form

$$\mathbf{v} = \begin{bmatrix} 4 \\ 2 \end{bmatrix}.$$

Now determine which is greater in absolute value, a or b. Since $|-2| < |4|$, we choose

$$\mathbf{p} = \begin{bmatrix} 0 \\ 4/4 \end{bmatrix} = \begin{bmatrix} 0 \\ 1 \end{bmatrix}.$$

The parametric equation is

$$\mathbf{l}(t) = \begin{bmatrix} 0 \\ 1 \end{bmatrix} + t \begin{bmatrix} 4 \\ 2 \end{bmatrix}.$$

The implicit and parametric forms both allow an infinite number of representations for the same line. In fact, in the example we just finished, the loop

$$\text{parametric} \rightarrow \text{implicit} \rightarrow \text{parametric}$$

produced two different parametric forms. We started with

$$\mathbf{l}(t) = \begin{bmatrix} 2 \\ 2 \end{bmatrix} + t \begin{bmatrix} 4 \\ 2 \end{bmatrix},$$

and ended with

$$\mathbf{l}(t) = \begin{bmatrix} 0 \\ 1 \end{bmatrix} + t \begin{bmatrix} 4 \\ 2 \end{bmatrix}.$$

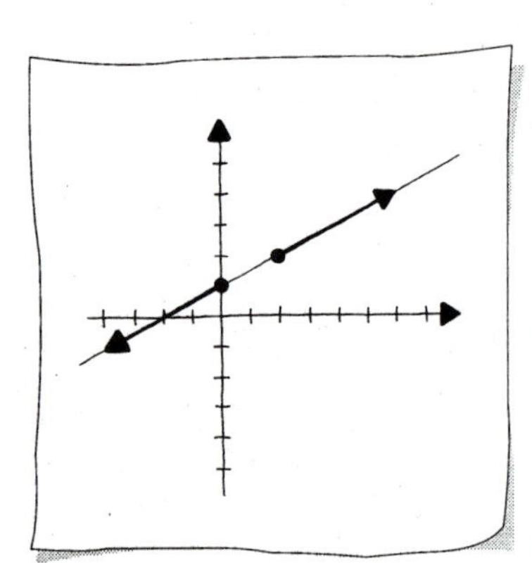

Sketch 36
Two parametric representations for the same line.

We could have just as easily generated the line

$$\mathbf{l}(t) = \begin{bmatrix} 0 \\ 1 \end{bmatrix} + t \begin{bmatrix} -4 \\ -2 \end{bmatrix},$$

if $\mathbf{v}$ was formed with the rule

$$\mathbf{v} = \begin{bmatrix} -b \\ a \end{bmatrix}.$$

Sketch 36 illustrates the first and third parametric representations of this line. These three parametric forms represent the same line! However, the manner in which the lines will be *traced* will differ. This is referred to as the *parametrization* of the line. We already encountered this concept in Section 3.2.

3.6 Distance of a Point to a Line

If you are given a point $\mathbf{r}$ and a line $\mathbf{l}$, how far is that point from the line? This problem arises frequently. For example, as in Figure 3.4, a line has been fit to point data. In order to measure how well this line *approximates* the data, it is necessary to check the distance of each point to the line. It should be intuitively clear that the distance $d(\mathbf{r}, \mathbf{l})$ of a point to a line is the *perpendicular distance*.

3.6.1 Starting with an Implicit Line. Suppose our problem is formulated as follows:
Given: a line $\mathbf{l}$ in implicit form, defined by the coefficients a, b, and c, and a point $\mathbf{r}$.
Find: $d(\mathbf{r}, \mathbf{l})$, or d for brevity.
Solution:

$$d = \frac{ax_1 + bx_2 + c}{\|\mathbf{a}\|},$$

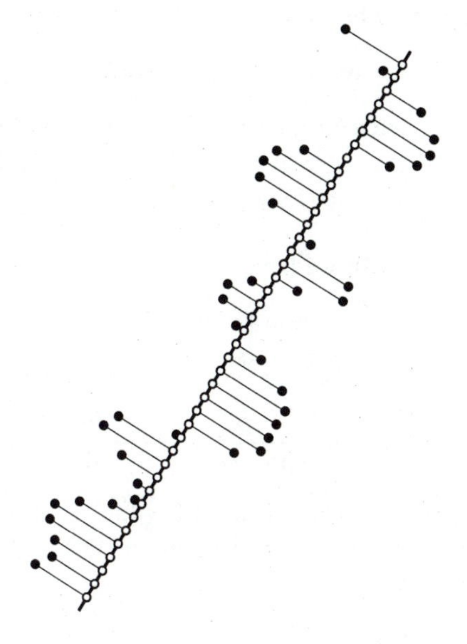

Figure 3.4.
Distance point to line: measuring the distance of each point to the line.

or in vector notation

$$d = \frac{\mathbf{a} \cdot (\mathbf{x} - \mathbf{p})}{\|\mathbf{a}\|}.$$

Let's investigate why this is so. Recall that the implicit equation of a line was derived through use of the dot product

$$\mathbf{a} \cdot (\mathbf{x} - \mathbf{p}) = 0,$$

as in (3.3); a line is given by a point $\mathbf{p}$ and a vector $\mathbf{a}$ normal to the line. Any point $\mathbf{x}$ on the line will satisfy this equality.

As in Sketch 37, we now consider a point $\mathbf{r}$ that is clearly not on the line. As a result, the equality will not be satisfied; however, let's assign a value v to the left-hand side:

$$v = \mathbf{a} \cdot (\mathbf{r} - \mathbf{p}).$$

To simplify, define $\mathbf{w} = \mathbf{r} - \mathbf{p}$, as in Sketch 37. Recall the definition of the dot product in (2.19) as

$$\mathbf{v} \cdot \mathbf{w} = \|\mathbf{v}\|\|\mathbf{w}\| \cos\theta.$$

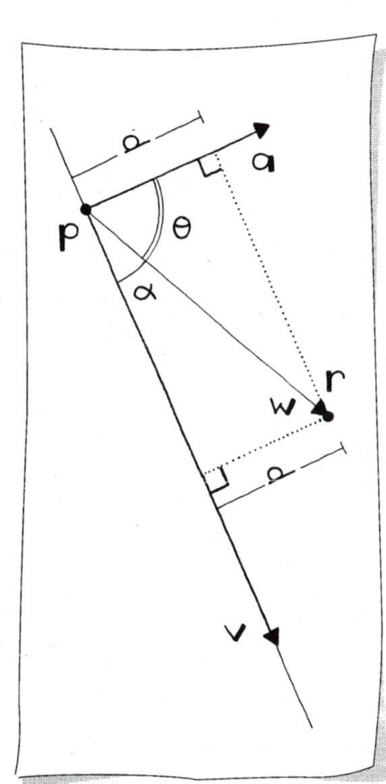

Sketch 37
Distance point to line.

Thus the expression for v becomes

$$v = \mathbf{a} \cdot \mathbf{w} = \|\mathbf{a}\| \|\mathbf{w}\| \cos(\theta). \tag{3.10}$$

The right triangle in Sketch 37 allows for an expression for $\cos(\theta)$ as

$$\cos(\theta) = \frac{d}{\|\mathbf{w}\|}.$$

Substituting this into (3.10), we have

$$v = \|\mathbf{a}\| d.$$

This indicates that the actual distance of $\mathbf{r}$ to the line is

$$d = \frac{v}{\|\mathbf{a}\|} = \frac{\mathbf{a} \cdot (\mathbf{r} - \mathbf{p})}{\|\mathbf{a}\|} = \frac{ax_1 + bx_2 + c}{\|\mathbf{a}\|}. \tag{3.11}$$

If many points will be checked against a line, it is advantageous to store the line in *point normal form.* This means that $\|\mathbf{a}\| = 1$. This will eliminate the division in (3.11).

Example 3.5 Start with the line

$$\mathbf{l} : 4x_1 + 2x_2 - 8 = 0$$

and the point

$$\mathbf{r} = \begin{bmatrix} 5 \\ 3 \end{bmatrix}.$$

Find the distance from $\mathbf{r}$ to the line. (Draw your own sketch for this example, similar to Sketch 37

First, calculate

$$\|\mathbf{a}\| = \sqrt{4^2 + 2^2} = 2\sqrt{5}.$$

Then the distance is

$$d(\mathbf{r}, \mathbf{l}) = \frac{4 \times 5 + 2 \times 3 - 8}{2\sqrt{5}} = \frac{9}{\sqrt{5}} \approx 4.02$$

As another exercise, let's rewrite the line in point normal form with coefficients

$$\hat{a} = \frac{4}{2\sqrt{5}} = \frac{2}{\sqrt{5}},$$

$$\hat{b} = \frac{2}{2\sqrt{5}} = \frac{1}{\sqrt{5}},$$

$$\hat{c} = \frac{c}{\|\mathbf{a}\|} = \frac{-8}{2\sqrt{5}} = -\frac{4}{\sqrt{5}},$$

thus making the point normal form of the line

$$\frac{2}{\sqrt{5}}x_1 + \frac{1}{\sqrt{5}}x_2 - \frac{4}{\sqrt{5}} = 0.$$

3.6.2 Starting with a Parametric Line. Alternatively, suppose our problem is formulated as follows:
Given: a line $\mathbf{l}$ in parametric form, defined by a point $\mathbf{p}$ and a vector $\mathbf{v}$, and a point $\mathbf{r}$.
Find: $d(\mathbf{r}, \mathbf{l})$, or d for brevity. Again, this is illustrated in Sketch 37.
Solution: Form the vector $\mathbf{w} = \mathbf{r} - \mathbf{p}$. Use the relationship

$$d = \|\mathbf{w}\| \sin(\alpha).$$

Later, in Section 10.2, we will see how to express $\sin(\alpha)$ directly in terms of $\mathbf{v}$ and $\mathbf{w}$; for now, we express it in terms of the cosine:

$$\sin(\alpha) = \sqrt{1 - \cos(\alpha)^2},$$

and as before

$$\cos(\alpha) = \frac{\mathbf{v} \cdot \mathbf{w}}{\|\mathbf{v}\|\|\mathbf{w}\|}.$$

Thus we have defined the distance d.

Example 3.6 We'll use the same line as in the previous example, but now it will be given in parametric form as

$$\mathbf{l}(t) = \begin{bmatrix} 0 \\ 4 \end{bmatrix} + t \begin{bmatrix} 2 \\ -4 \end{bmatrix}.$$

We'll also use the same point

$$\mathbf{r} = \begin{bmatrix} 5 \\ 3 \end{bmatrix}.$$

Add any new vectors for this example to the sketch you drew for the previous example.

First create the vector

$$\mathbf{w} = \begin{bmatrix} 5 \\ 3 \end{bmatrix} - \begin{bmatrix} 0 \\ 4 \end{bmatrix} = \begin{bmatrix} 5 \\ -1 \end{bmatrix}.$$

Next calculate $\|\mathbf{w}\| = \sqrt{26}$ and $\|\mathbf{v}\| = \sqrt{20}$. Compute

$$\cos(\alpha) = \frac{\begin{bmatrix} 2 \\ -4 \end{bmatrix} \cdot \begin{bmatrix} 5 \\ -1 \end{bmatrix}}{\sqrt{26}\sqrt{20}} \approx 0.614$$

Thus, the distance to the line becomes

$$d(\mathbf{r}, \mathbf{l}) \approx \sqrt{26}\sqrt{1 - (0.614)^2} \approx 4.02,$$

which rightly produces the same result as the previous example.

3.7 The Foot of a Point

Section 3.6 detailed how to calculate the distance of a point from a line. A new question arises: which point on the line is closest to the point? This point will be called the *foot* of the given point.

If you are given a line in implicit form, it is best to convert it to parametric form for this problem. This illustrates how the implicit form is handy for *testing* if a point is on the line, however it is not as handy for *finding* points on the line. The problem at hand is thus:

Given: a line $\mathbf{l}$ in parametric form, defined by a point $\mathbf{p}$ and a vector $\mathbf{v}$, and another point $\mathbf{r}$.
Find: the point $\mathbf{q}$ on the line which is closest to $\mathbf{r}$. See Sketch 38.
Solution: The point $\mathbf{q}$ can be defined as

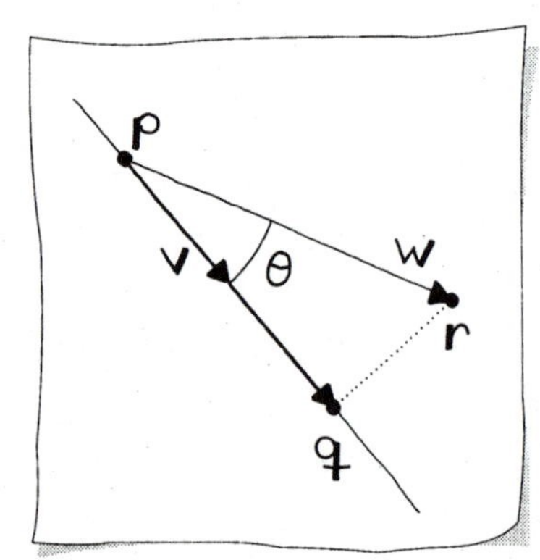

Sketch 38
Closest point $\mathbf{q}$ on line to point $\mathbf{r}$.

$$\mathbf{q} = \mathbf{p} + t\mathbf{v}, \tag{3.12}$$

so our problem is solved once we have found the scalar factor t. From Sketch 38, we see that

$$\cos(\theta) = \frac{\|t\mathbf{v}\|}{\|\mathbf{w}\|}.$$

Using

$$\cos(\theta) = \frac{\mathbf{v} \cdot \mathbf{w}}{\|\mathbf{v}\|\|\mathbf{w}\|},$$

we find

$$t = \frac{\mathbf{v} \cdot \mathbf{w}}{\|\mathbf{v}\|^2}.$$

Example 3.7

Given: the parametric line $\mathbf{l}$ defined as

$$\mathbf{l}(t) = \begin{bmatrix} 0 \\ 1 \end{bmatrix} + t \begin{bmatrix} 0 \\ 2 \end{bmatrix},$$

and point

$$\mathbf{r} = \begin{bmatrix} 3 \\ 4 \end{bmatrix}.$$

Find the point $\mathbf{q}$ on $\mathbf{l}$ that is closest to $\mathbf{r}$. This example is easy enough to find the answer by simply drawing a sketch, but let's go through the steps.

Define the vector

$$\mathbf{w} = \mathbf{r} - \mathbf{p} = \begin{bmatrix} 3 \\ 4 \end{bmatrix} - \begin{bmatrix} 0 \\ 1 \end{bmatrix} = \begin{bmatrix} 3 \\ 3 \end{bmatrix}.$$

Compute $\mathbf{v} \cdot \mathbf{w} = 6$ and $\|\mathbf{v}\| = 2$. Thus $t = 3/2$ and

$$\mathbf{q} = \begin{bmatrix} 0 \\ 4 \end{bmatrix}.$$

Try this example with $\mathbf{r} = \begin{bmatrix} 2 \\ -1 \end{bmatrix}$.

3.8 A Meeting Place — Computing Intersections

Finding a point in common between two lines is done many times over in a CAD or graphics package. Take for example Figure 3.5: the top part of the figure shows a great number of intersection lines. In order to color some of the areas, as in the bottom part of the figure, it is necessary to know the intersection points. Intersection problems arise in many other applications, and the first question to ask is what type of information do you want?

- Do you want to know merely whether the lines intersect?
- Do you want to know the point at which they intersect?
- Do you want a parameter value on one or both lines for the intersection point?

The particular question(s) you want to answer along with the line representation(s) will determine the best method for solving the intersection problem.

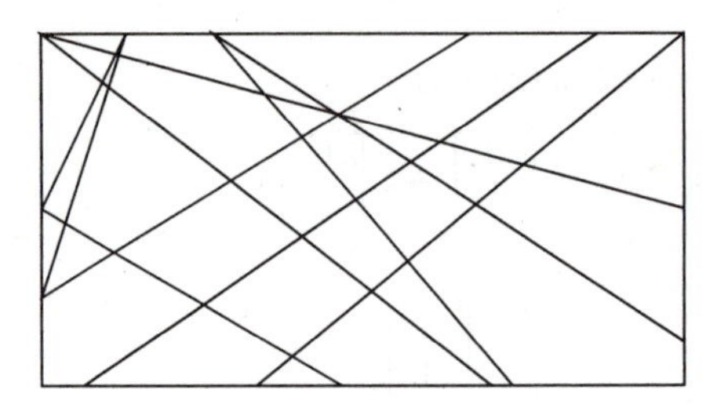

Figure 3.5.
Intersecting lines: the top figure may be drawn without knowing where the shown lines intersect. By finding line/line intersections (bottom), it is possible to color areas — creating an artistic image!

3.8.1 Parametric and Implicit. We then want to solve the following:
Given: two lines $\mathbf{l}_1$ and $\mathbf{l}_2$:

$$\begin{aligned} \mathbf{l}_1 : &\quad \mathbf{l}_1(t) = \mathbf{p} + t\mathbf{v} \\ \mathbf{l}_2 : &\quad ax_1 + bx_2 + c = 0 \end{aligned}$$

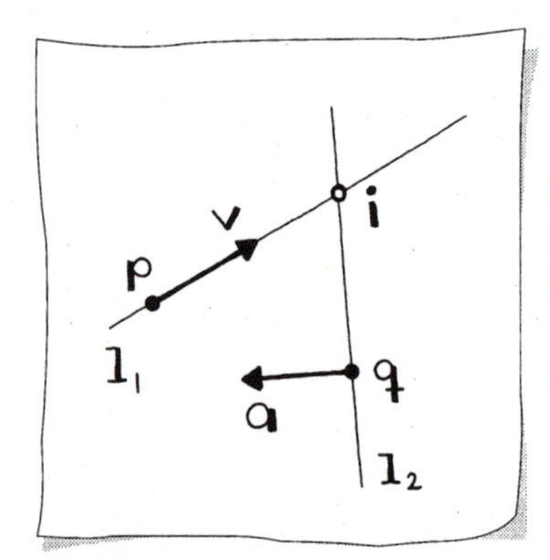

Sketch 39
Parametric and implicit line intersection.

Find: the intersection point $\mathbf{i}$. See Sketch 39 for an illustration.[1]
Solution: We will approach the problem by finding the specific parameter $\hat{t}$ with respect to $\mathbf{l}_1$ of the intersection point.

This intersection point, when inserted into the equation of $\mathbf{l}_2$, will cause the left-hand side to evaluate to zero:

$$a[p_1 + \hat{t}v_1] + b[p_2 + \hat{t}v_2] + c = 0.$$

This is one equation and one unknown! Just solve for $\hat{t}$,

$$\hat{t} = \frac{-c - ap_1 - bp_2}{av_1 + bv_2}, \tag{3.13}$$

then $\mathbf{i} = \mathbf{l}(\hat{t})$.

[1] In Section 3.3 we studied the conversion from the geometric elements of a point $\mathbf{q}$ and perpendicular vector $\mathbf{a}$ to the implicit line coefficients.

But wait — we must check if the denominator of (3.13) is zero before carrying out this calculation. Besides causing havoc numerically, what else does a zero denominator infer? The denominator

$$\text{denom} = av_1 + bv_2$$

can be rewritten as

$$\text{denom} = \mathbf{a} \cdot \mathbf{v}.$$

We know from (2.7) that a zero dot product implies that two vectors are perpendicular. Since $\mathbf{a}$ is perpendicular to the line $\mathbf{l}_2$ in implicit form, the lines are *parallel* if

$$\mathbf{a} \cdot \mathbf{v} = 0.$$

Of course, we always check for equality within a tolerance! A physically meaningful tolerance is best. Thus it is better to check the quantity

$$\cos(\theta) = \frac{\mathbf{a} \cdot \mathbf{v}}{\|\mathbf{a}\|\|\mathbf{v}\|}; \tag{3.14}$$

the tolerance will be the cosine of an angle. It usually suffices to use a tolerance between $\cos(0.1°)$ and $\cos(0.5°)$. Angle tolerances are particularly nice to have because they are *dimension independent.*[2]

If the test in (3.14) indicates the lines are parallel, then we might want to determine if the lines are identical. By simply plugging in the coordinates of $\mathbf{p}$ into the equation of $\mathbf{l}_2$, and computing, we get

$$d = \frac{ap_1 + bp_2 + c}{\|\mathbf{a}\|}.$$

If d is equal to zero (within tolerance), then the lines are identical.

Example 3.8

Given: two lines $\mathbf{l}_1$ and $\mathbf{l}_2$:

$$\mathbf{l}_1 : \quad \mathbf{l}_1(t) = \begin{bmatrix} 0 \\ 3 \end{bmatrix} + t \begin{bmatrix} -2 \\ -1 \end{bmatrix}$$
$$\mathbf{l}_2 : \quad 2x_1 + x_2 - 8 = 0$$

Find: the intersection point $\mathbf{i}$. Create your own sketch and try to predict what the answer should be!

Solution: Find the parameter $\hat{t}$ for $\mathbf{l}_1$ as given in (3.13). First check the denominator:

$$\text{denom} = 2 \times (-2) + 1 \times (-1) = -5.$$

[2]Note that we do not need to use the actual angle, just the cosine of the angle.

This is not zero, so we proceed to find

$$\hat{t} = \frac{8 - 2 \times 0 - 1 \times 3}{-5} = -1.$$

Plug this parameter value into $\mathbf{l}_1$ to find the intersection point:

$$\mathbf{l}_1(-1) = \begin{bmatrix} 0 \\ 3 \end{bmatrix} + -1 \begin{bmatrix} -2 \\ -1 \end{bmatrix} = \begin{bmatrix} 2 \\ 4 \end{bmatrix}.$$

3.8.2 Both Parametric. Another method for finding the intersection of two lines arises by using the parametric form for both, illustrated in Sketch 40.

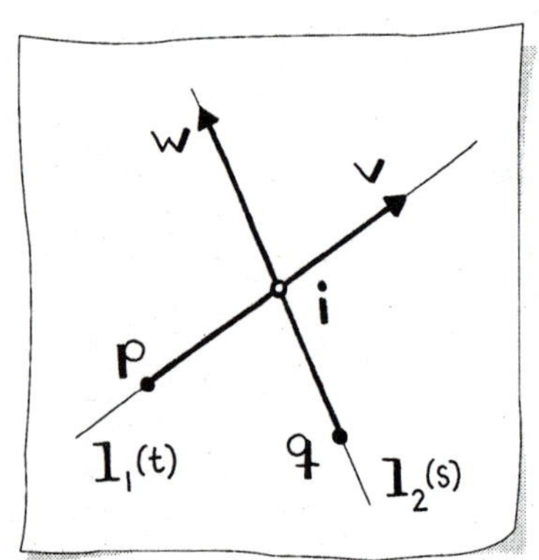

Sketch 40
Intersection of lines in parametric form.

Given: two lines in parametric form:

$$\begin{aligned} \mathbf{l}_1 &: \ \mathbf{l}_1(t) = \mathbf{p} + t\mathbf{v} \\ \mathbf{l}_2 &: \ \mathbf{l}_2(s) = \mathbf{q} + s\mathbf{w}. \end{aligned}$$

Note that we use two different parameters, t and s, here. This is because the lines are totally independent of each other.

Find: the intersection point $\mathbf{i}$.

Solution: we need two parameter values $\hat{t}$ and $\hat{s}$ such that

$$\mathbf{p} + \hat{t}\mathbf{v} = \mathbf{q} + \hat{s}\mathbf{w}.$$

This may be rewritten as

$$\hat{t}\mathbf{v} - \hat{s}\mathbf{w} = \mathbf{q} - \mathbf{p}. \tag{3.15}$$

We have two equations (one for each coordinate) and two unknowns $\hat{t}$ and $\hat{s}$. To solve for the unknowns, we could formulate an expression for $\hat{t}$ using the first equation, and substitute this expression into the second equation. This then generates a solution for $\hat{s}$. Use this solution in the expression for $\hat{t}$, and solve for $\hat{t}$. (Equations like this are systematically treated in Chapter 5.) Once we have $\hat{t}$ and $\hat{s}$, the intersection point is found by inserting one of these values into its respective parametric line equation.

If the vectors $\mathbf{v}$ and $\mathbf{w}$ are linearly dependent, as discussed in Section 2.6, then it will not be possible to find a unique $\hat{t}$ and $\hat{s}$. The lines are parallel and possibly identical.

Example 3.9
Given: two lines $\mathbf{l}_1$ and $\mathbf{l}_2$:

$$\mathbf{l}_1: \quad \mathbf{l}_1(t) = \begin{bmatrix} 0 \\ 3 \end{bmatrix} + t\begin{bmatrix} -2 \\ -1 \end{bmatrix}$$

$$\mathbf{l}_2: \quad \mathbf{l}_2(s) = \begin{bmatrix} 4 \\ 0 \end{bmatrix} + s\begin{bmatrix} -1 \\ 2 \end{bmatrix}$$

Find: the intersection point $\mathbf{i}$. This means that we need to find $\hat{t}$ and $\hat{s}$ such that $\mathbf{l}_1(\hat{t}) = \mathbf{l}_2(\hat{s})$.[3] Again, create your own sketch and try to predict what the answer should be!
Solution: Set up the two equations with two unknowns as in (3.15).

$$\hat{t}\begin{bmatrix} -2 \\ -1 \end{bmatrix} - \hat{s}\begin{bmatrix} -1 \\ 2 \end{bmatrix} = \begin{bmatrix} 4 \\ 0 \end{bmatrix} - \begin{bmatrix} 0 \\ 3 \end{bmatrix}.$$

Use the methods in Chapter 5 to solve these equations, resulting in $\hat{t} = -1$ and $\hat{s} = 2$. Plug these values into the line equations to verify the same intersection point is produced for each:

$$\mathbf{l}_1(-1) = \begin{bmatrix} 0 \\ 3 \end{bmatrix} + (-1)\begin{bmatrix} -2 \\ -1 \end{bmatrix} = \begin{bmatrix} 2 \\ 4 \end{bmatrix}$$

$$\mathbf{l}_2(2) = \begin{bmatrix} 4 \\ 0 \end{bmatrix} + 2\begin{bmatrix} -1 \\ 2 \end{bmatrix} = \begin{bmatrix} 2 \\ 4 \end{bmatrix}.$$

3.8.3 Both Implicit. And yet a third method:
Given: two lines in implicit form:

$$\mathbf{l}_1: \quad ax_1 + bx_2 + c = 0,$$
$$\mathbf{l}_2: \quad \bar{a}x_1 + \bar{b}x_2 + \bar{c} = 0.$$

As illustrated in Sketch 41, each line isgeometrically given in terms of a point and a vector perpendicular to the line.
Find: The intersection point

$$\mathbf{i} = \hat{\mathbf{x}} = \begin{bmatrix} \hat{x}_1 \\ \hat{x}_2 \end{bmatrix}$$

that simultaneously satisfies $\mathbf{l}_1$ and $\mathbf{l}_2$.
Solution: we have two equations

$$a\hat{x}_1 + b\hat{x}_2 = -c, \tag{3.16}$$
$$\bar{a}\hat{x}_1 + \bar{b}\hat{x}_2 = -\bar{c} \tag{3.17}$$

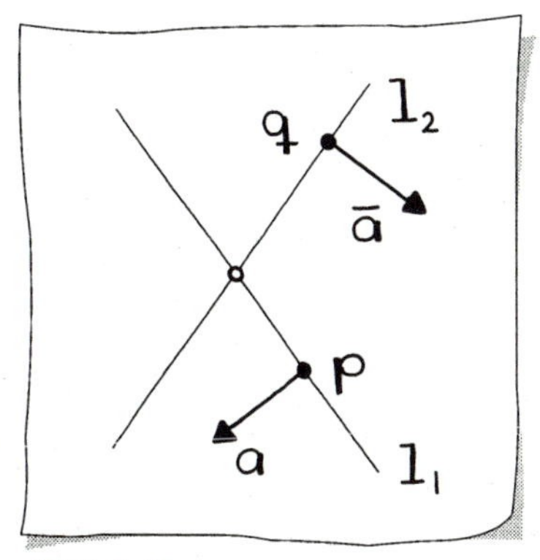

Sketch 41
Intersection of two lines in implicit form.

[3]Really we only need $\hat{t}$ or $\hat{s}$ to find the intersection point.

with two unknowns, $\hat{x}_1$ and $\hat{x}_2$. Equations like this are solved in Chapter 5.

If the lines are parallel then it will not be possible to find $\hat{\mathbf{x}}$. This means that $\mathbf{a}$ and $\bar{\mathbf{a}}$, as in Sketch 41, are linearly dependent.

Example 3.10

Given: two lines $\mathbf{l}_1$ and $\mathbf{l}_2$:

$$\begin{aligned} \mathbf{l}_1: &\quad x_1 - 2x_2 + 6 = 0 \\ \mathbf{l}_2: &\quad 2x_1 + x_2 - 8 = 0 \end{aligned}$$

Find: the intersection point $\hat{\mathbf{x}}$ as above. Create your own sketch and try to predict what the answer should be!

Solution: Reformulate the equations for $\mathbf{l}_1$ and $\mathbf{l}_2$ as in (3.16) and (3.17). Using the techniques in Chapter 5, you will find that

$$\hat{\mathbf{x}} = \begin{bmatrix} 2 \\ 4 \end{bmatrix}.$$

Plug this point into the equations for $\mathbf{l}_1$ and $\mathbf{l}_2$ to verify.

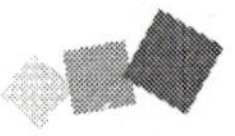

3.9 Exercises

1. Find the *computationally* best way to organize the evaluation of n equally spaced points on a line.
2. Find the equation for a line in implicit form which passes through the points
$$\mathbf{p} = \begin{bmatrix} -2 \\ 0 \end{bmatrix} \quad \text{and} \quad \mathbf{q} = \begin{bmatrix} 0 \\ -1 \end{bmatrix}.$$
3. Test if the following points lie on the line defined in Exercise 2.
$$\begin{bmatrix} 0 \\ 0 \end{bmatrix} \quad \begin{bmatrix} -4 \\ 1 \end{bmatrix} \quad \begin{bmatrix} 5 \\ 1 \end{bmatrix} \quad \begin{bmatrix} -3 \\ -1 \end{bmatrix}.$$
4. For the points in Exercise 3, if a point does not lie on the line, calculate the distance from the line.
5. Redefine the line in Exercise 2 using $-\mathbf{a}$. Recompute the distance of each point in Exercise 4 to the line.
6. Given two lines: $\mathbf{l}_1$ defined by points
$$\begin{bmatrix} 1 \\ 0 \end{bmatrix} \quad \text{and} \quad \begin{bmatrix} 0 \\ 3 \end{bmatrix},$$

and $\mathbf{l}_2$ defined by points

$$\begin{bmatrix} -1 \\ 6 \end{bmatrix} \quad \text{and} \quad \begin{bmatrix} -4 \\ 1 \end{bmatrix},$$

find the intersection point using each of the three methods in Section 3.8.

7. Find the intersection of the lines

$$\mathbf{l}_1: \quad \mathbf{l}_1(t) = \begin{bmatrix} 0 \\ -1 \end{bmatrix} + t \begin{bmatrix} 1 \\ 1 \end{bmatrix} \quad \text{and}$$
$$\mathbf{l}_2: \quad -x_1 + x_2 + 1 = 0.$$

8. Find the intersection of the lines

$$\mathbf{l}_1: \quad -x_1 + x_2 + 1 = 0 \quad \text{and}$$
$$\mathbf{l}_2: \quad \mathbf{l}(t) = \begin{bmatrix} 2 \\ 2 \end{bmatrix} + t \begin{bmatrix} 2 \\ 2 \end{bmatrix}.$$

9. Find the closest point on the line

$$\mathbf{l}(t) = \begin{bmatrix} 2 \\ 2 \end{bmatrix} + t \begin{bmatrix} 2 \\ 2 \end{bmatrix}.$$

to the point

$$\mathbf{r} = \begin{bmatrix} 0 \\ 2 \end{bmatrix}.$$

4 Linear Maps in 2D

Figure 4.1.
Linear maps in 2D: an interesting geometric figure constructed by applying 2D linear maps to a square.

Geometry always has two parts to it: one part is the description of the objects that can be generated; the other investigates how these objects can be changed (or transformed). Any object formed by several vectors may be mapped to an arbitrarily bizarre curved or distorted object — here, we are interested in those maps that map vectors to vectors and are "benign" in some well-defined sense. All these maps may be described using the tools of matrix operations. An interesting pattern is generated from a simple square in Figure 4.1 by such "benign" 2D linear maps.

4.1 Skew Target Boxes

In Section 1.1, we saw how to map an object from a unit square to a rectangular target box. We will now make two changes to that approach. The first is merely a matter of notation: instead of using a unit square in some $[\mathbf{d}_1, \mathbf{d}_2]$-system, we will now use the unit square in the $[\mathbf{e}_1, \mathbf{e}_2]$-system. In addition, the target box will now be allowed to be a *parallelogram*.

Instead of specifying two extreme points as for a rectangle, it is more convenient to describe a parallelogram target box by a point $\mathbf{p}$ and two vectors $\mathbf{a}_1, \mathbf{a}_2$. A point $\mathbf{x}$ is now mapped to a point $\mathbf{x}'$ by

$$\mathbf{x}' = \mathbf{p} + x_1\mathbf{a}_1 + x_2\mathbf{a}_2, \tag{4.1}$$

as illustrated by Sketch 42. This simply states that we duplicate the $[\mathbf{e}_1, \mathbf{e}_2]$–geometry in the $[\mathbf{a}_1, \mathbf{a}_2]$–system: $\mathbf{x}'$ has the same coordinates in the new system as $\mathbf{x}$ did in the old one.

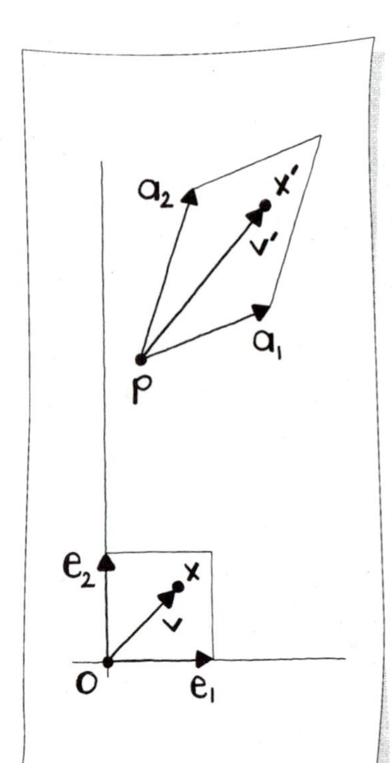

Sketch 42
A skew box.

Similarly, if we define a vector $\mathbf{v} = \mathbf{x} - \mathbf{o}$, i.e.,

$$\mathbf{v} = x_1\mathbf{e}_1 + x_2\mathbf{e}_2, \tag{4.2}$$

then it is mapped to

$$\mathbf{v}' = x_1\mathbf{a}_1 + x_2\mathbf{a}_2. \tag{4.3}$$

Reviewing a definition from Section 2.6, we recall that any combination $c\mathbf{u} + d\mathbf{v}$ of two vectors $\mathbf{u}$ and $\mathbf{v}$ is called a *linear combination*.

If $\mathbf{a}_1$ and $\mathbf{a}_2$ are lined up with the coordinate axes, i.e.,

$$\mathbf{a}_1 = \begin{bmatrix} a_{1,1} \\ 0 \end{bmatrix}, \quad \mathbf{a}_2 = \begin{bmatrix} 0 \\ a_{2,2} \end{bmatrix},$$

then we recover (1.1) and (1.2).

The components of a subscripted vector will be written with a double subscript as

$$\mathbf{a}_1 = \begin{bmatrix} a_{1,1} \\ a_{2,1} \end{bmatrix}.$$

The vector component index precedes the vector subscript. The next section will clarify the reason for this notation.

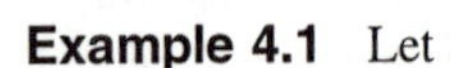

Example 4.1 Let

$$\mathbf{p} = \begin{bmatrix} 2 \\ 2 \end{bmatrix}, \quad \mathbf{a}_1 = \begin{bmatrix} 2 \\ 1 \end{bmatrix}, \quad \mathbf{a}_2 = \begin{bmatrix} -2 \\ 4 \end{bmatrix}$$

be the origin and two vectors of a new coordinate system, and let

$$\mathbf{x} = \begin{bmatrix} 2 \\ 1/2 \end{bmatrix}$$

be a point in the $[\mathbf{e}_1, \mathbf{e}_2]$-system. What can we say about the point $\mathbf{x}'$ in the $[\mathbf{a}_1, \mathbf{a}_2]$-system which has the same coordinates relative to it? We must be able to write $\mathbf{x}'$ as

$$\mathbf{x}' = \begin{bmatrix} 2 \\ 2 \end{bmatrix} + 2 \times \begin{bmatrix} 2 \\ 1 \end{bmatrix} + \frac{1}{2} \times \begin{bmatrix} -2 \\ 4 \end{bmatrix} = \begin{bmatrix} 5 \\ 6 \end{bmatrix}. \tag{4.4}$$

Thus $\mathbf{x}'$ has coordinates

$$\begin{bmatrix} 2 \\ 1/2 \end{bmatrix}$$

with respect to the $[\mathbf{a}_1, \mathbf{a}_2]$-system; with respect to the $[\mathbf{e}_1, \mathbf{e}_2]$-system, it has coordinates

$$\begin{bmatrix} 5 \\ 6 \end{bmatrix}.$$

See Sketch 43 for an illustration.

The example also shows that the vector $\mathbf{v} = \mathbf{x} - \mathbf{o}$ in the $[\mathbf{e}_1, \mathbf{e}_2]$-system is mapped to the vector $\mathbf{v}' = \mathbf{x}' - \mathbf{p}$ in the $[\mathbf{a}_1, \mathbf{a}_2]$-system.

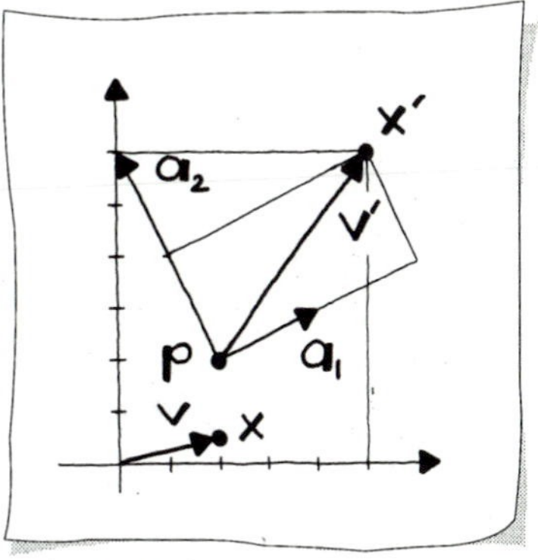

Sketch 43
Mapping a point and a vector.

4.2 The Matrix Form

The coordinates for $\mathbf{v}'$ in the $[\mathbf{a}_1, \mathbf{a}_2]$−system from the previous example are expressed as

$$\begin{bmatrix} 3 \\ 4 \end{bmatrix} = 2 \times \begin{bmatrix} 2 \\ 1 \end{bmatrix} + \frac{1}{2} \times \begin{bmatrix} -2 \\ 4 \end{bmatrix}. \tag{4.5}$$

This is strictly an equation between vectors. It invites a more concise notation using *matrix notation*:

$$\begin{bmatrix} 3 \\ 4 \end{bmatrix} = \begin{bmatrix} 2 & -2 \\ 1 & 4 \end{bmatrix} \begin{bmatrix} 2 \\ 1/2 \end{bmatrix}. \tag{4.6}$$

The 2×2 array in this equation is called a *matrix*. It has two columns, corresponding to the vectors $\mathbf{a}_1$ and $\mathbf{a}_2$. It also has two rows, namely the first row with entries $2, -2$ and the second one with $1, 4$.

In general, an equation like this one has the form

$$\mathbf{v}' = \begin{bmatrix} a_{1,1} & a_{1,2} \\ a_{2,1} & a_{2,2} \end{bmatrix} \begin{bmatrix} v_1 \\ v_2 \end{bmatrix}, \tag{4.7}$$

or,

$$\mathbf{v}' = A\mathbf{v}, \tag{4.8}$$

where A is the 2×2 matrix. The vector $\mathbf{v}'$ is called the *image* of $\mathbf{v}$. The *linear map* is described by the matrix A — we may think of A as being the map's coordinates. Just as we do for points and vectors, we will also refer to the linear map itself by A.

The elements $a_{1,1}$ and $a_{2,2}$ form the *diagonal* of the matrix. The product $A\mathbf{v}$ has two components, each of which is obtained as a dot product between the corresponding row of the matrix and $\mathbf{v}$. In full generality, we have

$$\begin{bmatrix} a_{1,1} & a_{1,2} \\ a_{2,1} & a_{2,2} \end{bmatrix} \begin{bmatrix} v_1 \\ v_2 \end{bmatrix} = \begin{bmatrix} v_1 a_{1,1} + v_2 a_{1,2} \\ v_1 a_{2,1} + v_2 a_{2,2} \end{bmatrix}.$$

For example,

$$\begin{bmatrix} 0 & -1 \\ 2 & 4 \end{bmatrix} \begin{bmatrix} -1 \\ 4 \end{bmatrix} = \begin{bmatrix} -4 \\ 14 \end{bmatrix}.$$

Another note on notation: coordinate systems, such as the $[\mathbf{e}_1, \mathbf{e}_2]$-system, can be interpreted as a matrix with columns $\mathbf{e}_1$ and $\mathbf{e}_2$. Thus

$$[\mathbf{e}_1, \mathbf{e}_2] \equiv \begin{bmatrix} 1 & 0 \\ 0 & 1 \end{bmatrix}.$$

There is a neat way to write the matrix-times-vector algebra in a way that facilitates manual computation. As explained above, every entry in the resulting vector is a dot product of the input vector and a row of the matrix. Let's arrange this as follows:

$$\begin{array}{cc|c} & & 2 \\ & & 1/2 \\ \hline 2 & -2 & 3 \\ 1 & 4 & 4 \end{array}$$

Each entry of the resulting vector is now at the intersection of the corresponding matrix row and the input vector, which is written as a column. As you multiply and then add the terms in your dot products, this scheme guides you to the correct position in the result automatically!

4.3 More about Matrices

Matrices were first introduced by H. Grassmann in 1844. They became the basis of *linear algebra*. Most of their properties can be studied by just considering the humble 2×2 case.

We will now demonstrate several matrix properties. For the sake of concreteness, we shall use the example

$$A = \begin{bmatrix} 2 & 1 \\ -1 & 3 \end{bmatrix}, \quad \mathbf{u} = \begin{bmatrix} 1 \\ 2 \end{bmatrix}, \quad \mathbf{v} \begin{bmatrix} -1 \\ 4 \end{bmatrix}.$$

We have already encountered 2×2 matrices. Sometimes it is convenient if we also think of the vector $\mathbf{v}$ as a matrix. It is a matrix with one column and two rows!

We may multiply all elements of a matrix by one factor; we then say that we have multiplied the matrix by that factor. Using our example, we may multiply the matrix A by a factor, say 2:

$$2 \times \begin{bmatrix} 2 & 1 \\ -1 & 3 \end{bmatrix} = \begin{bmatrix} 4 & 2 \\ -2 & 6 \end{bmatrix}.$$

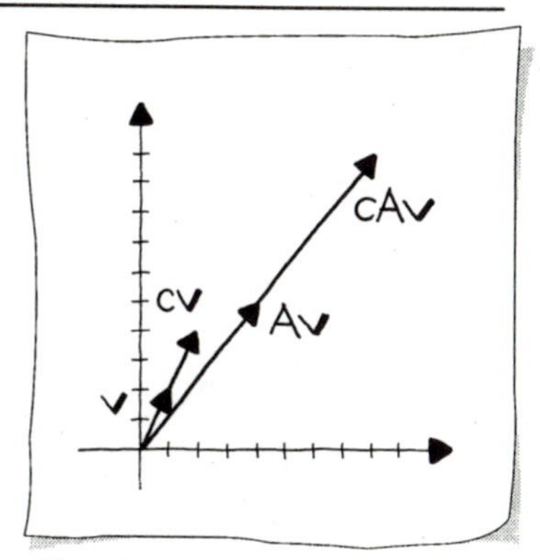

Sketch 44
Matrices preserve scalings.

Matrices are related to *linear* operations, i.e., multiplication by scalar factors or addition of vectors.

For example, if we scale a vector by a factor c, then its image will also be scaled by c:

$$A(c\mathbf{v}) = cA\mathbf{v}.$$

Sketch 44 shows this for an example with $c = 2$.

Example 4.2 Here are the computations that go along with Sketch 44.

$$\begin{bmatrix} 2 & 1 \\ -1 & 3 \end{bmatrix} \times 2 \times \begin{bmatrix} 1 \\ 2 \end{bmatrix} = 2 \times \begin{bmatrix} 2 & 1 \\ -1 & 3 \end{bmatrix} \begin{bmatrix} 1 \\ 2 \end{bmatrix} = \begin{bmatrix} 8 \\ 10 \end{bmatrix}.$$

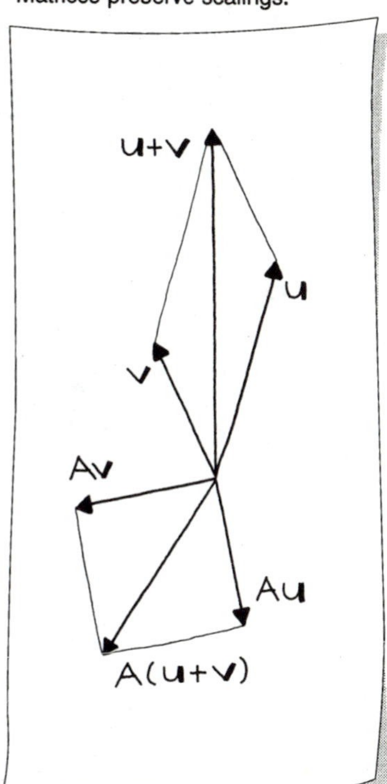

Sketch 45
Matrices preserve sums.

Matrices also preserve summations:

$$A(\mathbf{u} + \mathbf{v}) = A\mathbf{u} + A\mathbf{v},$$

see Sketch 45. This is also called the distributive law.

The last two properties taken together imply that matrices preserve *linear combinations*:

$$A(a\mathbf{u} + b\mathbf{v}) = aA\mathbf{u} + bA\mathbf{v}. \tag{4.9}$$

Example 4.3 The following illustrates that matrices preserve linear combinations:

$$\begin{bmatrix} -1 & 1/2 \\ 0 & -1/2 \end{bmatrix} \left(3 \begin{bmatrix} 1 \\ 2 \end{bmatrix} + 2 \begin{bmatrix} -1 \\ 4 \end{bmatrix} \right) =$$

$$3\begin{bmatrix} -1 & 1/2 \\ 0 & -1/2 \end{bmatrix}\begin{bmatrix} 1 \\ 2 \end{bmatrix} + 2\begin{bmatrix} -1 & 1/2 \\ 0 & -1/2 \end{bmatrix}\begin{bmatrix} -1 \\ 4 \end{bmatrix} = \begin{bmatrix} 6 \\ -7 \end{bmatrix}.$$

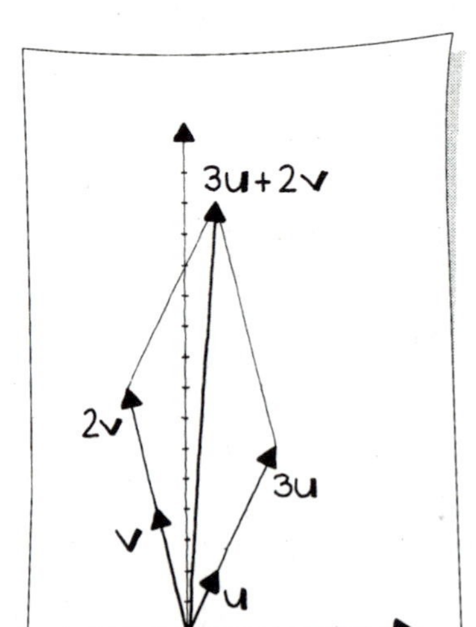

Sketch 46
Matrices preserve linear combinations.

Preservation of linear combinations is a key property of matrices — we will make substantial use of it throughout this book. Sketch 46 illustrates the example above.

Another matrix operation is *matrix addition.* Two matrices A and B may be added by adding corresponding elements:

$$\begin{bmatrix} a_{1,1} & a_{1,2} \\ a_{2,1} & a_{2,2} \end{bmatrix} + \begin{bmatrix} b_{1,1} & b_{1,2} \\ b_{2,1} & b_{2,2} \end{bmatrix} = \begin{bmatrix} a_{1,1}+b_{1,1} & a_{1,2}+b_{1,2} \\ a_{2,1}+b_{2,1} & a_{2,2}+b_{2,2} \end{bmatrix} \tag{4.10}$$

Notice that the matrices must be of the same dimensions; this is not true for matrix multiplication.

Using matrix addition, we may write

$$A\mathbf{v} + B\mathbf{v} = (A+B)\mathbf{v}.$$

This works because of the very simple definition of matrix addition. This is also called the distributive law.

Yet another matrix operation is forming the *transpose matrix.* It is denoted by A^{T} and is formed by interchanging the rows and columns of A: the first row of A^{T} is A's first column, and the second row of A^{T} is A's second column. For example, if

$$A = \begin{bmatrix} 1 & -2 \\ 3 & 5 \end{bmatrix}, \quad \text{then} \quad A^{\mathrm{T}} = \begin{bmatrix} 1 & 3 \\ -2 & 5 \end{bmatrix}.$$

Since we may think of a vector $\mathbf{v}$ as a matrix, we should be able to find $\mathbf{v}$'s transpose. Not very hard: it is a vector with one row and two columns:

$$\mathbf{v} = \begin{bmatrix} -1 \\ 4 \end{bmatrix}, \quad \text{and} \quad \mathbf{v}^{\mathrm{T}} = \begin{bmatrix} -1 & 4 \end{bmatrix}.$$

It is not hard to confirm that

$$[A+B]^{\mathrm{T}} = A^{\mathrm{T}} + B^{\mathrm{T}}. \tag{4.11}$$

Two more straightforward identities are:

$$A^{\mathrm{T}^{\mathrm{T}}} = A, \quad \text{and} \quad [cA]^{\mathrm{T}} = cA^{\mathrm{T}}. \tag{4.12}$$

A *symmetric matrix* is a special matrix that we will encounter many times. A matrix A is symmetric if $A = A^{\mathrm{T}}$, for example:

$$\begin{bmatrix} 5 & 8 \\ 8 & 5 \end{bmatrix}.$$

The columns of a matrix define an $[\mathbf{a}_1, \mathbf{a}_2]$-system. If the vectors $\mathbf{a}_1$ and $\mathbf{a}_2$ are linearly independent then the matrix is said to have *full rank*, or for the 2×2 case, the matrix has *rank* 2. If $\mathbf{a}_1$ and $\mathbf{a}_2$ are linearly dependent then the matrix has rank 1 (see Section 4.8). These two statements may be summarized as: the rank of a matrix equals the number of linearly independent column vectors.

4.4 Scalings

Consider the linear map given by

$$\mathbf{v}' = \begin{bmatrix} 1/2 & 0 \\ 0 & 1/2 \end{bmatrix} \mathbf{v} = \begin{bmatrix} v_1/2 \\ v_2/2 \end{bmatrix} \tag{4.13}$$

This map will "reduce" $\mathbf{v}$ since $\mathbf{v}' = 1/2\mathbf{v}$.

Its effect is illustrated in Figure 4.2. That figure — and more to follow — has two parts. The left part contains several vectors, and the right part shows what happens if we map each of them using the matrix from (4.13).

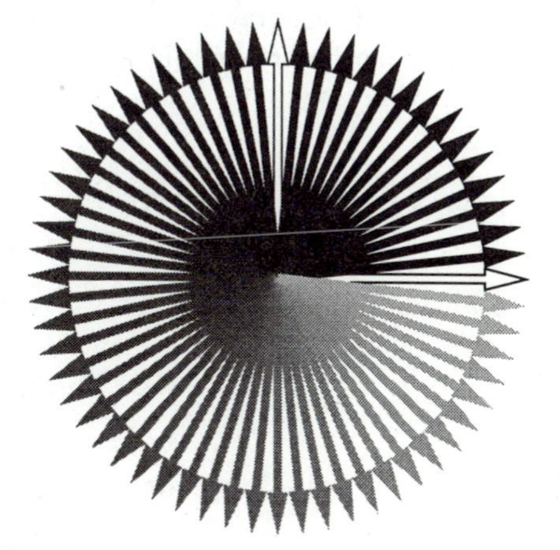

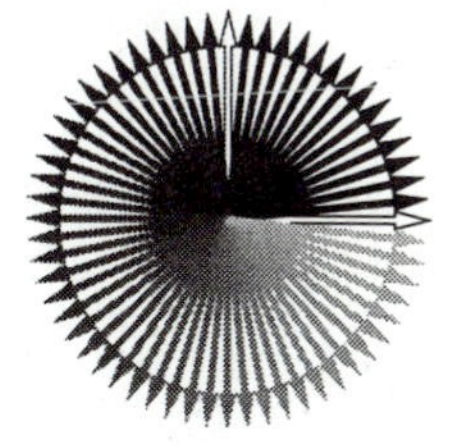

Figure 4.2.
Scaling: evenly scaling vectors.

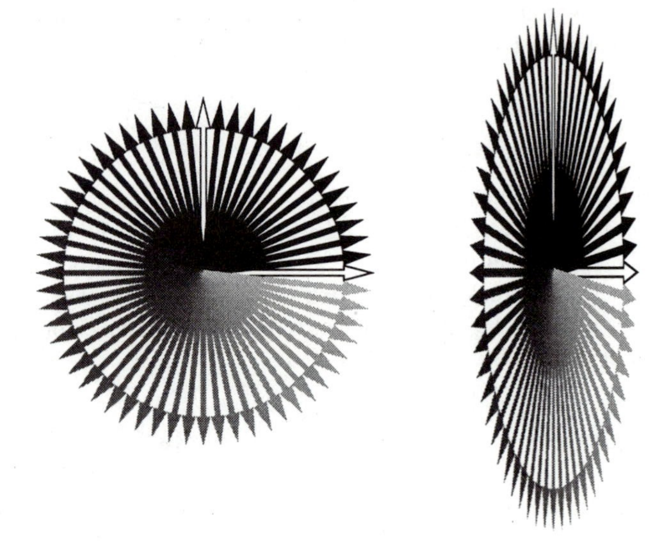

Figure 4.3.
Scaling: unevenly scaling vectors.

The two highlighted vectors in the right half are the vectors $\mathbf{a}_1$ and $\mathbf{a}_2$, corresponding to the unit vectors

$$\begin{bmatrix} 1 \\ 0 \end{bmatrix} \quad \text{and} \quad \begin{bmatrix} 0 \\ 1 \end{bmatrix}$$

in the left half. The vectors are shown in different gray shades to provide a sense of orientation. In this example, the linear map did not change orientation, but more complicated maps will.

Next, consider

$$\mathbf{v}' = \begin{bmatrix} 2 & 0 \\ 0 & 2 \end{bmatrix} \mathbf{v}.$$

Now, $\mathbf{v}$ will be "enlarged".

In general, a scaling is defined by the operation

$$\mathbf{v}' = \begin{bmatrix} s_{1,1} & 0 \\ 0 & s_{2,2} \end{bmatrix} \mathbf{v}, \tag{4.14}$$

thus allowing for uneven scalings in the $\mathbf{e}_1-$ and $\mathbf{e}_2$-direction. Figure 4.3 gives an example for $s_{1,1} = 1/2$ and $s_{2,2} = 2$.

A scaling affects the *area* of the object that is scaled. If we scale an object by $s_{1,1}$ in the $\mathbf{e}_1-$direction, then its area will be changed by a

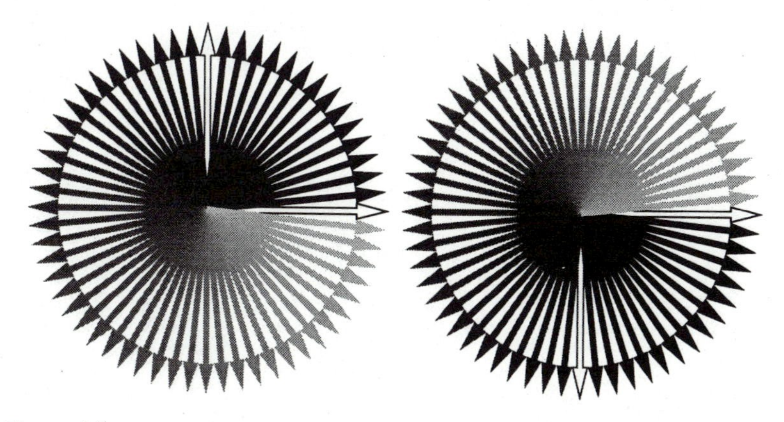

Figure 4.4.
Reflections: a reflection about the $\mathbf{e}_1$−axis.

factor $s_{1,1}$. Similarly, it will change by a factor of $s_{2,2}$ when we apply that scaling to the $\mathbf{e}_2$−direction. The total effect is thus a factor of $s_{1,1}s_{2,2}$.

You can see this from Figure 4.3 by mentally constructing the square spanned by $\mathbf{e}_1$ and $\mathbf{e}_2$, and comparing its area to the rectangle spanned by the image vectors.

4.5 Reflections

Consider the scaling

$$\mathbf{v}' = \begin{bmatrix} 1 & 0 \\ 0 & -1 \end{bmatrix} \mathbf{v}.$$

We may rewrite this as

$$\begin{bmatrix} v_1' \\ v_2' \end{bmatrix} = \begin{bmatrix} v_1 \\ -v_2 \end{bmatrix}.$$

The effect of this map is apparently a change in sign of the second component of $\mathbf{v}$, as shown in Figure 4.4. Geometrically, this means that the input vector $\mathbf{v}$ is reflected about the $\mathbf{e}_1$−axis.

We may also reflect about both axes:

$$\mathbf{v}' = \begin{bmatrix} -1 & 0 \\ 0 & -1 \end{bmatrix} \mathbf{v} \tag{4.15}$$

will do the job, as seen in Figure 4.5.

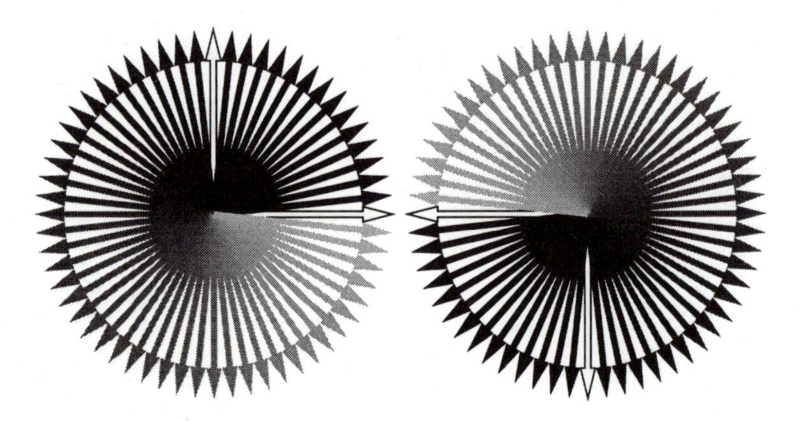

Figure 4.5.
Reflections: a reflection about both axes.

Obviously, reflections of this kind are just a special case of scalings — previously we simply had not given much thought to negative scaling factors.

A more involved reflection is achieved by

$$\mathbf{v}' = \begin{bmatrix} 0 & 1 \\ 1 & 0 \end{bmatrix} \mathbf{v}.$$

Its effect is shown in Figure 4.6.

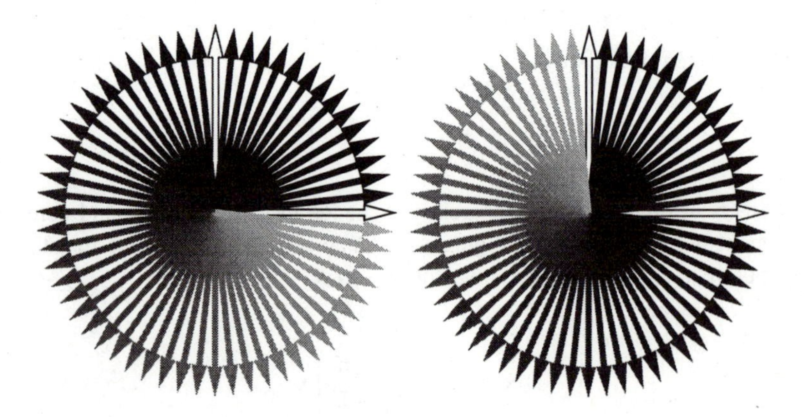

Figure 4.6.
Reflections: a reflection about the line $x_1 = x_2$.

The components of the input vector are interchanged:

$$\begin{bmatrix} 0 & 1 \\ 1 & 0 \end{bmatrix} \begin{bmatrix} v_1 \\ v_2 \end{bmatrix} = \begin{bmatrix} v_2 \\ v_1 \end{bmatrix}.$$

Geometrically, this is a reflection about the line $x_1 = x_2$. The map in (4.15) was similar: it reflected about the line $x_2 = 0$.

By inspection of the figures in this section, it appears that reflections do not change areas. But be careful — they may change the *sign* of an area. We will discuss this in more detail in Section 4.9.

4.6 Rotations

The notion of rotating a vector around the origin is intuitively clear — but a corresponding matrix takes a few moments to construct. To keep it easy at the beginning, let us rotate the unit vector

$$\mathbf{e} = \begin{bmatrix} 1 \\ 0 \end{bmatrix}$$

by α degrees, resulting in a new (rotated) vector[1]

$$\mathbf{e}' = \begin{bmatrix} \cos\alpha \\ \sin\alpha \end{bmatrix}.$$

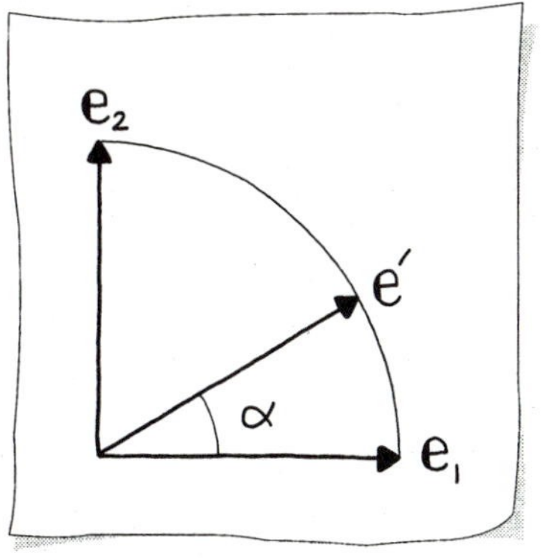

Sketch 47
Rotating a unit vector.

Consult Sketch 47 to convince yourself of this fact!

We thus need to find a matrix R which achieves

$$\begin{bmatrix} \cos\alpha \\ \sin\alpha \end{bmatrix} = \begin{bmatrix} r_{1,1} & r_{1,2} \\ r_{2,1} & r_{2,2} \end{bmatrix} \begin{bmatrix} 1 \\ 0 \end{bmatrix}.$$

Some trial and error should result in the correct *rotation matrix*; it is given by

$$R = \begin{bmatrix} \cos\alpha & -\sin\alpha \\ \sin\alpha & \cos\alpha \end{bmatrix}. \quad (4.16)$$

This does indeed rotate $\mathbf{e}$ to $\mathbf{e}'$ — what is not clear is that we have already found the solution to the general rotation problem!

Let $\mathbf{v}$ be an arbitrary vector. We claim that the matrix R from (4.16) will rotate it by α degrees to a new vector $\mathbf{v}'$. If this is so, then we must have

$$\mathbf{v} \cdot \mathbf{v}' = \|\mathbf{v}\|^2 \cos\alpha.$$

[1]Notice that $\cos^2\alpha + \sin^2\alpha = 1$, thus this is a rotation.

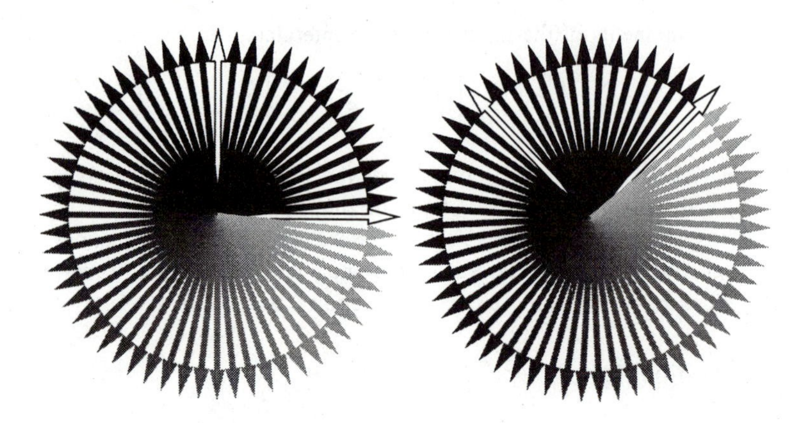

Figure 4.7.
Rotations: a rotation by 45 degrees.

according to the rules of dot products, see Section 2.7. Here, we made use of the fact that a rotation does not change the length of a vector, i.e., $\|\mathbf{v}\| = \|\mathbf{v}'\|$ and hence $\|\mathbf{v}\| \cdot \|\mathbf{v}'\| = \|\mathbf{v}\|^2$.

Since

$$\mathbf{v}' = \begin{bmatrix} v_1 \cos\alpha - v_2 \sin\alpha \\ v_1 \sin\alpha + v_2 \cos\alpha \end{bmatrix},$$

the dot product $\mathbf{v} \cdot \mathbf{v}'$ is given by

$$\begin{aligned} \mathbf{v} \cdot \mathbf{v}' &= v_1^2 \cos\alpha - v_1 v_2 \sin\alpha + v_1 v_2 \sin\alpha + v_2^2 \cos\alpha \\ &= (v_1^2 + v_2^2) \cos\alpha \\ &= \|\mathbf{v}\|^2 \cos\alpha, \end{aligned}$$

and all is shown! See Figure 4.7 for an illustration. There, $\alpha = 45°$, and the rotation matrix is thus given by

$$R = \begin{bmatrix} \sqrt{2}/2 & -\sqrt{2}/2 \\ \sqrt{2}/2 & \sqrt{2}/2 \end{bmatrix}$$

Rotations are in a special class of transformations; these are called *rigid body motions*. See Section 5.5 for more details. Finally, it should come without saying that rotations do not change areas.

4.7 Shears

What map takes a rectangle to a parallelogram? Pictorially, one such map is shown in Sketch 48.

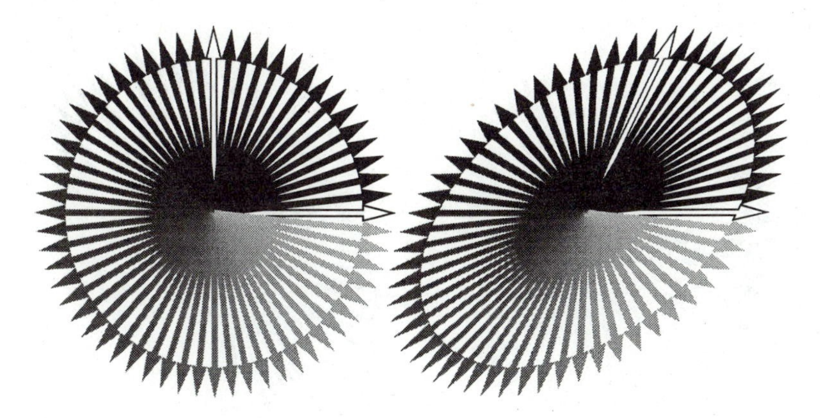

Figure 4.8.
Shears: shearing vectors parallel to the $\mathbf{e}_1$-axis.

In this example, we have a map:

$$\mathbf{v} = \begin{bmatrix} 0 \\ 1 \end{bmatrix} \longrightarrow \mathbf{v}' = \begin{bmatrix} d_1 \\ 1 \end{bmatrix}.$$

In matrix form, this is realized by

$$\begin{bmatrix} d_1 \\ 1 \end{bmatrix} = \begin{bmatrix} 1 & d_1 \\ 0 & 1 \end{bmatrix} \begin{bmatrix} 0 \\ 1 \end{bmatrix}. \tag{4.17}$$

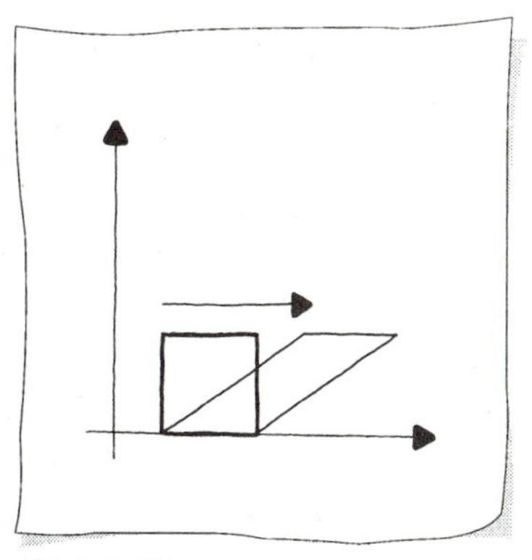

Sketch 48
A special shear.

Verify! The 2×2 matrix in this equation is called a *shear matrix*. It is the kind of matrix that is used when you generate italic fonts from standard ones.

A shear matrix may be applied to arbitrary vectors. If $\mathbf{v}$ is an input vector, then a shear maps it to $\mathbf{v}'$:

$$\mathbf{v}' = \begin{bmatrix} 1 & d_1 \\ 0 & 1 \end{bmatrix} \begin{bmatrix} v_1 \\ v_2 \end{bmatrix} = \begin{bmatrix} v_1 + v_2 d_1 \\ v_2 \end{bmatrix},$$

as illustrated in Figure 4.8.

We have so far restricted ourselves to shears along the $\mathbf{e}_1$-axis; we may also shear along the $\mathbf{e}_2$-axis. Then we would have

$$\mathbf{v}' = \begin{bmatrix} 1 & 0 \\ d_2 & 1 \end{bmatrix} \begin{bmatrix} v_1 \\ v_2 \end{bmatrix} = \begin{bmatrix} v_1 \\ v_1 d_2 + v_2 \end{bmatrix},$$

as illustrated in Figure 4.9.

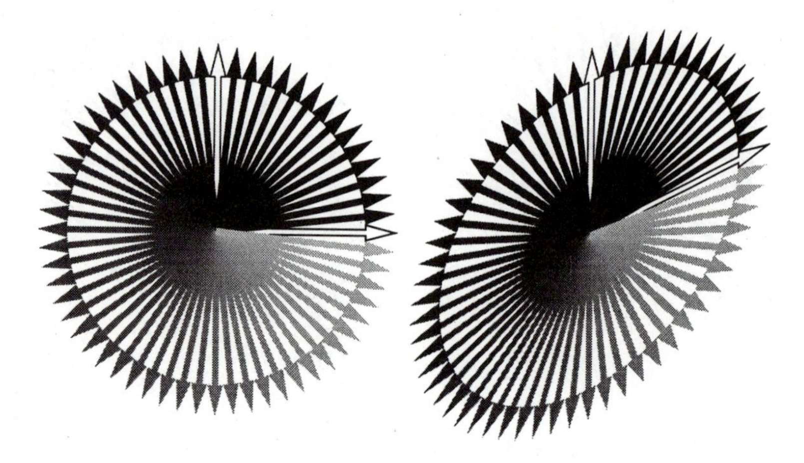

Figure 4.9.
Shears: shearing vectors parallel to the $\mathbf{e}_2$-axis.

Since it will be needed later, we look at the following: What is the shear that achieves

$$\mathbf{v} = \begin{bmatrix} v_1 \\ v_2 \end{bmatrix} \longrightarrow \mathbf{v}' = \begin{bmatrix} v_1 \\ 0 \end{bmatrix}?$$

It is obviously a shear parallel to the $\mathbf{e}_2$-saxis and is given by the map

$$\mathbf{v}' = \begin{bmatrix} v_1 \\ 0 \end{bmatrix} = \begin{bmatrix} 1 & 0 \\ -v_2/v_1 & 1 \end{bmatrix} \begin{bmatrix} v_1 \\ v_2 \end{bmatrix}. \tag{4.18}$$

Shears do not change areas. Consulting Sketch 48,we see that the rectangle and its image, a parallelogram, have the same area: both have the same base and the same height.

4.8 Projections

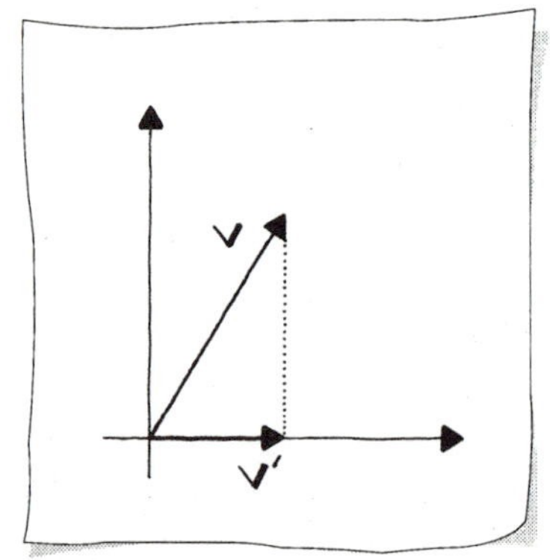

Sketch 49
A projection.

Projections — parallel projections, for our purposes — act like sunlight casting shadows. In 2D, this is modeled as follows: take any vector $\mathbf{v}$ and "flatten it out" onto the $\mathbf{e}_1$-axis. This simply means: set the v_2-coordinate of the vector to zero. For example, if we project the vector

$$\begin{bmatrix} 3 \\ 1 \end{bmatrix}$$

onto the $\mathbf{e}_1$-axis, it becomes

$$\begin{bmatrix} 3 \\ 0 \end{bmatrix},$$

as shown in Sketch 49.

Figure 4.10.
Projections: all vectors are "flattened out" onto the $\mathbf{e}_1$-axis.

What matrix achieves this map? That's easy:

$$\begin{bmatrix} 3 \\ 0 \end{bmatrix} = \begin{bmatrix} 1 & 0 \\ 0 & 0 \end{bmatrix} \begin{bmatrix} 3 \\ 1 \end{bmatrix}.$$

This matrix will not only project the vector

$$\begin{bmatrix} 3 \\ 1 \end{bmatrix}$$

onto the $\mathbf{e}_1$-axis, but in fact *every vector*! This is so since

$$\begin{bmatrix} v_1 \\ 0 \end{bmatrix} = \begin{bmatrix} 1 & 0 \\ 0 & 0 \end{bmatrix} \begin{bmatrix} v_1 \\ v_2 \end{bmatrix}.$$

While a somewhat trivial example of "real" projections, we see that this projection does indeed feature the main property of a projection: it *reduces dimensionality*. Every vector from 2D space is mapped into 1D space, namely onto the $\mathbf{e}_1$-axis. Figure 4.10 illustrates this property. The analogous case, projecting onto the $\mathbf{e}_2$-axis, is not more difficult; it is given by

$$\begin{bmatrix} 0 \\ v_2 \end{bmatrix} = \begin{bmatrix} 0 & 0 \\ 0 & 1 \end{bmatrix} \begin{bmatrix} v_1 \\ v_2 \end{bmatrix}.$$

Parallel projections are characterized by the fact that all vectors are projected in the *same* direction. This is a technique commonly used in computer graphics.[2] In 2D, all vectors are projected onto a line. If the

[2] Although it is more relevant in 3D.

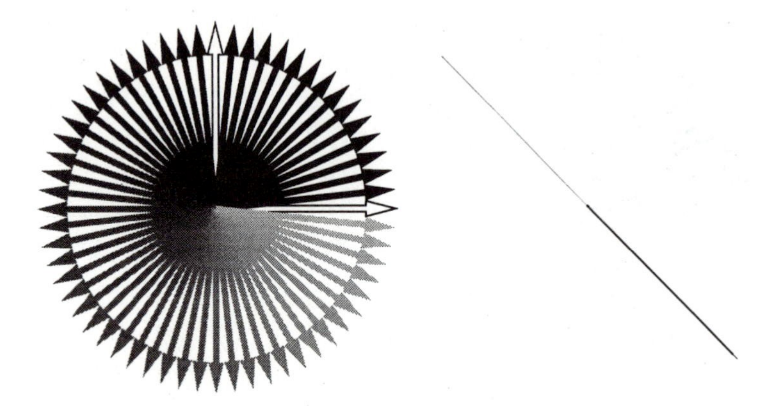

Figure 4.11.
Projections: all vectors are "flattened out" in one direction.

direction is perpendicular to the line, then the parallel projection is called *orthographic*; otherwise the projection is called *oblique*.

More general projections are easily obtained: just take $\mathbf{a}_1$ and $\mathbf{a}_2$ to be parallel. Since any mapped vector is of the form $\mathbf{v}' = v_1\mathbf{a}_1 + v_2\mathbf{a}_2$, it follows that now all vectors $\mathbf{v}'$ are also a multiple of $\mathbf{a}_1$ and $\mathbf{a}_2$. Figure 4.11 shows the effect of the matrix

$$\begin{bmatrix} 0.2 & 0.8 \\ -0.2 & -0.8 \end{bmatrix}.$$

The figure does not illustrate the fact that this is not a parallel projection. Try sketching the projection of a few vectors yourself to get a feel for how this projection works.

With the relationship that $\mathbf{a}_2 = c\mathbf{a}_1$, the *projection matrix* takes the form

$$\begin{bmatrix} a_{1,1} & ca_{1,1} \\ a_{2,1} & ca_{2,1} \end{bmatrix}.$$

As far as areas are concerned, projections take a lean approach: whatever an area was before the map, it is zero afterwards.

4.9 Areas and Linear Maps: Determinants

As you might have noticed, we discussed one particular aspect of linear maps for each type: how areas are changed. We will now discuss this aspect for an arbitrary linear map. Such a map takes the two vectors $[\mathbf{e}_1, \mathbf{e}_2]$ to the two vectors $[\mathbf{a}_1, \mathbf{a}_2]$. The area of the square spanned by $[\mathbf{e}_1, \mathbf{e}_2]$ is 1, that is $\text{area}(\mathbf{e}_1, \mathbf{e}_2) = 1$. If we knew the area of the parallelogram spanned by $[\mathbf{a}_1, \mathbf{a}_2]$, then we could say how the linear map affects areas.

How do we find the area P of a parallelogram spanned by two vectors $\mathbf{a}_1$ and $\mathbf{a}_2$? Referring to Sketch 50, let us first determine the area T of the triangle formed by $\mathbf{a}_1$ and $\mathbf{a}_2$. We see that

$$T = a_{1,1}a_{2,2} - T_1 - T_2 - T_3.$$

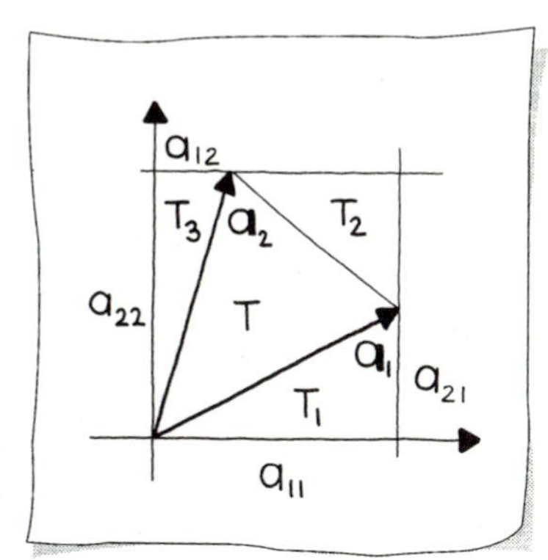

Sketch 50
Area of a parallelogram.

We then observe that

$$T_1 = \frac{1}{2}a_{1,1}a_{2,1}, \tag{4.19}$$

$$T_2 = \frac{1}{2}(a_{1,1} - a_{1,2})(a_{2,2} - a_{2,1}), \tag{4.20}$$

$$T_3 = \frac{1}{2}a_{1,2}a_{2,2}. \tag{4.21}$$

Working out the algebra, we arrive at

$$T = \frac{1}{2}a_{1,1}a_{2,2} - \frac{1}{2}a_{1,2}a_{2,1}.$$

Our aim was not really T, but the parallelogram area P. Clearly (see Sketch 51),

$$P = 2T,$$

and we have our desired area.

It is customary to use the term *determinant* for the area of the parallelogram spanned by $[\mathbf{a}_1, \mathbf{a}_2]$. Since the two vectors $\mathbf{a}_1$ and $\mathbf{a}_2$ form the columns of the matrix A, we also speak of the determinant of the matrix A, and denote it by $\det A$ or $|A|$:

$$|A| = \begin{vmatrix} a_{1,1} & a_{1,2} \\ a_{2,1} & a_{2,2} \end{vmatrix} = a_{1,1}a_{2,2} - a_{1,2}a_{2,1}. \tag{4.22}$$

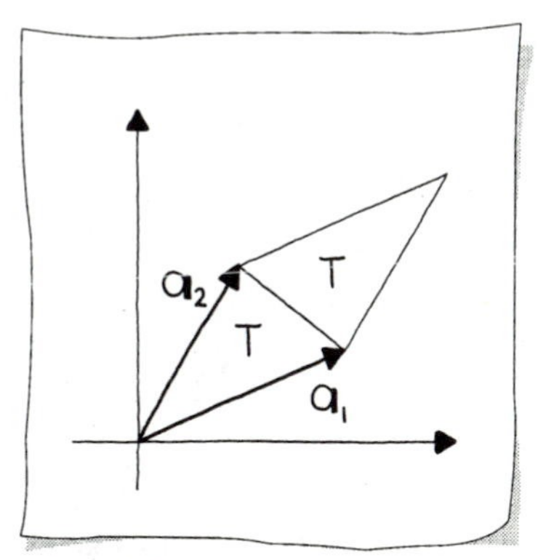

Sketch 51
Parallelogram and triangles.

Since A maps a square with area one onto a parallelogram with area $|A|$, the determinant of a matrix characterizes it as follows: If $|A| = 1$, then the corresponding linear map does not change areas; if $0 \leq |A| < 1$, the corresponding linear map shrinks areas, and if $|A| > 1$, it expands areas.

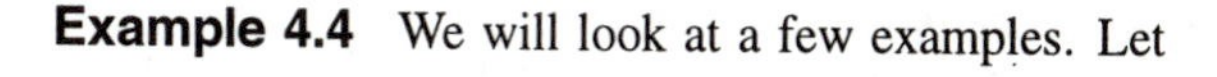

Example 4.4 We will look at a few examples. Let

$$A = \begin{bmatrix} 1 & 5 \\ 0 & 1 \end{bmatrix},$$

then

$$|A| = \begin{vmatrix} 1 & 5 \\ 0 & 1 \end{vmatrix} = 1.$$

Since A represents a *shear*, we see again that those maps do not change areas.

For another example, let

$$A = \begin{bmatrix} 1 & 0 \\ 0 & -1 \end{bmatrix}.$$

Then

$$|A| = \begin{vmatrix} 1 & 0 \\ 0 & -1 \end{vmatrix} = -1.$$

This matrix corresponds to a *reflection*, and it leaves areas unchanged, except for a *sign change*.

Finally, let

$$A = \begin{bmatrix} .2 & .8 \\ -.2 & -.8 \end{bmatrix}.$$

Then

$$|A| = \begin{vmatrix} .2 & .8 \\ -.2 & -.8 \end{vmatrix} = 0.$$

This matrix corresponds to the *projection* from Section 4.8. In that example, we saw that projections collapse any object onto a straight line, i.e., to a zero area.

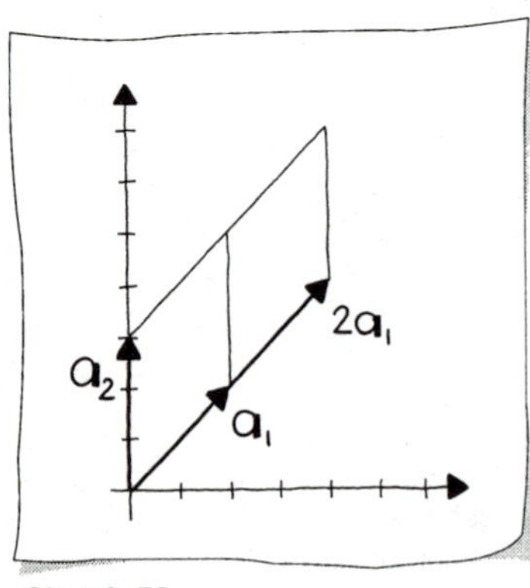

Sketch 52
Resulting area after scaling one column of A.

There are some rules for working with determinants:

If $A = [\mathbf{a}_1, \mathbf{a}_2]$, then

$$|c\mathbf{a}_1, \mathbf{a}_2| = c|\mathbf{a}_1, \mathbf{a}_2| = c|A|.$$

In other words, if one of the columns of A is scaled by a factor c, then A's determinant is also scaled by c. Verify that this is true from the definition of the determinant of A! Sketch 52 illustrates this for the example, $c = 2$.

If $|A|$ is positive and c is negative, then replacing $\mathbf{a}_1$ by $c\mathbf{a}_1$ will cause $c|A|$, the area formed by $c\mathbf{a}_1$ and $\mathbf{a}_2$, to become negative. The notion of a negative area is very useful computationally.

The area also changes sign when we *interchange* the columns of A:

$$|\mathbf{a}_1, \mathbf{a}_2| = -|\mathbf{a}_2, \mathbf{a}_1|. \tag{4.23}$$

This fact is easily computed using the definition of a determinant:

$$|\mathbf{a}_2, \mathbf{a}_1| = a_{1,2}a_{2,1} - a_{2,2}a_{1,1}.$$

Two 2D vectors whose determinant is positive are called *right-handed.* The standard example are the two vectors $\mathbf{e}_1$ and $\mathbf{e}_2$. If their determinant is negative, then they are called *left-handed.*[3] Sketch 53 shows a right-handed pair of vectors (top) and a pair of left-handed ones (bottom).

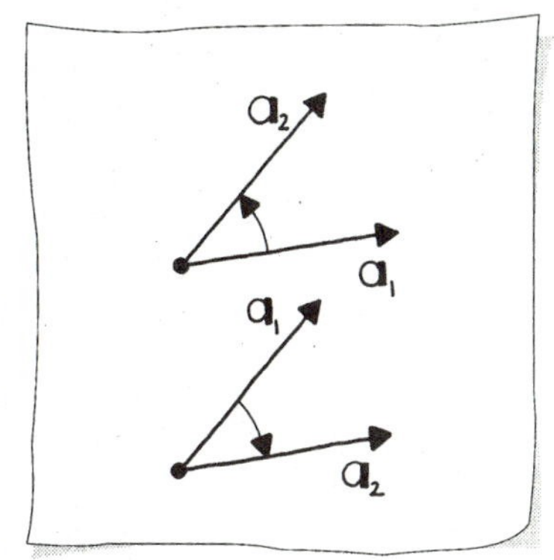

Sketch 53
Right-handed and left-handed vectors.

Our definition of positive and negative area is not totally arbitrary: the triangle formed by vectors $\mathbf{a}_1$ and $\mathbf{a}_2$ has area $1/2 \times \sin(\alpha)\|\mathbf{a}_1\|\|\mathbf{a}_2\|$. Here, the angle α indicates by how much we have to rotate $\mathbf{a}_1$ in order to line up with $\mathbf{a}_2$. If we interchange the two vectors, the sign of α and hence of $\sin(\alpha)$ also changes!

If *both* columns of A are scaled by c, then the determinant is scaled by c^2:

$$|c\mathbf{a}_1, c\mathbf{a}_2| = c^2|\mathbf{a}_1, \mathbf{a}_2| = c^2|A|.$$

Sketch 54 illustrates for the example, $c = 1/2$.

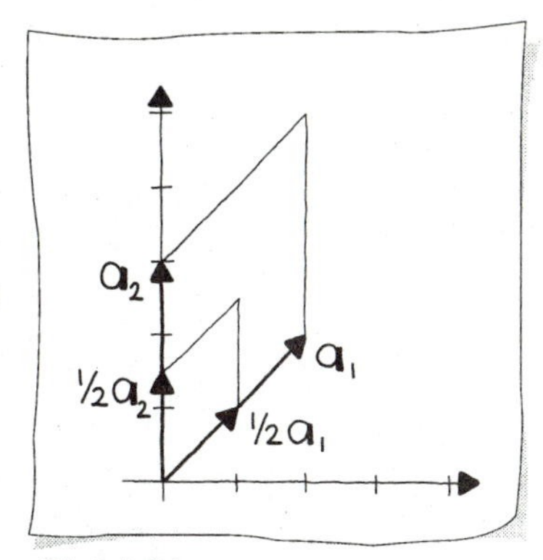

Sketch 54
Resulting area after scaling both columns of A.

4.10 Composing Linear Maps

Suppose you have mapped a vector $\mathbf{v}$ to $\mathbf{v}'$ using a matrix A. Next, you want to map $\mathbf{v}'$ to $\mathbf{v}''$ using a matrix B. We start out with

$$\mathbf{v}' = \begin{bmatrix} a_{1,1} & a_{1,2} \\ a_{2,1} & a_{2,2} \end{bmatrix} \begin{bmatrix} v_1 \\ v_2 \end{bmatrix} = \begin{bmatrix} a_{1,1}v_1 + a_{1,2}v_2 \\ a_{2,1}v_1 + a_{2,2}v_2 \end{bmatrix}.$$

Next, we have

$$\mathbf{v}'' = \begin{bmatrix} b_{1,1} & b_{1,2} \\ b_{2,1} & b_{2,2} \end{bmatrix} \begin{bmatrix} a_{1,1}v_1 + a_{1,2}v_2 \\ a_{2,1}v_1 + a_{2,2}v_2 \end{bmatrix} =$$

$$\begin{bmatrix} b_{1,1}(a_{1,1}v_1 + a_{1,2}v_2) + b_{1,2}(a_{2,1}v_1 + a_{2,2}v_2) \\ b_{2,1}(a_{1,1}v_1 + a_{1,2}v_2) + b_{2,2}(a_{2,1}v_1 + a_{2,2}v_2) \end{bmatrix}.$$

Collecting the terms in v_1 and v_2, we get

$$\mathbf{v}'' = \begin{bmatrix} b_{1,1}a_{1,1} + b_{1,2}a_{2,1} & b_{1,1}a_{1,2} + b_{1,2}a_{2,2} \\ b_{2,1}a_{1,1} + b_{2,2}a_{2,1} & b_{2,1}a_{1,2} + b_{2,2}a_{2,2} \end{bmatrix} \begin{bmatrix} v_1 \\ v_2 \end{bmatrix}.$$

The matrix that we have created here, let's call it C, is called the *product matrix* of B and A:

$$B \cdot A = C.$$

[3]The reason for this terminology will become transparent when we revisit these definitions for the 3D case, see Section 10.2.

In more detail,

$$\begin{bmatrix} b_{1,1} & b_{1,2} \\ b_{2,1} & b_{2,2} \end{bmatrix} \begin{bmatrix} a_{1,1} & a_{1,2} \\ a_{2,1} & a_{2,2} \end{bmatrix} = \begin{bmatrix} b_{1,1}a_{1,1} + b_{1,2}a_{2,1} & b_{1,1}a_{1,2} + b_{1,2}a_{2,2} \\ b_{2,1}a_{1,1} + b_{2,2}a_{2,1} & b_{2,1}a_{1,2} + b_{2,2}a_{2,2} \end{bmatrix} \tag{4.24}$$

This looks messy, but a simple rule puts order into chaos: the element $c_{i,j}$ is computed as the dot product of B's i^{th} row and A's j^{th} column.

We can use this product to describe the composite map:

$$\mathbf{v}'' = B\mathbf{v}' = B[A\mathbf{v}] = BA\mathbf{v}.$$

Example 4.5 Let

$$\mathbf{v} = \begin{bmatrix} 2 \\ -1 \end{bmatrix}, \quad A = \begin{bmatrix} -1 & 2 \\ 0 & 3 \end{bmatrix}, \quad B = \begin{bmatrix} 0 & -2 \\ -3 & 1 \end{bmatrix}.$$

Then

$$\mathbf{v}' = \begin{bmatrix} -1 & 2 \\ 0 & 3 \end{bmatrix} \begin{bmatrix} 2 \\ -1 \end{bmatrix} = \begin{bmatrix} -4 \\ -3 \end{bmatrix}$$

and

$$\mathbf{v}'' = \begin{bmatrix} 0 & -2 \\ -3 & 1 \end{bmatrix} \begin{bmatrix} -4 \\ -3 \end{bmatrix} = \begin{bmatrix} 6 \\ 9 \end{bmatrix}.$$

We can also compute $\mathbf{v}''$ using the matrix product BA:

$$C = BA = \begin{bmatrix} 0 & -2 \\ -3 & 1 \end{bmatrix} \begin{bmatrix} -1 & 2 \\ 0 & 3 \end{bmatrix} = \begin{bmatrix} 0 & -6 \\ 3 & -3 \end{bmatrix}.$$

Verify for yourself that $\mathbf{v}'' = C\mathbf{v}$!

There is a neat way to arrange two matrices when forming their product for manual computation (yes, that is still encountered!), analogous to the matrix/vector product from Section 4.2. Using the above example, and highlighting the computation of $c_{2,1}$, we write

$$\begin{array}{rr|rr} & & \mathbf{-1} & 2 \\ & & \mathbf{0} & 3 \\ \hline 0 & -2 & & \\ \mathbf{-3} & \mathbf{1} & \mathbf{3} & \end{array}$$

You see how $c_{2,1}$ is at the intersection of column one of the "top" matrix and row two of the "left" matrix.

The complete multiplication scheme is then arranged like this:

		−1	2
		0	3
0	−2	0	−6
−3	1	3	−3

While we use the term "product" for BA, it is very important to realize that this kind of product differs significantly from products of real numbers: it is not *commutative*. That is, in general

$$AB \neq BA.$$

Example 4.6 Let us take two very simple matrices to illustrate this. A rotates by 90 degrees, and B shears parallel to the $\mathbf{e}_2$−axis:

$$A = \begin{bmatrix} 0 & -1 \\ 1 & 0 \end{bmatrix} \quad B = \begin{bmatrix} 1 & 0.5 \\ 0 & 1 \end{bmatrix}$$

We first form AB:

$$AB = \begin{bmatrix} 0 & -1 \\ 1 & 0 \end{bmatrix} \begin{bmatrix} 1 & 0.5 \\ 0 & 1 \end{bmatrix} = \begin{bmatrix} 0 & -1 \\ 1 & 0.5 \end{bmatrix}.$$

Next, BA :

$$BA = \begin{bmatrix} 1 & 0.5 \\ 0 & 1 \end{bmatrix} \begin{bmatrix} 0 & -1 \\ 1 & 0 \end{bmatrix} = \begin{bmatrix} 0.5 & -1 \\ 1 & 0 \end{bmatrix}.$$

Clearly these are not the same!

Matrix products correspond to linear map compositions — since the products are not commutative, it follows that it matters in which order we carry out linear maps. *Linear map composition is order dependent.* Figure 4.12 gives an example.

Of course, *some* maps *do* commute: for example, the rotations. It does not matter if we rotate by α first and then by β or the other way around. In either case, we have rotated by $\alpha + \beta$. In terms of matrices:

$$\begin{bmatrix} \cos\alpha & -\sin\alpha \\ \sin\alpha & \cos\alpha \end{bmatrix} \begin{bmatrix} \cos\beta & -\sin\beta \\ \sin\beta & \cos\beta \end{bmatrix} = \tag{4.25}$$

$$\begin{bmatrix} \cos\alpha\cos\beta - \sin\alpha\sin\beta & -\cos\alpha\sin\beta - \sin\alpha\cos\beta \\ \sin\alpha\cos\beta + \cos\alpha\sin\beta & -\sin\alpha\sin\beta + \cos\alpha\cos\beta \end{bmatrix}. \tag{4.26}$$

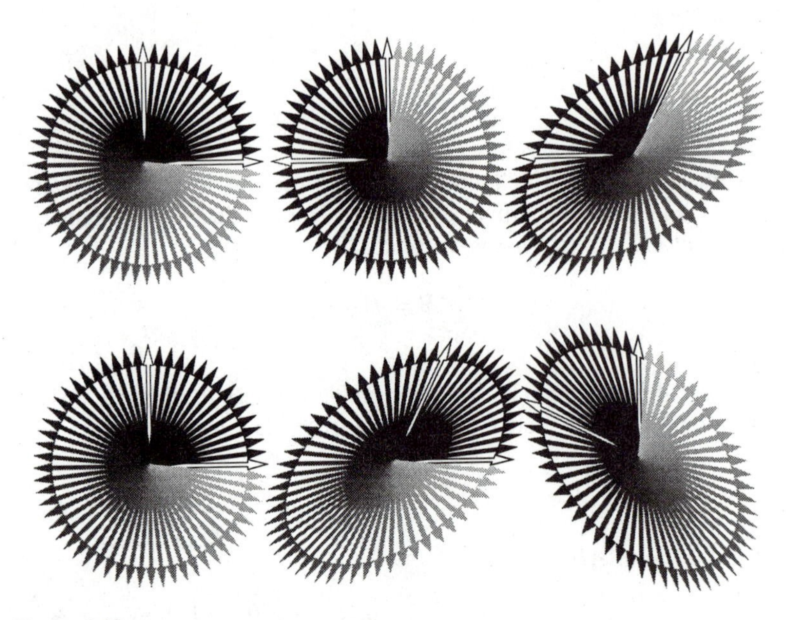

Figure 4.12.
Linear map composition is order dependent. Top: rotate, then shear. Bottom: shear, then rotate.

Check for yourself that the other alternative gives the same result! The product matrix must correspond to a rotation by $\alpha + \beta$, and thus it must equal to

$$\begin{bmatrix} \cos(\alpha+\beta) & -\sin(\alpha+\beta) \\ \sin(\alpha+\beta) & \cos(\alpha+\beta) \end{bmatrix}.$$

We have thus shown that

$$\begin{bmatrix} \cos(\alpha+\beta) & -\sin(\alpha+\beta) \\ \sin(\alpha+\beta) & \cos(\alpha+\beta) \end{bmatrix} = \tag{4.27}$$

$$\begin{bmatrix} \cos\alpha\cos\beta - \sin\alpha\sin\beta & -\cos\alpha\sin\beta - \sin\alpha\cos\beta \\ \sin\alpha\cos\beta + \cos\alpha\sin\beta & -\sin\alpha\sin\beta + \cos\alpha\cos\beta \end{bmatrix}, \tag{4.28}$$

and so we have, without much effort, proved two trig identities!

4.11 More on Matrix Multiplication

Matrix multiplication is not limited to the product of 2×2 matrices. In fact, we are constantly using a different kind of matrix multiplication; when we multiply a matrix by a vector, we follow the rules of matrix multiplication!

If $\mathbf{v}' = A\mathbf{v}$, then the first component of $\mathbf{v}'$ is the dot product of A's first row and $\mathbf{v}$; the second component of $\mathbf{v}'$ is the dot product of A's second row and $\mathbf{v}$.

We may even write the dot product of two vectors in the form of matrix multiplication, as an example should show:

$$\begin{bmatrix} 3 & 4 \end{bmatrix} \cdot \begin{bmatrix} -3 \\ 6 \end{bmatrix} = 15.$$

Usually, we write $\mathbf{u} \cdot \mathbf{v}$ for the dot product of $\mathbf{u}$ and $\mathbf{v}$, but sometimes the above form $\mathbf{u}^{\mathrm{T}}\mathbf{v}$ is useful as well.

In Section 4.3, we introduced the transpose A^{T} of a matrix A. We saw that addition of matrices is "well-behaved" under transposition; matrix multiplication is not that straightforward. We have

$$[AB]^{\mathrm{T}} = B^{\mathrm{T}}A^{\mathrm{T}}. \tag{4.29}$$

To see why this is true, recall that each element of a product matrix is obtained as a dot product. How do dot products react to transposition? If we have a product $\mathbf{u}^{\mathrm{T}}\mathbf{v}$, what is $[\mathbf{u}^{\mathrm{T}}\mathbf{v}]^{\mathrm{T}}$? This has an easy answer:

$$[\mathbf{u}^{\mathrm{T}}\mathbf{v}]^{\mathrm{T}} = \mathbf{v}^{\mathrm{T}}\mathbf{u},$$

as an example will clarify.

Example 4.7

$$\begin{bmatrix} 3 & 4 \end{bmatrix} \begin{bmatrix} -3 \\ 6 \end{bmatrix} = \begin{bmatrix} -3 & 6 \end{bmatrix} \begin{bmatrix} 3 \\ 4 \end{bmatrix} = 15.$$

Since matrix multiplication is just the application of several dot products, we see that (4.29) does make sense.

What is the *determinant* of a product matrix? If $C = AB$ denotes a matrix product, then we know that B scales objects by $|B|$, and A scales objects by $|A|$. It is clear then that the composition of the maps scales by the product of the individual scales:

$$|AB| = |A| \cdot |B|. \tag{4.30}$$

Example 4.8 As a simple example, take two scalings:

$$A = \begin{bmatrix} 1/2 & 0 \\ 0 & 1/2 \end{bmatrix}, \quad B = \begin{bmatrix} 4 & 0 \\ 0 & 4 \end{bmatrix}.$$

We have $|A| = 1/4$ and $|B| = 16$. Thus A scales down, and B scales up, but the effect of B's scaling is greater than that of A's. The product:

$$AB = \begin{bmatrix} 2 & 0 \\ 0 & 2 \end{bmatrix}$$

thus scales up: $|AB| = |A| \cdot |B| = 4$.

Just as for real numbers, we can define *exponents* for matrices:

$$A^r = \underbrace{A \cdot \ldots \cdots A}_{r \text{ times}}.$$

Here are some rules:

$$A^{r+s} = A^r A^s$$
$$A^{rs} = (A^r)^s$$
$$A^0 = I.$$

4.12 Working with Matrices

Yet more rules of matrix arithmetic! We encountered many of these rules throughout this chapter in terms of matrix and vector multiplication: a vector is simply a special matrix.

For A, B, C 2×2 matrices:

Commutative law for addition	$A + B = B + A$
Associative law for addition	$A + (B + C) = (A + B) + C$
Associative law for multiplication	$A(BC) = (AB)C$
Distributive law	$A(B + C) = AB + AC$
Distributive law	$(B + C)A = BA + CA$.

And some rules involving scalars:

$$a(B + C) = aB + aC$$
$$(a + b)C = aC + bC$$
$$(ab)C = a(bC)$$
$$a(BC) = (aB)C = B(aC).$$

Although the focus of this chapter is on 2×2 matrices, these rules apply to matrices of any size. Of course it is assumed in the rules above that the sizes of the matrices are such that the operations can be performed. See Chapter 14 for more information on matrices larger than 2×2.

4.13 Exercises

For the following exercises, let

$$A = \begin{bmatrix} 0 & -1 \\ 1 & 0 \end{bmatrix}, \quad B = \begin{bmatrix} 1 & -1 \\ -1 & 1/2 \end{bmatrix}, \quad \mathbf{v} = \begin{bmatrix} 2 \\ 3 \end{bmatrix}.$$

1. Describe geometrically the effect of A and B. (You may do this analytically or by suitably modifying the PostScript file `Rotate.ps` in the web site.)[4]
2. Compute $A\mathbf{v}$ and $B\mathbf{v}$.
3. Compute $A + B$. Show that $A\mathbf{v} + B\mathbf{v} = (A + B)\mathbf{v}$.
4. Compute $AB\mathbf{v}$ and $BA\mathbf{v}$.
5. Compute $B^{\mathrm{T}}A$.
6. What is the shear matrix that maps $\mathbf{v}$ onto the $\mathbf{e}_2$−axis?
7. What is the determinant of A?

PS1 Modify the ps-file `Nocomm.ps` to show that $AB \neq BA$. The whole PostScript program may look somewhat involved – all you have to do is change the definition of the input matrices `mat1` and `mat2` and watch what happens!

PS2 Modify the ps-file `Rotate.ps` such that a rotation of 100 degrees is achieved.

[4] The web site for *The Geometry Toolbox* is http://eros.cagd.eas.asu.edu/˜farin/gbook/gbook_home.html.

2×2 Linear Systems

5

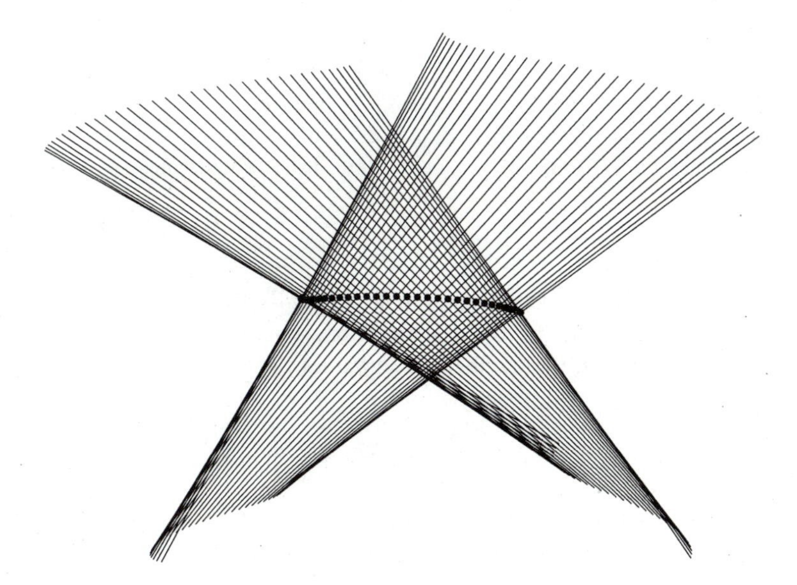

Figure 5.1.
Intersections of lines: two families of lines are shown; the intersections of corresponding line pairs are marked. For each intersection, a 2×2 linear system has to be solved.

Just about anybody can solve two equations in two unknowns by somehow manipulating the equations. In this chapter, we will develop a systematic way for finding the solution, simply by checking the underlying geometry. This approach will later enable us to solve much larger systems of equations. Figure 5.1 illustrates repeatedly, the intersection of two lines: a problem that can be formulated as a 2×2 linear system.

5.1 Coordinate Transformations

In our standard $[\mathbf{e}_1, \mathbf{e}_2]$-coordinate system, suppose we are given a point $\mathbf{p}$ and two vectors $\mathbf{a}_1$ and $\mathbf{a}_2$. We may use $\mathbf{p}, \mathbf{a}_1, \mathbf{a}_2$ as a new coordinate

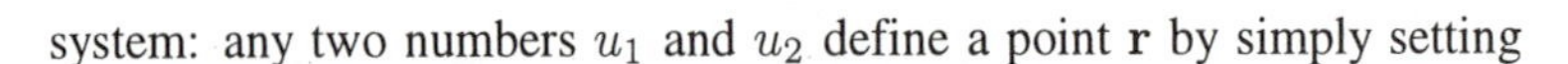

system: any two numbers u_1 and u_2 define a point $\mathbf{r}$ by simply setting

$$\mathbf{r} = \mathbf{p} + u_1\mathbf{a}_1 + u_2\mathbf{a}_2. \tag{5.1}$$

This is illustrated in Sketch 55.

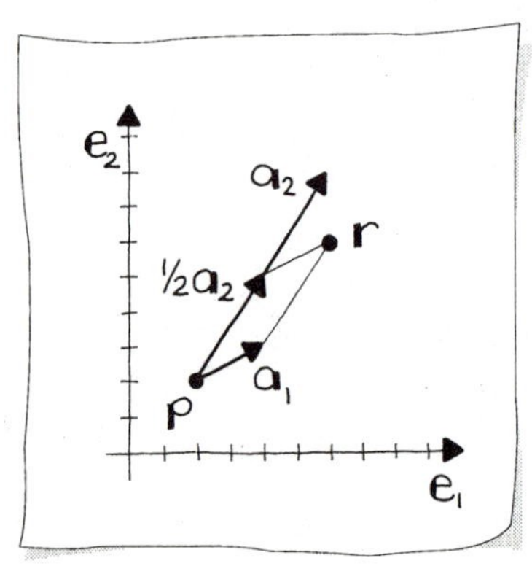

Sketch 55
A new coordinate system.

Example 5.1 Before we proceed further, we give an example. Let

$$\mathbf{p} = \begin{bmatrix} 2 \\ 2 \end{bmatrix}, \quad \mathbf{a}_1 = \begin{bmatrix} 2 \\ 1 \end{bmatrix}, \quad \mathbf{a}_2 = \begin{bmatrix} 4 \\ 6 \end{bmatrix},$$

and $u_1 = 1, u_2 = 1/2$. Thus

$$\mathbf{r} = \begin{bmatrix} 2 \\ 2 \end{bmatrix} + 1 \times \begin{bmatrix} 2 \\ 1 \end{bmatrix} + \frac{1}{2} \times \begin{bmatrix} 4 \\ 6 \end{bmatrix} = \begin{bmatrix} 6 \\ 6 \end{bmatrix}.$$

This example is illustrated in Sketch 55. In the $[\mathbf{a}_1, \mathbf{a}_2]$-system, $\mathbf{r}$ has coordinates $(1, 1/2)$. In the $[\mathbf{e}_1, \mathbf{e}_2]$-system, it has coordinates $(6, 6)$.

If we were given $\mathbf{r}$'s coordinates in the $[\mathbf{e}_1, \mathbf{e}_2]$-system, how can we find its coordinates in the $[\mathbf{a}_1, \mathbf{a}_2]$-system? Let's return to our example, and pretend that we don't know the correct values for u_1 and u_2. If we want to find them, we could write

$$\begin{bmatrix} 2 \\ 2 \end{bmatrix} + u_1 \times \begin{bmatrix} 2 \\ 1 \end{bmatrix} + u_2 \times \begin{bmatrix} 4 \\ 6 \end{bmatrix} = \begin{bmatrix} 6 \\ 6 \end{bmatrix}.$$

What we have here are really two equations in the two unknowns u_1 and u_2, which we see by expanding the point/vector equations into

$$\begin{aligned} 2 + 2u_1 + 4u_2 &= 6 \\ 2 + u_1 + 6u_2 &= 6. \end{aligned}$$

This quickly becomes

$$\begin{aligned} 2u_1 + 4u_2 &= 4 \\ u_1 + 6u_2 &= 4. \end{aligned} \tag{5.2}$$

These two equations in two unknowns have the solution $u_1 = 1$ and $u_2 = 1/2$, as is seen by inserting these values for u_1 and u_2 into the equations.

Being able to solve two simultaneous sets of equations allows us to switch back and forth between different coordinate systems. The rest of this chapter is dedicated to a detailed discussion of how to solve these equations.

5.2 The Matrix Form

The two equations in (5.2) are also called a *linear system*. It can be written more compactly if we use matrix notation:

$$\begin{bmatrix} 2 & 4 \\ 1 & 6 \end{bmatrix} \begin{bmatrix} u_1 \\ u_2 \end{bmatrix} = \begin{bmatrix} 4 \\ 4 \end{bmatrix}. \tag{5.3}$$

In general, a 2×2 linear system looks like this:

$$\begin{bmatrix} a_{1,1} & a_{1,2} \\ a_{2,1} & a_{2,2} \end{bmatrix} \begin{bmatrix} u_1 \\ u_2 \end{bmatrix} = \begin{bmatrix} b_1 \\ b_2 \end{bmatrix}. \tag{5.4}$$

It is shorthand notation for the equations

$$a_{1,1}u_1 + a_{1,2}u_2 = b_1$$
$$a_{2,1}u_1 + a_{2,2}u_2 = b_2.$$

We sometimes write it even shorter, using a matrix A:

$$A\mathbf{u} = \mathbf{b}, \tag{5.5}$$

where

$$A = \begin{bmatrix} a_{1,1} & a_{1,2} \\ a_{2,1} & a_{2,2} \end{bmatrix}, \quad \mathbf{u} = \begin{bmatrix} u_1 \\ u_2 \end{bmatrix}, \quad \mathbf{b} = \begin{bmatrix} b_1 \\ b_1 \end{bmatrix}.$$

Both $\mathbf{u}$ and $\mathbf{b}$ represent vectors, not points! See Sketch 56 for an illustration.

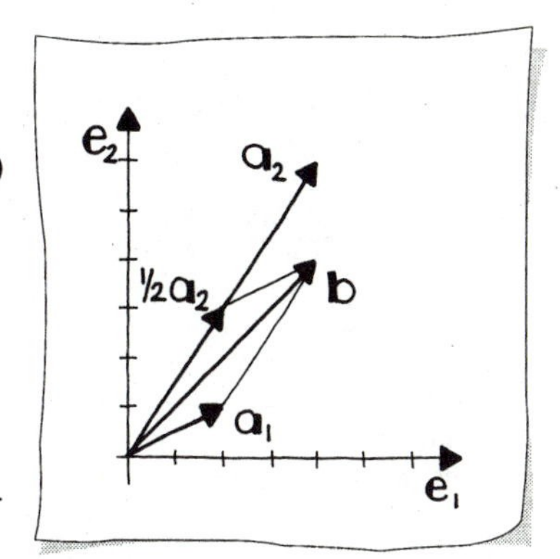

Sketch 56
Geometry of a 2x2 system.

While the savings of this notation is not completely obvious in the 2×2 case, it will save a lot of work for more complicated cases with more equations and unknowns.

The columns of the matrix A correspond to the vectors $\mathbf{a}_1$ and $\mathbf{a}_2$. We could then rewrite our linear system as

$$u_1\mathbf{a}_1 + u_2\mathbf{a}_2 = \mathbf{b}.$$

Geometrically, we are trying to express the given vector $\mathbf{b}$ as a linear combination of the given vectors $\mathbf{a}_1$ and $\mathbf{a}_2$; we need to determine the factors u_1 and u_2.

5.3 A Direct Approach: Cramer's Rule

Sketch 57 offers a direct solution to our linear system. By simply inspecting the areas of the parallelograms in the sketch, we see that

$$u_1 = \frac{\text{area}(\mathbf{b}, \mathbf{a}_2)}{\text{area}(\mathbf{a}_1, \mathbf{a}_2)},$$
$$u_2 = \frac{\text{area}(\mathbf{a}_1, \mathbf{b})}{\text{area}(\mathbf{a}_1, \mathbf{a}_2)}.$$

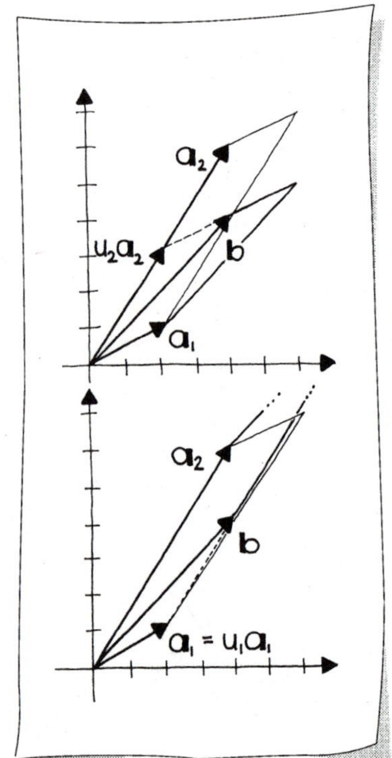

Sketch 57
Cramer's rule.

The area of a parallelogram is given by the determinant of the two vectors spanning it. Recall from Section 4.9 that this is a signed area. This method of solving for the solution of a linear system is called *Cramer's rule.*

Example 5.2 Applying Cramer's rule to the linear system in (5.3), we get

$$u_1 = \frac{\begin{vmatrix} 4 & 4 \\ 4 & 6 \end{vmatrix}}{\begin{vmatrix} 2 & 4 \\ 1 & 6 \end{vmatrix}} = \frac{8}{8},$$

$$u_2 = \frac{\begin{vmatrix} 2 & 4 \\ 1 & 4 \end{vmatrix}}{\begin{vmatrix} 2 & 4 \\ 1 & 6 \end{vmatrix}} = \frac{4}{8}.$$

Examining the determinant in the numerator, notice that $\mathbf{b}$ replaces $\mathbf{a}_1$ in the solution for u_1 and then $\mathbf{b}$ replaces $\mathbf{a}_2$ in the solution for u_2.

Cramer's rule is primarily of theoretical importance. For larger systems, Cramer's rule is both expensive and numerically unstable. Hence, we now study a more effective method.

5.4 Gauss Elimination

Let's consider a special 2 × 2 linear system:

$$\begin{bmatrix} a_{1,1} & a_{1,2} \\ 0 & a_{2,2} \end{bmatrix} \mathbf{u} = \mathbf{b}. \tag{5.6}$$

This situation is shown in Sketch 58. The matrix is called *upper triangular* because all elements below the diagonal are zero, forming a triangle of numbers above the diagonal.

We can solve this system without much work. Examining the last equation, we see it is possible to solve for

$$u_2 = b_2/a_{2,2}.$$

With u_2 in hand, we can solve the first equation for

$$u_1 = \frac{1}{a_{1,1}}(b_1 - u_2 a_{1,2}).$$

This technique of solving the equations from the bottom up is called *back substitution.*

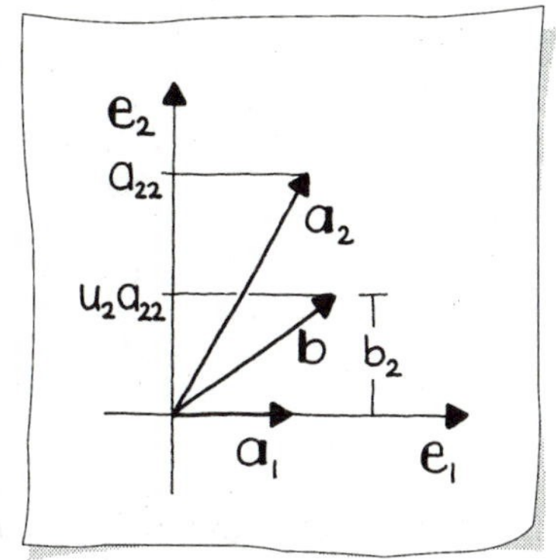

Sketch 58
A special linear system.

In general, we will not be so lucky to encounter an upper triangular system as in (5.6). But any (almost any, at least) linear system may be *transformed* to this simple case, as we shall see by re-examining the system in (5.3). We write it as

$$u_1 \begin{bmatrix} 2 \\ 1 \end{bmatrix} + u_2 \begin{bmatrix} 4 \\ 6 \end{bmatrix} = \begin{bmatrix} 4 \\ 4 \end{bmatrix}.$$

This situation is shown in Sketch 59. Clearly, $\mathbf{a}_1$ is not on the $\mathbf{e}_1$-axis as we would like, but we can apply a stepwise procedure so that it will become just that. This systematic, stepwise procedure is called *Gauss elimination.*

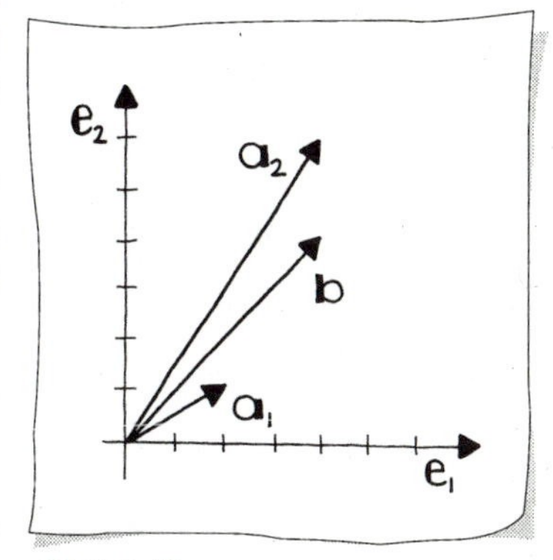

Sketch 59
The geometry of a linear system.

Recall one key fact from Chapter 4: *linear maps do not change linear combinations.* That means if we apply the same linear map to all vectors in our system, then the factors u_1 and u_2 won't change. If the map is given by a matrix S, then

$$S\left[u_1 \begin{bmatrix} 2 \\ 1 \end{bmatrix} + u_2 \begin{bmatrix} 4 \\ 6 \end{bmatrix}\right] = S \begin{bmatrix} 4 \\ 4 \end{bmatrix}.$$

In order to get $\mathbf{a}_1$ to line up with the $\mathbf{e}_1$-axis, we will employ a *shear* parallel to the $\mathbf{e}_2$-axis, such that

$$\begin{bmatrix} 2 \\ 1 \end{bmatrix} \quad \text{is mapped to} \quad \begin{bmatrix} 2 \\ 0 \end{bmatrix}.$$

That shear (see Section 4.7) is given by the matrix

$$S_1 = \begin{bmatrix} 1 & 0 \\ -1/2 & 1 \end{bmatrix}.$$

We apply S_1 to all vectors involved in our system:

$$\begin{bmatrix} 1 & 0 \\ -1/2 & 1 \end{bmatrix} \begin{bmatrix} 2 \\ 1 \end{bmatrix} = \begin{bmatrix} 2 \\ 0 \end{bmatrix}, \quad \begin{bmatrix} 1 & 0 \\ -1/2 & 1 \end{bmatrix} \begin{bmatrix} 4 \\ 6 \end{bmatrix} = \begin{bmatrix} 4 \\ 4 \end{bmatrix},$$

$$\begin{bmatrix} 1 & 0 \\ -1/2 & 1 \end{bmatrix} \begin{bmatrix} 4 \\ 4 \end{bmatrix} = \begin{bmatrix} 4 \\ 2 \end{bmatrix}.$$

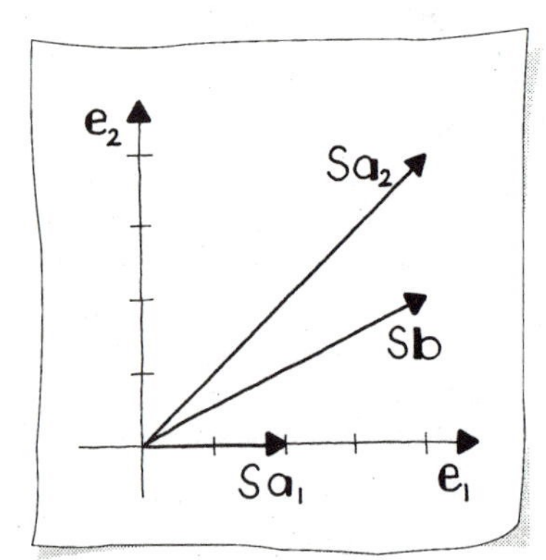

Sketch 60
Shearing the vectors in a linear system.

The effect of this map is shown in Sketch 60.

Our transformed system now reads

$$\begin{bmatrix} 2 & 4 \\ 0 & 4 \end{bmatrix} \begin{bmatrix} u_1 \\ u_2 \end{bmatrix} = \begin{bmatrix} 4 \\ 2 \end{bmatrix}.$$

Now we can employ *back substitution* to find:

$$u_2 = 2/4 = 1/2$$
$$u_1 = \frac{1}{2}(4 - 4 \times \frac{1}{2}) = 1.$$

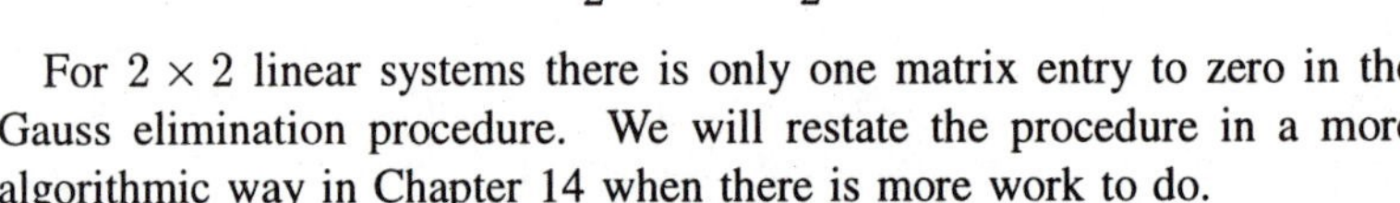

For 2×2 linear systems there is only one matrix entry to zero in the Gauss elimination procedure. We will restate the procedure in a more algorithmic way in Chapter 14 when there is more work to do.

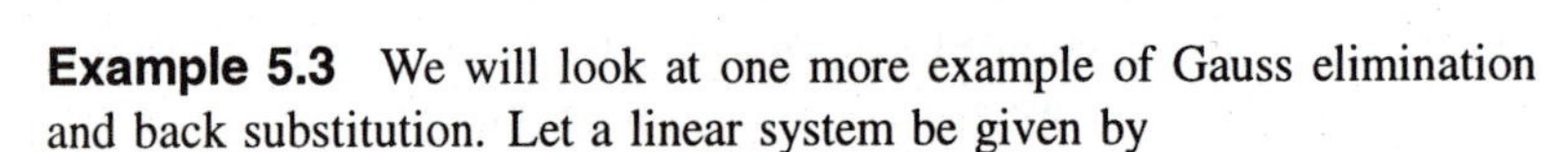

Example 5.3 We will look at one more example of Gauss elimination and back substitution. Let a linear system be given by

$$\begin{bmatrix} -1 & 4 \\ 2 & 2 \end{bmatrix} \begin{bmatrix} u_1 \\ u_2 \end{bmatrix} = \begin{bmatrix} 0 \\ 2 \end{bmatrix}.$$

The shear that takes $\mathbf{a}_1$ to the $\mathbf{e}_1$-axis is given by

$$S_1 = \begin{bmatrix} 1 & 0 \\ 2 & 1 \end{bmatrix},$$

and it transforms the system to

$$\begin{bmatrix} -1 & 4 \\ 0 & 10 \end{bmatrix} \begin{bmatrix} u_1 \\ u_2 \end{bmatrix} = \begin{bmatrix} 0 \\ 2 \end{bmatrix}.$$

Draw your own sketch to understand the geometry.

Using back substitution, the solution is now easily found as $u_1 = 8/10$ and $u_2 = 2/10$.

5.5 Undoing Maps: Inverse Matrices

In this section, we will see how to *undo* a linear map. Reconsider the linear system

$$A\mathbf{u} = \mathbf{b}.$$

The matrix A maps $\mathbf{u}$ to $\mathbf{b}$. Now that we know $\mathbf{u}$, what is the matrix B that maps $\mathbf{b}$ back to $\mathbf{u}$,

$$\mathbf{u} = B\mathbf{b}? \tag{5.7}$$

Defining B — the inverse map — is the purpose of this section.

In solving the original linear system, we applied shears to the column vectors of A and to $\mathbf{b}$. After the first shear, we had

$$S_1 A\mathbf{u} = S_1\mathbf{b}.$$

This demonstrated how shears can be used to zero elements of the matrix. Let's return to the example linear system in (5.3). After applying S_1 the system became

$$\begin{bmatrix} 2 & 4 \\ 0 & 4 \end{bmatrix} \begin{bmatrix} u_1 \\ u_2 \end{bmatrix} = \begin{bmatrix} 4 \\ 2 \end{bmatrix}.$$

Let's use another shear to zero the upper right element. Geometrically, this corresponds to constructing a shear that will map the new $\mathbf{a}_2$ to the $\mathbf{e}_2$−axis. It is given by the matrix

$$S_2 = \begin{bmatrix} 1 & -1 \\ 0 & 1 \end{bmatrix}.$$

Applying it to all vectors gives the new system

$$\begin{bmatrix} 2 & 0 \\ 0 & 4 \end{bmatrix} \begin{bmatrix} u_1 \\ u_2 \end{bmatrix} = \begin{bmatrix} 2 \\ 2 \end{bmatrix}.$$

After the second shear, our linear system has been changed to

$$S_2 S_1 A\mathbf{u} = S_2 S_1\mathbf{b}.$$

We are not limited to applying shears — any affine map will do. Next apply a non-uniform scaling S_3 in the $\mathbf{e}_1$ and $\mathbf{e}_2$ directions that will map the latest $\mathbf{a}_1$ and $\mathbf{a}_2$ onto the vectors $\mathbf{e}_1$ and $\mathbf{e}_2$. For our current example,

$$S_3 = \begin{bmatrix} 1/2 & 0 \\ 0 & 1/4 \end{bmatrix}.$$

The new system becomes

$$\begin{bmatrix} 1 & 0 \\ 0 & 1 \end{bmatrix} \begin{bmatrix} u_1 \\ u_2 \end{bmatrix} = \begin{bmatrix} 1 \\ 1/2 \end{bmatrix},$$

which corresponds to

$$S_3 S_2 S_1 A\mathbf{u} = S_3 S_2 S_1 \mathbf{b}.$$

This is a very special system. First of all, to solve for $\mathbf{u}$ is now trivial because A has been transformed into the *unit matrix* or *identity matrix* I:

$$I = \begin{bmatrix} 1 & 0 \\ 0 & 1 \end{bmatrix}. \tag{5.8}$$

This process of transforming A until it becomes the identity is theoretically equivalent to the back substitution process of Section 5.4. Hoewever, back substitution uses fewer operations and thus is the method of choice for solving linear systems.

Yet we have now found the matrix B in (5.7)! The two shears and scaling transformed A into the identity matrix I:

$$S_3 S_2 S_1 A = I; \tag{5.9}$$

thus the solution of the system is

$$\mathbf{u} = S_3 S_2 S_1 \mathbf{b}. \tag{5.10}$$

This leads to the definition of the *inverse matrix* A^{-1} of a matrix A:

$$A^{-1} = S_3 S_2 S_1. \tag{5.11}$$

The matrix A^{-1} *undoes* the effect of matrix the A: the vector $\mathbf{u}$ was mapped to $\mathbf{b}$ by A, and $\mathbf{b}$ is mapped back to $\mathbf{u}$ by A^{-1}. Thus we can now write (5.10) as

$$\mathbf{u} = A^{-1}\mathbf{b}.$$

If we combine (5.9) and (5.11), we immediately get

$$A^{-1}A = I. \tag{5.12}$$

This makes intuitive sense, since the actions of a map and its inverse should cancel out, i.e., not change anything — that is what I does!

The inverse of the identity is the identity:

$$I^{-1} = I.$$

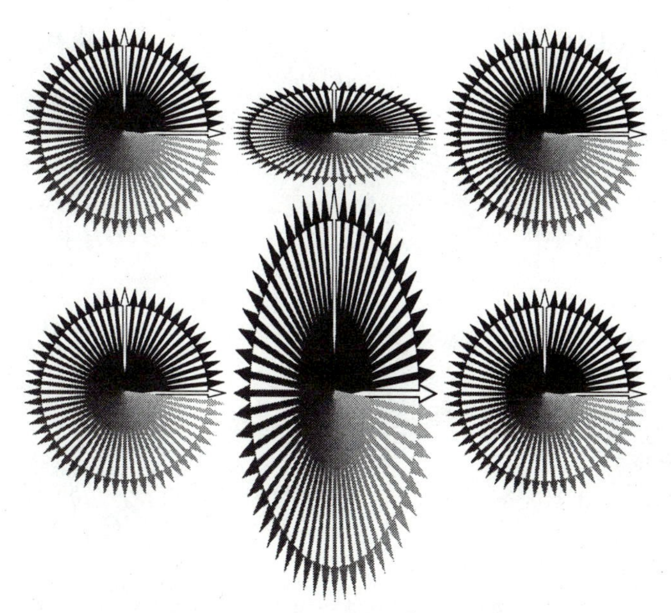

Figure 5.2.
Inverse matrices: top, a set of unit vectors, the result of applying a scale, then the result of the inverse scale. Bottom: a set of unit vectors, the result of applying the inverse scale, then the result of the original scale.

The inverse of a scaling is given by:

$$\begin{bmatrix} s & 0 \\ 0 & t \end{bmatrix}^{-1} = \begin{bmatrix} 1/s & 0 \\ 0 & 1/t \end{bmatrix}.$$

Multiply this out to convince yourself!

Figure 5.2 shows the effects of a matrix and its inverse for the scaling

$$\begin{bmatrix} 1 & 0 \\ 0 & 0.5 \end{bmatrix}.$$

Figure 5.3 shows the effects of a matrix and its inverse for the shear

$$\begin{bmatrix} 1 & 1 \\ 0 & 1 \end{bmatrix}.$$

We consider the inverse of a rotation as follows: if R_α rotates by α degrees counterclockwise, then $R_{-\alpha}$ rotates by α degrees clockwise, or

$$R_\alpha^{-1} = R_\alpha^{\mathrm{T}},$$

as you can see from the definition of a rotation matrix (4.16).

Figure 5.3.
Inverse matrices: top, a set of unit vectors, the result of applying a shear, then the result of the inverse shear. Bottom: a set of unit vectors, the result of applying the inverse shear, then the result of the original shear.

The rotation matrix is an example of an *orthogonal matrix.* An orthogonal matrix A is characterized by the fact that

$$A^{-1} = A^{\mathrm{T}}.$$

The column vectors $\mathbf{a}_1$ and $\mathbf{a}_2$ of an orthogonal matrix satisfy $\|\mathbf{a}_1\| = 1$, $\|\mathbf{a}_2\| = 1$ and $\mathbf{a}_1 \cdot \mathbf{a}_2 = 0$. In words, the column vectors are *orthonormal.* The row vectors are orthonormal as well. Those transformations that are described by orthogonal matrices are called *rigid body motions*. The determinant of an orthogonal matrix is ± 1.

We add without proof two fairly obvious identities:

$$A^{-1^{-1}} = A, \tag{5.13}$$

which should be obvious from Figures 5.2 and 5.3.

Next:

$$A^{-1^{\mathrm{T}}} = A^{\mathrm{T}^{-1}}. \tag{5.14}$$

Figure 5.4 illustrates this for

$$A = \begin{bmatrix} 1 & 0 \\ 1 & 0.5 \end{bmatrix}.$$

Figure 5.4.
Inverse matrices: top illustrates $I, A^{-1}, A^{-1^{\mathrm{T}}}$ and bottom illustrates $I, A^{\mathrm{T}}, A^{\mathrm{T}^{-1}}$.

Given a matrix A, how do we compute its inverse? Let us start with

$$AA^{-1} = I. \tag{5.15}$$

If we denote the two (unknown) columns of A^{-1} by $\bar{\mathbf{a}}_1$ and $\bar{\mathbf{a}}_2$, and those of I by $\mathbf{e}_1$ and $\mathbf{e}_2$, then (5.15) may be written as

$$A\begin{bmatrix} \bar{\mathbf{a}}_1 & \bar{\mathbf{a}}_2 \end{bmatrix} = \begin{bmatrix} \mathbf{e}_1 & \mathbf{e}_2 \end{bmatrix}.$$

This is really short for two linear systems:

$$A\bar{\mathbf{a}}_1 = \mathbf{e}_1 \quad \text{and} \quad A\bar{\mathbf{a}}_2 = \mathbf{e}_2.$$

Both systems have the same matrix A and can thus be solved *simultaneously*. All we have to do is to apply the familiar shears and scale — those that transform A to I — to both $\mathbf{e}_1$ and $\mathbf{e}_2$.

Example 5.4 Let's revisit an example from above with

$$A = \begin{bmatrix} -1 & 4 \\ 2 & 2 \end{bmatrix}.$$

Our two simultaneous equations are:

$$\begin{bmatrix} -1 & 4 \\ 2 & 2 \end{bmatrix}\begin{bmatrix} \bar{\mathbf{a}}_1 & \bar{\mathbf{a}}_2 \end{bmatrix} = \begin{bmatrix} 1 & 0 \\ 0 & 1 \end{bmatrix}.$$

The first shear takes this to

$$\begin{bmatrix} -1 & 4 \\ 0 & 10 \end{bmatrix} \begin{bmatrix} \bar{\mathbf{a}}_1 & \bar{\mathbf{a}}_2 \end{bmatrix} = \begin{bmatrix} 1 & 0 \\ 2 & 1 \end{bmatrix}.$$

The second shear yields

$$\begin{bmatrix} -1 & 0 \\ 0 & 10 \end{bmatrix} \begin{bmatrix} \bar{\mathbf{a}}_1 & \bar{\mathbf{a}}_2 \end{bmatrix} = \begin{bmatrix} 2/10 & -4/10 \\ -2 & 1 \end{bmatrix}.$$

Finally the scaling produces

$$\begin{bmatrix} 1 & 0 \\ 0 & 1 \end{bmatrix} \begin{bmatrix} \bar{\mathbf{a}}_1 & \bar{\mathbf{a}}_2 \end{bmatrix} = \begin{bmatrix} -2/10 & 4/10 \\ -2/10 & 1/10 \end{bmatrix}.$$

Thus the inverse matrix

$$A^{-1} = \begin{bmatrix} -2/10 & 4/10 \\ -2/10 & 1/10 \end{bmatrix}.$$

5.6 Unsolvable Systems

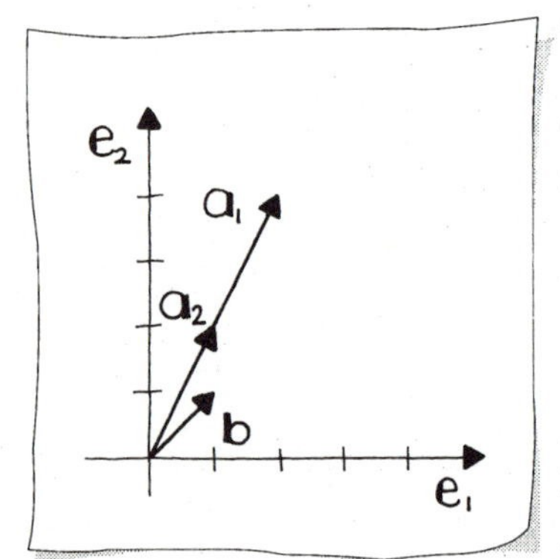

Sketch 61
An unsolvable linear system.

Consider the situation shown in Sketch 61. The two vectors $\mathbf{a}_1$ and $\mathbf{a}_2$ are multiples of each other. In other words, they are *linearly dependent.*

The corresponding linear system is

$$\begin{bmatrix} 2 & 1 \\ 4 & 2 \end{bmatrix} \begin{bmatrix} u_1 \\ u_2 \end{bmatrix} = \begin{bmatrix} 1 \\ 1 \end{bmatrix}.$$

It is obvious from the sketch that we have a problem here, but let's just blindly apply Gauss elimination; apply a shear such that $\mathbf{a}_1$ is mapped to the $\mathbf{e}_1$-axis. The resulting system is

$$\begin{bmatrix} 2 & 1 \\ 0 & 0 \end{bmatrix} \begin{bmatrix} u_1 \\ u_2 \end{bmatrix} = \begin{bmatrix} 1 \\ -1 \end{bmatrix}.$$

But the last equation reads $0 = -1$, and now we really are in trouble! This means that our system does not have a solution.

5.7 Underdetermined Systems

Consider the system

$$\begin{bmatrix} 2 & 1 \\ 4 & 2 \end{bmatrix} \begin{bmatrix} u_1 \\ u_2 \end{bmatrix} = \begin{bmatrix} 3 \\ 6 \end{bmatrix},$$

shown in Sketch 62.

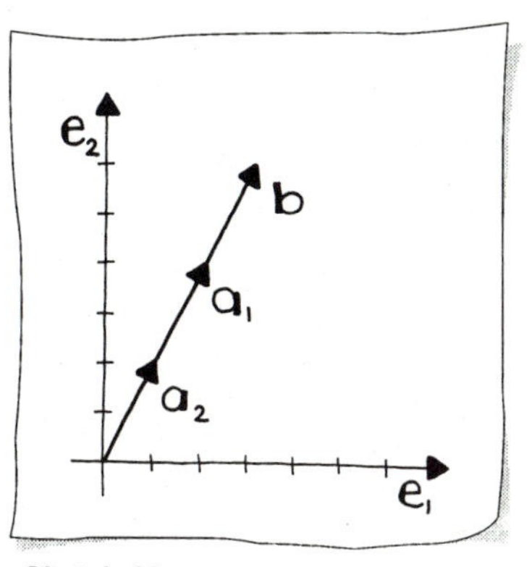

Sketch 62
An underdetermined linear system.

Again, we shear $\mathbf{a}_1$ onto the $\mathbf{e}_1$−axis, and obtain

$$\begin{bmatrix} 1 & 3 \\ 0 & 0 \end{bmatrix} \begin{bmatrix} u_1 \\ u_2 \end{bmatrix} = \begin{bmatrix} 0 \\ 0 \end{bmatrix}.$$

Now the last equation reads $0 = 0$ — true, but a bit trivial! In reality, our system is just one equation written down twice in slightly different forms. This is also clear from the sketch: $\mathbf{b}$ may be written as a multiple of either $\mathbf{a}_1$ or $\mathbf{a}_2$.

5.8 Homogeneous Systems

A system of the form

$$A\mathbf{u} = \mathbf{0}, \tag{5.16}$$

i.e., one where the right-hand side consists of the zero vector, is called *homogeneous*. If it has a solution $\mathbf{u}$, then clearly all multiples $c\mathbf{u}$ are also solutions: we multiply both sides of the equations by a common factor c. One obvious solution is the zero vector itself; this is called the *trivial solution* and is usually of little interest.

Not all homogeneous systems do have a solution, however. Equation (5.16) may be read as follows: What vector $\mathbf{u}$, when mapped by A, has the zero vector as its image? The only maps capable of achieving this are *projections*. They are characterized by the fact that their two columns $\mathbf{a}_1$ and $\mathbf{a}_2$ are parallel, or linearly dependent.

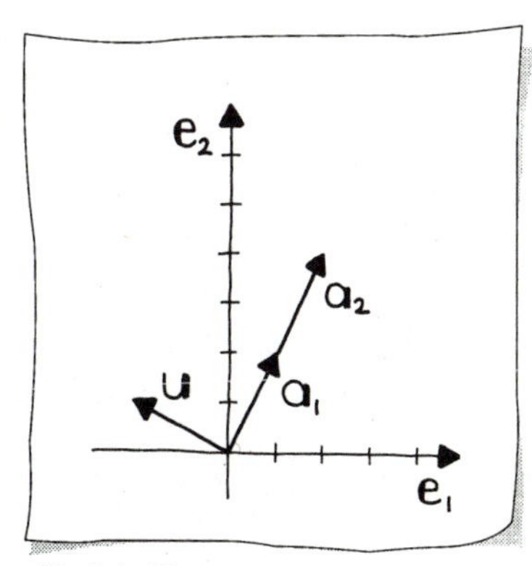

Sketch 63
A homogeneous system.

Example 5.5 An example, illustrated in Sketch 63, should help. Let our homogeneous system be

$$\begin{bmatrix} 1 & 2 \\ 2 & 4 \end{bmatrix} \mathbf{u} = \begin{bmatrix} 0 \\ 0 \end{bmatrix}.$$

Clearly, $\mathbf{a}_2 = 2\mathbf{a}_1$; the matrix A projects all vectorsonto this direction. In this example, because A is symmetric, any vector $\mathbf{u}$ which is perpendicular to this direction will be projected to the zero vector:

$$A[c\mathbf{u}] = \mathbf{0}.$$

An easy check reveals that

$$\mathbf{u} = \begin{bmatrix} -2 \\ 1 \end{bmatrix}$$

is a solution to the system; so is any multiple of it. Also check that $\mathbf{a}_1 \cdot \mathbf{u} = 0$, so they are in fact perpendicular.

Example 5.6 We now consider an example of a homogeneous system that does not have a solution:

$$\begin{bmatrix} 1 & 2 \\ 2 & 1 \end{bmatrix} \mathbf{u} = \begin{bmatrix} 0 \\ 0 \end{bmatrix}.$$

The two columns of A are not linearly dependent; therefore, A is not a projection. Then it cannot map any (nonzero) vector $\mathbf{u}$ to the zero vector!

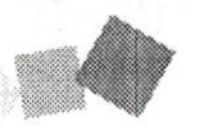

In general, we may state that a homogeneous system has nonzero solutions only if the columns of the matrix are linearly dependent.

5.9 Numerical Strategies: Pivoting

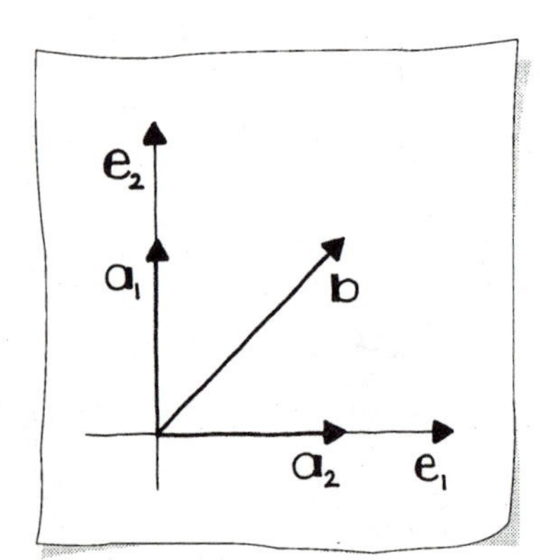

Sketch 64
A linear system that needs pivoting.

Consider the system

$$\begin{bmatrix} 0 & 1 \\ 1 & 0 \end{bmatrix} \begin{bmatrix} u_1 \\ u_2 \end{bmatrix} = \begin{bmatrix} 1 \\ 1 \end{bmatrix},$$

illustrated in Sketch 64.

Our standard approach, shearing $\mathbf{a}_1$ onto the $\mathbf{e}_1$-axis, will not work here; there is no shear that takes

$$\begin{bmatrix} 0 \\ 1 \end{bmatrix}$$

onto the $\mathbf{e}_1$−axis. However, there is no problem if we simply interchange the two equations! Then we have

$$\begin{bmatrix} 1 & 0 \\ 0 & 1 \end{bmatrix} \begin{bmatrix} u_1 \\ u_2 \end{bmatrix} = \begin{bmatrix} 1 \\ 1 \end{bmatrix},$$

and thus $u_1 = u_2 = 1$. So we cannot blindly apply a shear to $\mathbf{a}_1$; we must first check that one exists. If it does not — i.e., if $a_{1,1} = 0$ — interchange the equations.

As a rule of thumb, if a method fails because some number equals zero, then it will work poorly if that number is small. It is thus advisable to interchange the two equations anytime we have $|a_{1,1}| < |a_{1,2}|$. The absolute value is used here since we are interested in the magnitude of the involved numbers, not their sign. The process of interchanging equations is called *pivoting*, and it is used to improve numerical stability.

Example 5.7 Let's study a somewhat realistic example taken from [15]:

$$\begin{bmatrix} 0.0001 & 1 \\ 1 & 1 \end{bmatrix} \begin{bmatrix} u_1 \\ u_2 \end{bmatrix} = \begin{bmatrix} 1 \\ 2 \end{bmatrix},$$

If we shear $\mathbf{a}_1$ onto the $\mathbf{e}_1$-axis, the new system reads

$$\begin{bmatrix} 0.0001 & 1 \\ 0 & -9999 \end{bmatrix} \begin{bmatrix} u_1 \\ u_2 \end{bmatrix} = \begin{bmatrix} 1 \\ -9998 \end{bmatrix},$$

Note how "far out" the new $\mathbf{a}_2$ and $\mathbf{b}$ are relative to $\mathbf{a}_1$! This is the type of behavior that causes numerical problems. Numbers that differ greatly in magnitude tend to need more digits for calculation. On a finite precision machine these extra digits are not always available, thus round-off can take us far from the true solution.

Suppose we have a machine which only stores three digits, although it calculates with six digits. Due to round-off, the system above would be stored as

$$\begin{bmatrix} 0.0001 & 1 \\ 0 & -10000 \end{bmatrix} \begin{bmatrix} u_1 \\ u_2 \end{bmatrix} = \begin{bmatrix} 1 \\ -10000 \end{bmatrix},$$

which would result in a solution of $u_2 = 1$ and $u_1 = 0$, which is not close to the true solution (rounded to five digits) of $u_2 = 1.0001$ and $u_1 = 0.99990$.

Luckily, pivoting is a tool to damper the effects of round-off. Now employ pivoting by interchanging the rows, yielding the system

$$\begin{bmatrix} 1 & 1 \\ 0.0001 & 1 \end{bmatrix} \begin{bmatrix} u_1 \\ u_2 \end{bmatrix} = \begin{bmatrix} 2 \\ 1 \end{bmatrix}.$$

Shear $\mathbf{a}_1$ onto the $\mathbf{e}_1$-axis, and the new system reads

$$\begin{bmatrix} 1 & 1 \\ 0 & 0.9999 \end{bmatrix} \begin{bmatrix} u_1 \\ u_2 \end{bmatrix} = \begin{bmatrix} 2 \\ 0.9998 \end{bmatrix}.$$

Notice that the vectors are all within the same range. Even with the three digit machine, this system will allow us to compute a result which is "close" to the true solution, because the effects of round-off have been minimized.

5.10 Defining a Map

Matrices map vectors to vectors. If we know the result of such a map, namely that two vectors $\mathbf{v}_1$ and $\mathbf{v}_2$ were mapped to $\mathbf{v}_1'$ and $\mathbf{v}_2'$, can we find the matrix that did it?

Suppose some matrix A was responsible for the map. We would then have the two equations

$$A\mathbf{v}_1 = \mathbf{v}_1' \quad \text{and} \quad A\mathbf{v}_2 = \mathbf{v}_2'.$$

Combining them, we can write

$$A[\mathbf{v}_1, \mathbf{v}_2] = [\mathbf{v}_1', \mathbf{v}_2'],$$

or, even shorter:

$$AV = V'. \tag{5.17}$$

To define A, we simply use shears, S_1 and S_2, and a scaling S_3 to transform V into the identity matrix as follows:

$$\begin{aligned} AVS_1S_2S_3 &= V'S_1S_2S_3 \\ AVV^{-1} &= V'V^{-1} \\ A &= V'V^{-1}. \end{aligned}$$

5.11 Exercises

1. Using the matrix form, write down the linear system to find $\begin{bmatrix} 6 \\ 3 \end{bmatrix}$ in terms of the local coordinate system defined by

$$\mathbf{p} = \begin{bmatrix} 3 \\ 3 \end{bmatrix}, \quad \mathbf{a}_1 = \begin{bmatrix} 2 \\ -3 \end{bmatrix}, \quad \mathbf{a}_2 = \begin{bmatrix} 6 \\ 0 \end{bmatrix}.$$

2. Use Cramer's rule to solve the system in Exercise 1.
3. Give an example of an upper triangular matrix.
4. Use Gauss elimination and back substitution to solve the system in Exercise 1.

5. Find the inverse of the matrix in Exercise 1.

6. What is the inverse of the matrix

$$\begin{bmatrix} 10 & 0 \\ 0 & 0.5 \end{bmatrix}?$$

7. What is the inverse of the matrix

$$\begin{bmatrix} \cos 30^\circ & -\sin 30^\circ \\ \sin 30^\circ & \cos 30^\circ \end{bmatrix}?$$

8. Give an example along with a sketch of an unsolvable system. Do the same for an underdetermined system.

9. Under what conditions can a nontrivial solution be found to a homogeneous system?

10. Re-solve the system in Exercise 1 with Gauss elimination with pivoting.

11. Define the matrix A that maps

$$\begin{bmatrix} 1 \\ 0 \end{bmatrix} \rightarrow \begin{bmatrix} 1 \\ 0 \end{bmatrix} \quad \text{and} \quad \begin{bmatrix} 1 \\ 1 \end{bmatrix} \rightarrow \begin{bmatrix} 1 \\ -1 \end{bmatrix}.$$

Moving Things Around: Affine Maps

6

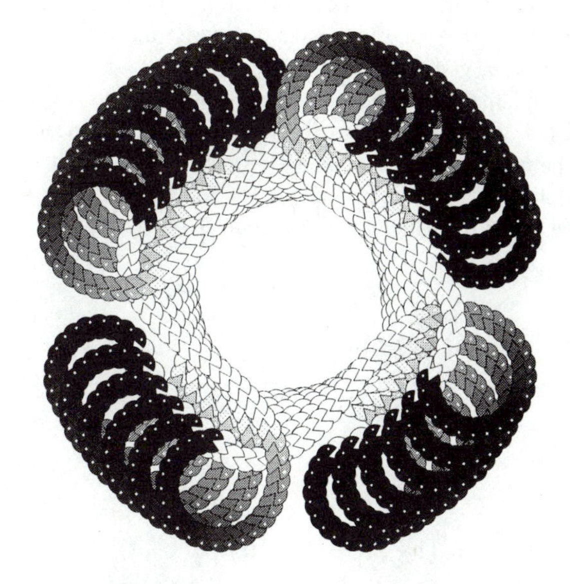

Figure 6.1.
Moving things around: affine maps in 2D applied to a familiar video game character.

Imagine playing a video game. As you press a button, figures and objects on the screen start moving around; they shift their position, they rotate, they zoom in or out. As you see this kind of motion, the video software must carry out quite a few computations. Such computations have been applied to the familiar face in Figure 6.1. These computations are implementations of *affine maps*, the subject of this chapter.

6.1 Affine and Linear Maps

A map of the form $\mathbf{v}' = A\mathbf{v}$ is called a *linear map* because it preserves linear combinations of vectors, see (4.9) or Sketch 46.

While linear maps take vectors to vectors, *affine maps* take points $\mathbf{x}$ to points $\mathbf{x}'$ and are of the form

$$\mathbf{x}' = \mathbf{p} + A(\mathbf{x} - \mathbf{o}), \tag{6.1}$$

where $\mathbf{o}$ is the origin of $\mathbf{x}$'s coordinate system. In most cases, we will have the familiar

$$\mathbf{o} = \begin{bmatrix} 0 \\ 0 \end{bmatrix},$$

and then we will simply drop the "$-\mathbf{o}$" part.

Affine maps are the basic tool to move and orient objects. All are of the form given in (6.1) and thus have two parts: a *translation*, given by $\mathbf{p}$ and a *linear map*, given by A.

A very fundamental property of linear maps has to do with *ratios* (see Section 2.4). What happens to the ratio of three collinear points when we map them by an affine map? The answer to this question is fairly fundamental to all of geometry, and it is: nothing. In other words, affine maps leave ratios unchanged, or *invariant*. To see this, let

$$\mathbf{p}_2 = (1-t)\mathbf{p}_1 + t\mathbf{p}_3$$

and let an affine map be defined by

$$\mathbf{x}' = A\mathbf{x} + \mathbf{p}.$$

We now have

$$\begin{aligned} \mathbf{p}_2' &= A((1-t)\mathbf{p}_1 + t\mathbf{p}_3) + \mathbf{p} \\ &= (1-t)A\mathbf{p}_1 + tA\mathbf{p}_3 + [(1-t)+t]\mathbf{p} \\ &= (1-t)[A\mathbf{p}_1 + \mathbf{p}] + t[A\mathbf{p}_3 + \mathbf{p}] \\ &= (1-t)\mathbf{p}_1' + t\mathbf{p}_3'. \end{aligned}$$

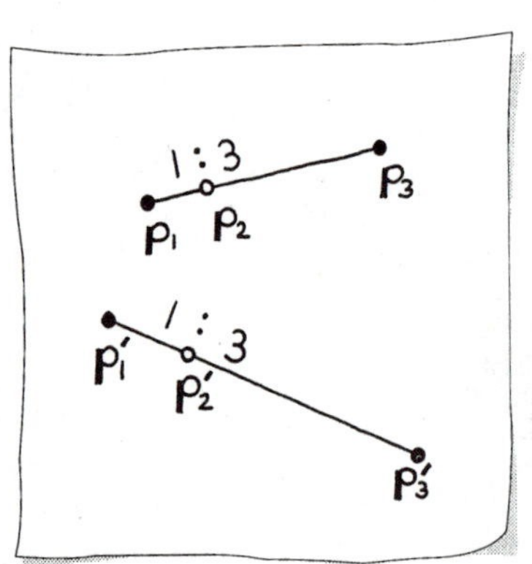

Sketch 65
Midpoints go to midpoints.

The step from the first to the second equation may seem a bit contrived; yet it is the one that makes crucial use of the fact that we are combining points using *barycentric combinations*: $(1-t)+t=1$.

The last equation shows that the linear $(1-t), t-$ relationship between three points is not changed by affine maps — meaning that their ratio is invariant. In particular, the midpoint of two points will be mapped to the midpoint of the image points (see Sketch 65).

The other basic property of affine maps is this: they map parallel lines to parallel lines. If two lines do not intersect before they are mapped, then they will not intersect afterwards either. Conversely, two lines that intersect before the map will also do so afterwards. Figure 6.2 shows how

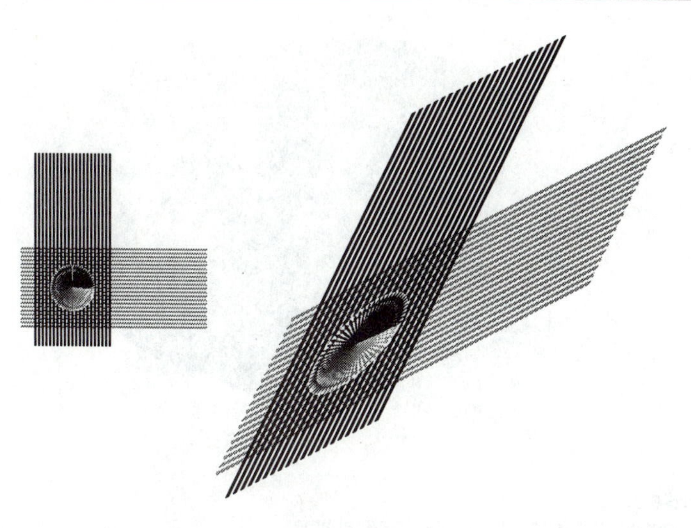

Figure 6.2.
Affine maps: parallel lines are mapped to parallel lines.

two families of parallel lines are mapped to two families of parallel lines. The two families intersect before and after the affine map. The map uses the matrix

$$A = \begin{bmatrix} 1 & 2 \\ 2 & 1 \end{bmatrix}.$$

6.2 Translations

If an object is moved without changing its orientation, then it is *translated*. See Figure 6.3 for several translations of the letter **D**.[1] How is this action covered by the general affine map in (6.1)? Recall the the *identity matrix* from Section 5.5 which has no effect whatsoever on any vector: we always have

$$I\mathbf{x} = \mathbf{x},$$

which you should be able to verify without effort.

A translation is thus written in the context of (6.1) as

$$\mathbf{x}' = \mathbf{p} + I\mathbf{x}.$$

One property of translations is that they do not change areas; all **D**'s in Figure 6.3 have the same area. A translation causes a *rigid body motion*. Recall that rotations are also of this type.

[1]We rendered the **D** darker as it was translated more.

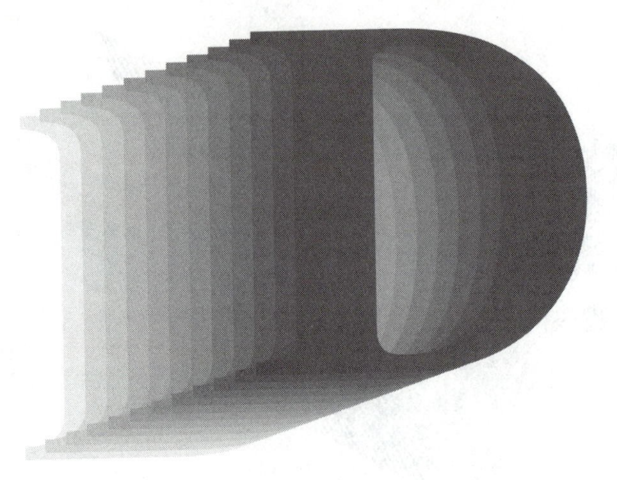

Figure 6.3.
Translations: the letter **D** is moved several times.

6.3 More General Affine Maps

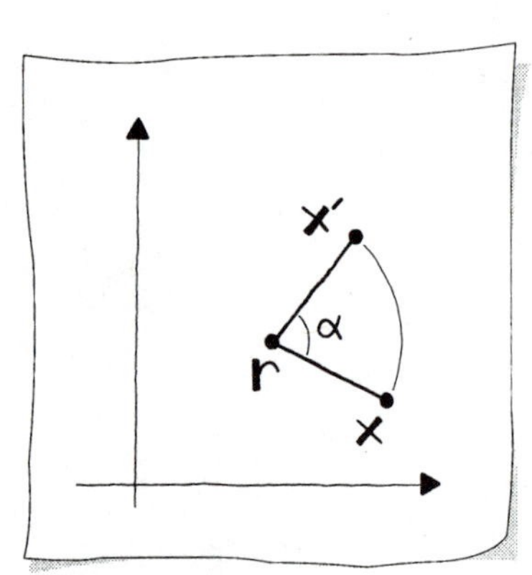

Sketch 66
Rotating a point about another point.

It is one thing to say "every affine map is of the form $A\mathbf{x}+\mathbf{p}$," but it is not always clear what A and $\mathbf{p}$ should be for a given problem.

Problem 1. let $\mathbf{r}$ be some point around which you would like to *rotate* some other point $\mathbf{x}$ by α degrees, as shown in Sketch 66. Let $\mathbf{x}'$ be the rotated point.

Rotations have only been defined around the origin, not around arbitrary points. Hence, we translate our given geometry (the two points $\mathbf{r}$ and $\mathbf{x}$) such that $\mathbf{r}$ moves to the origin. This is easy:

$$\bar{\mathbf{r}} = \mathbf{r} - \mathbf{r} = \mathbf{0}, \quad \bar{\mathbf{x}} = \mathbf{x} - \mathbf{r}.$$

Now we rotate the vector $\bar{\mathbf{x}}$ around the origin by α degrees:

$$\bar{\bar{\mathbf{x}}} = A\bar{\mathbf{x}}.$$

The matrix A would be taken directly from (4.16). Finally, we translate $A\bar{\mathbf{x}}$ back to the center $\mathbf{r}$ of rotation:

$$\mathbf{x}' = A\bar{\mathbf{x}} + \mathbf{r}.$$

This is the solution, but it is not in the standard form for an affine map. This is achieved by replacing $\bar{\mathbf{x}}$ by its definition:

$$\mathbf{x}' = A\mathbf{x} - A\mathbf{r} + \mathbf{r}$$

or

$$\mathbf{x}' = A(\mathbf{x} - \mathbf{r}) + \mathbf{r}. \tag{6.2}$$

Example 6.1 Let

$$\mathbf{r} = \begin{bmatrix} 2 \\ 1 \end{bmatrix}, \quad \mathbf{x} = \begin{bmatrix} 3 \\ 0 \end{bmatrix}$$

and $\alpha = 90°$. We obtain

$$\mathbf{x}' = \begin{bmatrix} 0 & -1 \\ 1 & 0 \end{bmatrix} \begin{bmatrix} 1 \\ -1 \end{bmatrix} + \begin{bmatrix} 2 \\ 1 \end{bmatrix} = \begin{bmatrix} 3 \\ 2 \end{bmatrix}.$$

See Sketch 67 for an illustration.

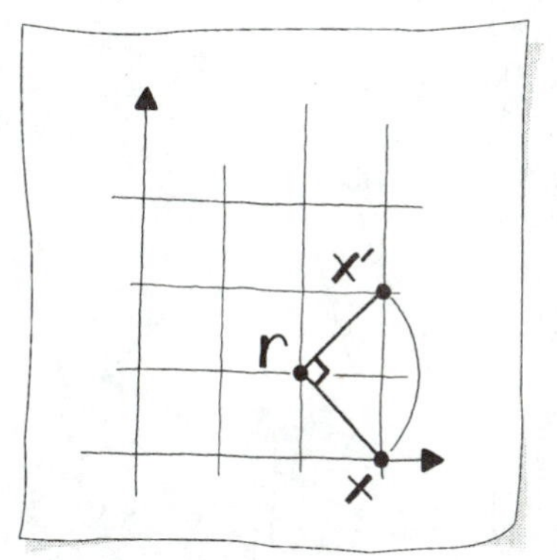

Sketch 67
Rotate 90°.

Problem 2. let $\mathbf{l}$ be a line and $\mathbf{x}$ be a point. You want to *reflect* $\mathbf{x}$ across $\mathbf{l}$, with result $\mathbf{x}'$, as shown in Sketch 68.

This problem could be solved using affine maps; find the intersection $\mathbf{r}$ of $\mathbf{l}$ with the $\mathbf{e}_1$−axis, find the cosine of the angle between $\mathbf{l}$ and $\mathbf{e}_1$, rotate $\mathbf{x}$ around $\mathbf{r}$ such that $\mathbf{l}$ is mapped to the $\mathbf{e}_1$−axis, reflect the rotated $\mathbf{x}$ across the $\mathbf{e}_1$−axis, and finally undo the rotation. Complicated!

It is much easier to employ the 'closest point on a line to a point' algorithm from Section 3.7. It will compute the point $\mathbf{p}$ on $\mathbf{l}$ that is closest to $\mathbf{x}$. Then $\mathbf{p}$ must be the midpoint of $\mathbf{x}$ and $\mathbf{x}'$:

$$\mathbf{p} = \frac{1}{2}\mathbf{x} + \frac{1}{2}\mathbf{x}',$$

from which we conclude

$$\mathbf{x}' = 2\mathbf{p} - 1\mathbf{x}.$$

While this does not have the standard affine map form, it is equivalent to it, yet computationally much less complex.

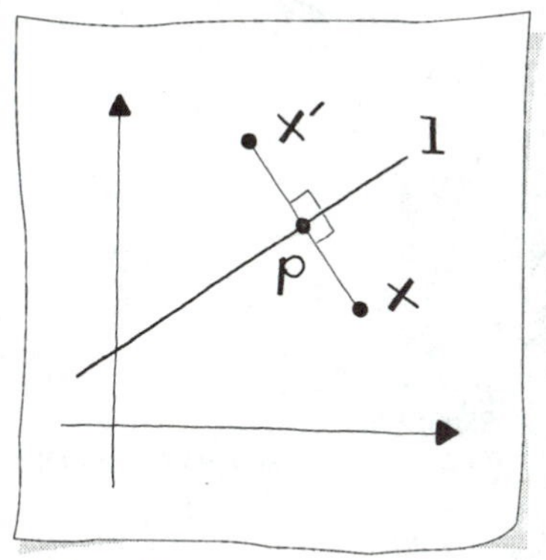

Sketch 68
Reflect a point across a line.

6.4 Mapping Triangles to Triangles

Affine maps may be viewed as combinations of the maps discussed above. Another flavor of affine maps is described in this section; it draws from concepts in Chapter 5.

This other flavor arises like this: given a (source) triangle T with vertices $\mathbf{a}_1, \mathbf{a}_2, \mathbf{a}_3$, and a (target) triangle T' with vertices $\mathbf{a}'_1, \mathbf{a}'_2, \mathbf{a}'_3$, what affine maps takes T to T'? More precisely, if $\mathbf{x}$ is a point inside T, it will be mapped to a point $\mathbf{x}'$ inside T': how do we find $\mathbf{x}'$? For starters, see Sketch 69.

Our desired affine map will be of the form

$$\mathbf{x}' = A[\mathbf{x} - \mathbf{a}_1] + \mathbf{a}'_1,$$

thus we need to find the matrix A. We define (see Sketch 69)

$$\mathbf{v}_2 = \mathbf{a}_2 - \mathbf{a}_1, \qquad \mathbf{v}_3 = \mathbf{a}_3 - \mathbf{a}_1$$

and

$$\mathbf{v}'_2 = \mathbf{a}'_2 - \mathbf{a}'_1, \qquad \mathbf{v}'_3 = \mathbf{a}'_3 - \mathbf{a}'_1.$$

We know

$$A\mathbf{v}_2 = \mathbf{v}'_2,$$
$$A\mathbf{v}_3 = \mathbf{v}'_3.$$

These two vector equations may be combined into one matrix equation:

$$A\begin{bmatrix} \mathbf{v}_2 & \mathbf{v}_3 \end{bmatrix} = \begin{bmatrix} \mathbf{v}'_2 & \mathbf{v}'_3 \end{bmatrix},$$

which we abbreviate as

$$AV = V'.$$

We multiply both sides of this equation by V's inverse V^{-1}, see Chapter 5, and obtain A as

$$A = V'V^{-1}.$$

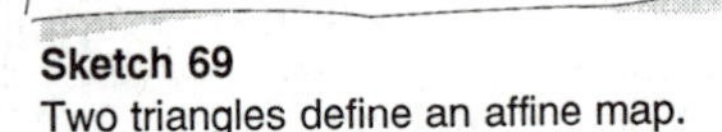

Sketch 69
Two triangles define an affine map.

Example 6.2 Triangle T is defined by the vertices

$$\mathbf{a}_1 = \begin{bmatrix} 0 \\ 1 \end{bmatrix} \quad \mathbf{a}_2 = \begin{bmatrix} -1 \\ -1 \end{bmatrix} \quad \mathbf{a}_3 = \begin{bmatrix} 1 \\ -1 \end{bmatrix},$$

and triangle T' is defined by the vertices

$$\mathbf{a}'_1 = \begin{bmatrix} 0 \\ 1 \end{bmatrix} \quad \mathbf{a}'_2 = \begin{bmatrix} 1 \\ 3 \end{bmatrix} \quad \mathbf{a}'_3 = \begin{bmatrix} -1 \\ 3 \end{bmatrix}.$$

The matrices V and V' are then defined as

$$V = \begin{bmatrix} -1 & 1 \\ -2 & -2 \end{bmatrix} \quad V' = \begin{bmatrix} 1 & -1 \\ 2 & 2 \end{bmatrix}.$$

The inverse of the matrix V is

$$V^{-1} = \begin{bmatrix} -1/2 & -1/4 \\ 1/2 & -1/4 \end{bmatrix},$$

thus the linear map A is defined as

$$A = \begin{bmatrix} -1 & 0 \\ 0 & -1 \end{bmatrix}.$$

Do you recognize the map?

Let's try a sample point $\mathbf{x} = \begin{bmatrix} 0 \\ -1/3 \end{bmatrix}$ in T. This point is mapped to

$$\mathbf{x}' = \begin{bmatrix} -1 & 0 \\ 0 & -1 \end{bmatrix} \left(\begin{bmatrix} 0 \\ -1/3 \end{bmatrix} - \begin{bmatrix} 0 \\ 1 \end{bmatrix} \right) + \begin{bmatrix} 0 \\ 1 \end{bmatrix} = \begin{bmatrix} 0 \\ 7/3 \end{bmatrix}$$

in T'.

Note that V's inverse V^{-1} might not exist; this is the case when $\mathbf{v}_2$ and $\mathbf{v}_3$ are collinear and thus $\det V = 0$.

6.5 Composing Affine Maps

Linear maps are an important theoretical tool, but ultimately we are interested in affine maps; they map objects which are defined by points to other such objects.

If an affine map is given by

$$\mathbf{x}' = \mathbf{p} + A(\mathbf{x} - \mathbf{o}),$$

nothing keeps us from applying it twice, resulting in $\mathbf{x}''$:

$$\mathbf{x}'' = \mathbf{p} + A(\mathbf{x}' - \mathbf{o}).$$

This may be repeated several times — for interesting choices of A and $\mathbf{p}$, interesting images will result. (We added extra gray scales to our figures to help distinguish between the individual maps.)

Our test objects will be the letters **D** and **S**. If we apply a sequence of scalings, we get a result as in Figure 6.4. As the iteration of maps proceeds, we darkened the corresponding letters. We have moved the origin to **D**'s center. The affine map is defined by

$$A = \begin{bmatrix} 1.1 & 0 \\ 0 & 1.1 \end{bmatrix} \quad \text{and} \quad \mathbf{p} = \begin{bmatrix} 0 \\ 0 \end{bmatrix}.$$

In affine space, we can introduce a translation, and obtain Figure 6.5. This was achieved by setting

$$\mathbf{p} = \begin{bmatrix} 2 \\ 2 \end{bmatrix}.$$

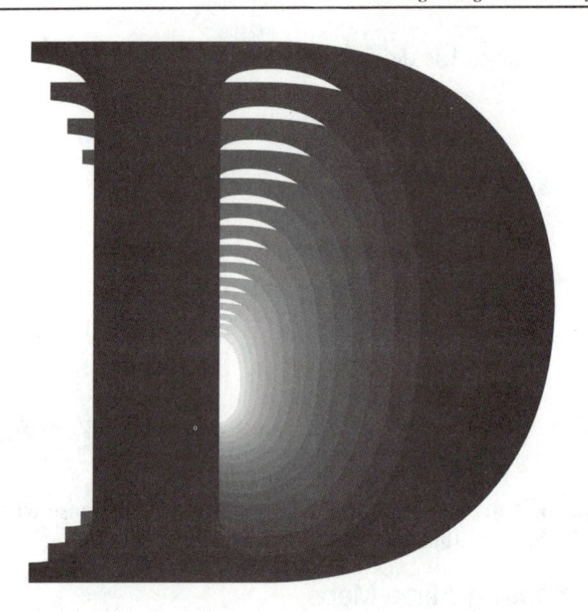

Figure 6.4.
Scaling: the letter **D** is scaled several times; the origin is at its center.

Figure 6.5.
Scaling: the letter **D** is scaled several times; a translation was applied at each step as well.

Figure 6.6.
Rotations: the letter **S** is rotated several times; the origin is at the lower left.

Figure 6.7.
Rotations: the letter **S** is rotated several times; scalings and translations are also applied.

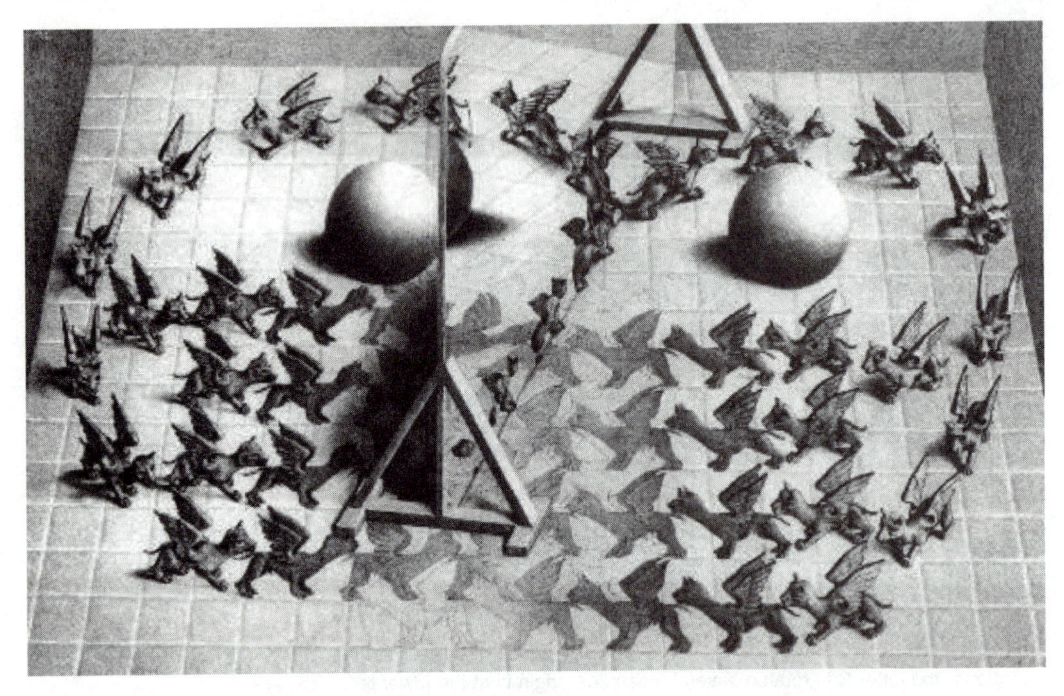

Figure 6.8.
M.C. Escher: Magic Mirror (1949).

Rotations can also be made more interesting. In Figure 6.6, you see the letter **S** rotated several times around the origin, which is near the lower left of the letter.

Adding scaling and rotation results in Figure 6.7.

The basic affine map for this case is given by

$$\mathbf{x}' = S[R\mathbf{x} + \mathbf{p}]$$

where R rotates by $-20°$, S scales nonuniformly, and $\mathbf{p}$ translates:

$$R = \begin{bmatrix} \cos(-20) & -\sin(-20) \\ \sin(-20) & \cos(-20) \end{bmatrix}, \quad S = \begin{bmatrix} 1.25 & 0 \\ 0 & 1.1 \end{bmatrix} \quad \mathbf{p} = \begin{bmatrix} 5 \\ 5 \end{bmatrix}.$$

We finish this chapter with Figure 6.8 by the Dutch artist, M.C. Escher, who in a very unique way mixed complex geometric issues with a unique style. See [4], http://homepage.seas.upenn.edu/˜dalig/escher.html, or http://www.bxscience.edu/˜saltykov/art/escher/escher.html.

Figure 6.8 is itself a 2D object, and so may be subjected to affine maps.

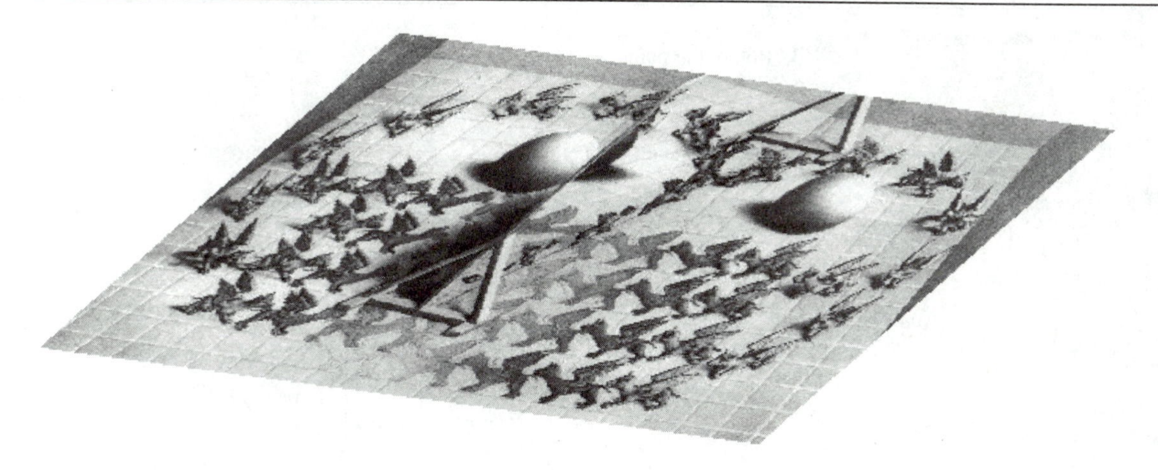

Figure 6.9.
M.C. Escher: Magic Mirror (1949); affine map applied.

Figure 6.9 gives an example. The matrix used here is

$$A = \begin{bmatrix} 1 & 0.5 \\ -0.2 & 0.7 \end{bmatrix}. \tag{6.3}$$

6.6 Exercises

For exercises 1 and 2 let

$$A = \begin{bmatrix} 2 & 1 \\ 1 & 2 \end{bmatrix} \quad \text{and} \quad \mathbf{p} = \begin{bmatrix} 2 \\ 2 \end{bmatrix}.$$

1. Let

$$\mathbf{r} = \begin{bmatrix} 0 \\ 1 \end{bmatrix} \quad \text{and} \quad \mathbf{s} = \begin{bmatrix} 1 \\ 3/2 \end{bmatrix},$$

and $\mathbf{q} = 1/3\mathbf{r} + 2/3\mathbf{s}$. Compute $\mathbf{r}', \mathbf{s}', \mathbf{q}'$; e.g., $\mathbf{r} = A\mathbf{r} + \mathbf{p}$. Show that $\mathbf{q}' = 1/3\mathbf{r}' + 2/3\mathbf{s}'$.

2. Let

$$\mathbf{t} = \begin{bmatrix} 0 \\ 1 \end{bmatrix} \quad \text{and} \quad \mathbf{m} = \begin{bmatrix} 2 \\ 1 \end{bmatrix}.$$

Compute $\mathbf{t}'$ and $\mathbf{m}'$. Sketch the lines defined by $\mathbf{t}, \mathbf{m}$ and $\mathbf{t}', \mathbf{m}'$. Do the same for $\mathbf{r}$ and $\mathbf{s}$ from the previous exercise. What does this illustrate?

3. Rotate the point
$$\mathbf{x} = \begin{bmatrix} -2 \\ -2 \end{bmatrix}$$
by 90° around the point
$$\mathbf{r} = \begin{bmatrix} -2 \\ 2 \end{bmatrix}.$$
Define A and $\mathbf{p}$ of the affine map.

4. Reflect the point
$$\mathbf{x} = \begin{bmatrix} 0 \\ 2 \end{bmatrix}$$
about the line
$$\mathbf{l}(t) = \begin{bmatrix} 0 \\ 0 \end{bmatrix} + t \begin{bmatrix} 1 \\ 2 \end{bmatrix}.$$

5. Given a triangle T with vertices
$$\mathbf{a}_1 = \begin{bmatrix} 2 \\ 0 \end{bmatrix} \quad \mathbf{a}_2 = \begin{bmatrix} 0 \\ 1 \end{bmatrix} \quad \mathbf{a}_3 = \begin{bmatrix} -2 \\ 0 \end{bmatrix},$$
and T' with vertices
$$\mathbf{a}'_1 = \begin{bmatrix} 2 \\ 0 \end{bmatrix} \quad \mathbf{a}'_2 = \begin{bmatrix} 0 \\ -1 \end{bmatrix} \quad \mathbf{a}'_3 = \begin{bmatrix} -2 \\ 0 \end{bmatrix}$$
Suppose the triangle T has been mapped to T' via an affine map. What are the coordinates of the point $\mathbf{x}'$ corresponding to
$$\mathbf{x} = \begin{bmatrix} 0 \\ 0 \end{bmatrix}?$$

PS1 Experiment with the file S_rotran.ps by changing some of the parameters in the for loop.

PS2 Experiment with the file Escher_aff.ps. Just before the unreadable part, you see the line

/matrix[1 – .2 0.5 0.7 0 0]def.

This is PostScript's way of defining the matrix from (6.3). The last two zeroes are meaningless here. Change some of the other elements and see what happens.

7 Eigen Things

Figure 7.1.
The Tacoma Narrows bridge: a view from the approach shortly before collapsing.

A linear map is described by a matrix, but that does not say much about its geometric properties. When you look at the 2D linear map figures from Chapter 4, you see that they all map a circle to some ellipse, thereby stretching and rotating the circle. This stretching and rotating is the geometry of a linear map; it is captured by its eigenvectors and eigenvalues, the subject of this chapter.

Eigenvalues and eigenvectors play an important role in the analysis of mechanical structures. If a bridge starts to sway because of strong winds, then this may be described in terms of certain eigenvalues associated with the bridge's mathematical model. Figures 7.1 and 7.2 shows how the Tacoma Narrows bridge swayed violently during mere 42-mile-per-hour winds on Nov. 7, 1940. It collapsed seconds later. Today, a careful eigen-

Figure 7.2.
The Tacoma Narrows bridge: a view from shore shortly before collapsing.

value analysis is carried before any bridge is built! For more images, see http://www.nwwf.com/wa003a.htm.

The essentials of all eigen-theory are already present in the humble 2D case, the subject of this chapter. A brief discussion of the higher-dimensional case is given at the end.

7.1 Fixed Directions

Consider Figure 4.2 in Section 4.4. You see that the $\mathbf{e}_1$−axis is mapped to itself; so is the $\mathbf{e}_2$−axis. This means that any vector of the form $c\mathbf{e}_1$ or $d\mathbf{e}_2$ is mapped to some multiple of itself. Similarly, in Sketch 4.7 in Section 4.7, you see that all vectors of the form $c\mathbf{e}_1$ are mapped to multiples of each other.

The directions defined by those vectors are called *fixed directions*, for the reason that those directions are not changed by the map. All vectors in the fixed directions change only in length. The fixed directions need not be the coordinate axes.

If a matrix A takes a (nonzero) vector $\mathbf{r}$ to a multiple of itself, then this may be written as

$$A\mathbf{r} = \lambda\mathbf{r} \tag{7.1}$$

with some real number λ. Then A will treat any multiple of $\mathbf{r}$ in this way as well. Given a matrix A, one might then ask which vectors it treats in this special way. It turns out that there are at most two directions (in 2D), and they are orthogonal to each other, as in the case of a scaling. These special vectors are called the *eigenvectors* of A, from the German, "*eigen*", meaning special or proper. An eigenvector is mapped to a multiple of itself, and the corresponding factor λ is called its *eigenvalue*. The eigenvalues and eigenvectors of a matrix are the key to understanding its geometry.

7.2 Eigenvalues

We now develop a way to find the eigenvalues of a matrix A. First, we rewrite (7.1) as

$$A\mathbf{r} = \lambda\mathbf{r}I,$$

with I being the identity matrix. We may change this to

$$[A - \lambda I]\mathbf{r} = \mathbf{0}. \tag{7.2}$$

This means that the matrix $[A - \lambda I]$ maps a nonzero vector $\mathbf{r}$ to the zero vector; $[A - \lambda I]$ must be a projection. Then $[A - \lambda I]$'s determinant vanishes:

$$\det[A - \lambda I] = 0. \tag{7.3}$$

This, as you will see, is a quadratic equation in λ, called the *characteristic equation* of A.

Example 7.1 Before we proceed further, we will look at an example. Let

$$A = \begin{bmatrix} 2 & 1 \\ 1 & 2 \end{bmatrix}.$$

Its action is shown in Figure 7.3.

So let's write out (7.3). It is

$$\begin{vmatrix} 2-\lambda & 1 \\ 1 & 2-\lambda \end{vmatrix} = 0.$$

Figure 7.3.
Action of a matrix: an example.

If we expand the determinant, we get the simple expression — the characteristic equation —

$$(2-\lambda)^2 - 1 = 0.$$

Expanding and gathering terms, we have a quadratic equation in λ:

$$\lambda^2 - 4\lambda + 3 = 0$$

with the solutions[1]

$$\lambda_1 = 3, \qquad \lambda_2 = 1.$$

Thus the eigenvalues of a 2×2 matrix are nothing but the zeroes of a quadratic equation. The eigenvectors are next.

7.3 Eigenvectors

Continuing the above example, we would still like to know the corresponding eigenvectors. We know that one of them will be mapped to three times

[1] Recall that a quadratic equation $a\lambda^2 + b\lambda + c = 0$ has the solutions $\lambda_1 = \frac{-b+\sqrt{b^2-4ac}}{2a}$ and $\lambda_2 = \frac{-b-\sqrt{b^2-4ac}}{2a}$.

itself, the other one to itself. Let's call the corresponding eigenvectors $\mathbf{r}_1$ and $\mathbf{r}_2$. The eigenvector $\mathbf{r}_1$ satisfies

$$\begin{bmatrix} 2-3 & 1 \\ 1 & 2-3 \end{bmatrix} \mathbf{r}_1 = \mathbf{0},$$

or

$$\begin{bmatrix} -1 & 1 \\ 1 & -1 \end{bmatrix} \mathbf{r}_1 = \mathbf{0}.$$

This is a *homogeneous* system, as discussed in Section 5.8. Such systems either have none or infinitely many solutions. In our case, since the matrix has rank 1, there are infinitely many solutions. Any vector of the form

$$\mathbf{r}_1 = \begin{bmatrix} c \\ c \end{bmatrix}$$

will do. And indeed, Figure 7.4 indicates that

$$\begin{bmatrix} 1 \\ 1 \end{bmatrix}$$

is stretched by a factor of three. Of course

$$\begin{bmatrix} -1 \\ -1 \end{bmatrix}$$

is also stretched by a factor of three.

Next, we determine $\mathbf{r}_2$. We get the linear system

$$\begin{bmatrix} 1 & 1 \\ 1 & 1 \end{bmatrix} \mathbf{r}_2 = \mathbf{0}.$$

Again, we have a homogeneous system with infinitely many solutions. They are all of the form

$$\mathbf{r}_2 = \begin{bmatrix} c \\ -c \end{bmatrix}.$$

Now recheck Figure 7.4; you see that the vector

$$\begin{bmatrix} 1 \\ -1 \end{bmatrix}$$

is indeed mapped to itself!

Typically, eigenvectors are normalized to achieve a degree of uniqueness, and we then have

$$\mathbf{r}_1 = \frac{1}{\sqrt{2}} \begin{bmatrix} 1 \\ 1 \end{bmatrix} \quad \mathbf{r}_2 = \frac{1}{\sqrt{2}} \begin{bmatrix} 1 \\ -1 \end{bmatrix}.$$

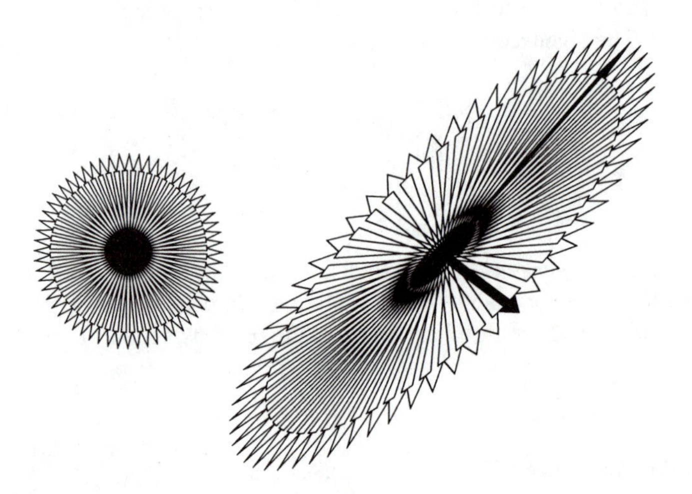

Figure 7.4.
Eigenvectors: the action of a matrix and its eigenvectors, scaled by their corresponding eigenvalues.

Let us return to the general case. The expression $\det[A - \lambda I] = 0$ is a quadratic polynomial in λ, and its zeroes λ_1 and λ_2 are A's eigenvalues. To find the corresponding eigenvectors, we set up the linear systems $[A - \lambda_1 I] = \mathbf{0}$ and $[A - \lambda_2 I] = \mathbf{0}$. Both are homogeneous linear systems with infinitely many solutions, corresponding to the eigenvectors $\mathbf{r}_1$ and $\mathbf{r}_2$.

7.4 Special Cases

Not all quadratic polynomials have zeroes which are real. As you might recall from calculus, there are either no, one, or two real zeroes of a quadratic polynomial,[2] illustrated in Figure 7.5. If there are no zeroes, then the corresponding matrix A has no fixed directions. We know one example — rotations. They rotate every vector, leaving no direction unchanged. Let's look at the familiar rotation by -90 degrees, given by

$$\begin{bmatrix} 0 & 1 \\ -1 & 0 \end{bmatrix}.$$

Its characteristic equation is

$$\begin{vmatrix} -\lambda & 1 \\ -1 & -\lambda \end{vmatrix} = 0$$

[2]Actually, every quadratic polynomial has two zeroes, but they may be complex numbers.

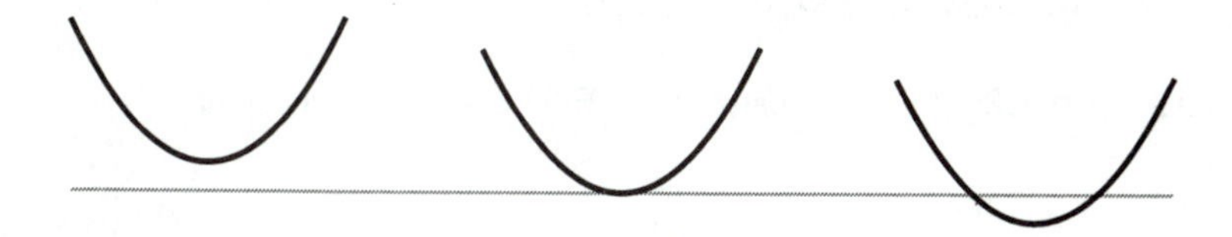

Figure 7.5.
Quadratic polynomials: from left to right, no zero, one zero, two zeroes.

or

$$\lambda^2 + 1 = 0.$$

This has no real solutions, as expected.

A quadratic equation may also have one double root; then there is only one fixed direction. A shear in the $\mathbf{e}_1$-direction provides an example — it maps all vectors in the $\mathbf{e}_1$-direction to themselves. An example is:

$$A = \begin{bmatrix} 1 & 1/2 \\ 0 & 1 \end{bmatrix}.$$

The action of this shear is illustrated in Figure 4.8 in Section 4.7. You clearly see that the $\mathbf{e}_1$-axis is not changed.

The characteristic equation for A is

$$\begin{vmatrix} 1-\lambda & 1/2 \\ 0 & 1-\lambda \end{vmatrix} = 0$$

or

$$(1-\lambda)^2 = 0.$$

It has the double root $\lambda_1 = \lambda_2 = 1$. For the corresponding eigenvector, we have to solve

$$\begin{bmatrix} 0 & 1/2 \\ 0 & 0 \end{bmatrix} \mathbf{r} = \mathbf{0}.$$

While this may look strange, it is nothing but a homogeneous system. All vectors of the form

$$\mathbf{r} = \begin{bmatrix} c \\ 0 \end{bmatrix}$$

are solutions. This is quite as expected; those vectors line up along the $\mathbf{e}_1$-direction.

Is there an easy way to decide if a matrix has real eigenvalues or not? In general, no. But there is one important special case: *every symmetric matrix has real eigenvalues.* We will skip the proof but note that these matrices do arise quite often in "real life." See Section 7.5 for more details.

The last special case to be covered is that of a *zero eigenvalue.*

Example 7.2 Take the matrix from Figure 4.11. It was given by

$$A = \begin{bmatrix} 0.2 & 0.8 \\ -0.2 & -0.8 \end{bmatrix}.$$

The characteristic equation is

$$\lambda(\lambda + 0.4) = 0,$$

and you can immediately check that one eigenvalue is $\lambda_1 = 0$. The corresponding eigenvector is found by solving

$$\begin{bmatrix} 0.2 & 0.8 \\ -0.2 & -0.8 \end{bmatrix} \begin{bmatrix} v_1 \\ v_2 \end{bmatrix} = \begin{bmatrix} 0 \\ 0 \end{bmatrix},$$

yet another homogeneous linear system. Its solutions are of the form

$$\mathbf{r}_1 = c \begin{bmatrix} -4 \\ 1 \end{bmatrix},$$

as you should convince yourself! Since this matrix maps nonzero vectors (multiples of λ_1's eigenvector) to the zero vector, it reduces dimensionality, and thus has rank one.

As a general statement, we may say that a 2×2 matrix with one zero eigenvalue has rank one. A matrix with two zero eigenvalues has rank zero, and thus must be the zero matrix.

7.5 The Geometry of Symmetric Matrices

With symmetric matrices, we don't have to worry about complex eigenvalues, and symmetric matrices come up frequently, as shown in Chapter 9. We know the two basic equations for eigenvalues and eigenvectors of a symmetric matrix A:

$$A\mathbf{r}_1 = \lambda_1 \mathbf{r}_1, \tag{7.4}$$

$$A\mathbf{r}_2 = \lambda_2 \mathbf{r}_2. \tag{7.5}$$

Since A is symmetric, we may use (7.4) to write the following:

$$\mathbf{r}_1^{\mathrm{T}} A \mathbf{r}_2 = \lambda_1 \mathbf{r}_1^{\mathrm{T}} \mathbf{r}_2. \tag{7.6}$$

Using (7.5), we obtain

$$\mathbf{r}_1^{\mathrm{T}} A \mathbf{r}_2 = \lambda_2 \mathbf{r}_1^{\mathrm{T}} \mathbf{r}_2 \tag{7.7}$$

and thus

$$\lambda_1 \mathbf{r}_1^{\mathrm{T}} \mathbf{r}_2 = \lambda_2 \mathbf{r}_1^{\mathrm{T}} \mathbf{r}_2$$

or

$$(\lambda_1 - \lambda_2) \mathbf{r}_1^{\mathrm{T}} \mathbf{r}_2 = 0.$$

If $\lambda_1 \neq \lambda_2$ (the standard case), then we conclude that $\mathbf{r}_1^{\mathrm{T}} \mathbf{r}_2 = 0$; in other words, A's two eigenvectors are *orthogonal*. Check this for the example of Section 7.3!

If we define (using the capital Greek Lambda: Λ)

$$\Lambda = \begin{bmatrix} \lambda_1 & 0 \\ 0 & \lambda_2 \end{bmatrix} \quad \text{and} \quad R = \begin{bmatrix} \mathbf{r}_1 & \mathbf{r}_2 \end{bmatrix},$$

then we may condense (7.4) and (7.5) into one matrix equation:

$$AR = R\Lambda. \tag{7.8}$$

Example 7.3 For the example of Section 7.3, $AR = R\Lambda$ becomes

$$\begin{bmatrix} 2 & 1 \\ 1 & 2 \end{bmatrix} \begin{bmatrix} \sqrt{2}/2 & \sqrt{2}/2 \\ \sqrt{2}/2 & -\sqrt{2}/2 \end{bmatrix} = \begin{bmatrix} \sqrt{2}/2 & \sqrt{2}/2 \\ -\sqrt{2}/2 & \sqrt{2}/2 \end{bmatrix} \begin{bmatrix} 3 & 0 \\ 0 & 1 \end{bmatrix}.$$

Verify this identity!

We know $\mathbf{r}_1^{\mathrm{T}} \mathbf{r}_1 = 1$ and $\mathbf{r}_2^{\mathrm{T}} \mathbf{r}_2 = 1$ (since we assume eigenvectors are normalized) and also $\mathbf{r}_1^{\mathrm{T}} \mathbf{r}_2 = \mathbf{r}_2^{\mathrm{T}} \mathbf{r}_1 = 0$ (since they are orthogonal). These four equations may also be written in matrix form

$$R^{\mathrm{T}} R = I, \tag{7.9}$$

with I the identity matrix. Thus

$$R^{-1} = R^{\mathrm{T}}$$

and R is an orthogonal matrix.

Now (7.8) becomes

$$A = R\Lambda R^{\mathrm{T}};$$

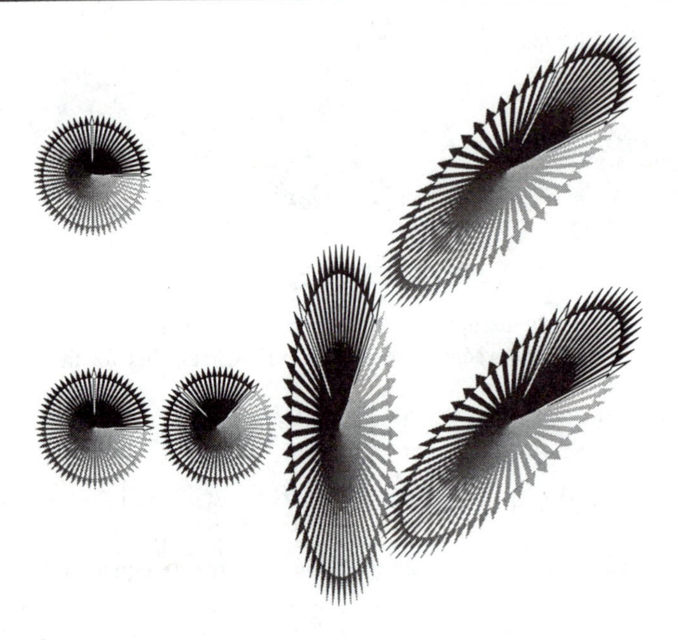

Figure 7.6.
Symmetric matrices: top, the matrix from Figure 7.4. Bottom: decomposed into rotation, scale, and inverse rotation.

after transposing, it is

$$A = R^{\mathrm{T}} \Lambda R. \tag{7.10}$$

What does this mean geometrically? Since R is an orthogonal matrix, it is a *rotation*. Its inverse, R^{T}, is also a rotation, just in the opposite direction. The diagonal matrix Λ is a scaling along each of the coordinate axes. Thus (7.10) states that the action of every symmetric matrix may be obtained by applying a rotation, then a scaling, and then undoing the rotation. Figure 7.6 gives an example.

7.6 Repeating Maps

When we studied matrices, we saw that they always map the unit circle to an ellipse. Nothing keeps us from now mapping the ellipse again, using the same map. We can then repeat again, and so on. Figures 7.7 and 7.8 show two such examples.

Figure 7.7.
Repetitions: a matrix is applied several times.

Figure 7.7 corresponds to the matrix

$$A = \begin{bmatrix} 1 & 0.3 \\ 0.3 & 1 \end{bmatrix}.$$

Being symmetric, it has two real eigenvalues and eigenvectors. As the map is repeated several times, the resulting ellipses become more and more stretched: they are elongated in the direction $\mathbf{r}_1$ by $\lambda_1 = 1.3$ and compacted in the direction of $\mathbf{r}_2$ by a factor of $\lambda_2 = 0.7$, with

$$\mathbf{r}_1 = \begin{bmatrix} 1 \\ 1 \end{bmatrix}, \quad \mathbf{r}_2 = \begin{bmatrix} -1 \\ 1 \end{bmatrix}.$$

To get some more insight into this phenomenon, consider applying A twice to $\mathbf{r}_1$. We get

$$AA\mathbf{r}_1 = A\lambda_1\mathbf{r}_1 = \lambda_1^2\mathbf{r}_1.$$

In general,

$$A^n\mathbf{r}_1 = \lambda^n\mathbf{r}_1.$$

The same holds for $\mathbf{r}_2$ and λ_2, of course. So you see that once a matrix has real eigenvectors, they play a more and more prominent role as the matrix is applied repeatedly.

By contrast, the matrix corresponding to Figure 7.8 is given by

$$A = \begin{bmatrix} 0.7 & 0.3 \\ -1 & 1 \end{bmatrix}.$$

As you should verify for yourself, this matrix does not have real eigenvalues. In that sense, it is related to a rotation matrix. If you study Figure 7.8,

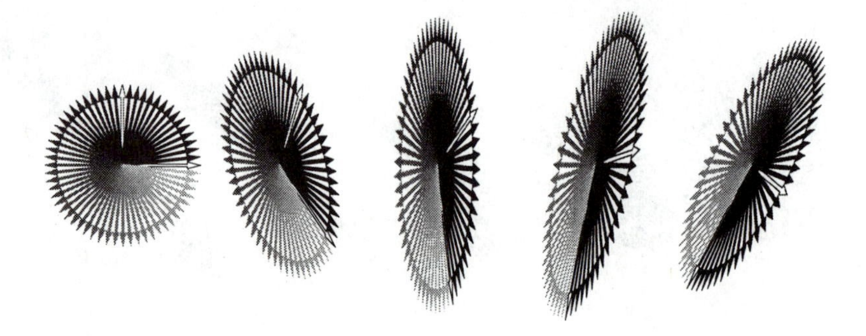

Figure 7.8.
Repetitions: a matrix is applied several times.

you will notice a rotation component as we progress — the ellipses do not line up along any fixed directions.

7.7 The Condition of a Map

In most of our figures about affine maps, we have mapped a circle (formed by many unit vectors) to an ellipse. This ellipse is evidently closely related to the geometry of the map, and indeed to its eigenvalues.

A unit circle is given by the equation

$$r_1^2 + r_2^2 = 1.$$

Using matrix notation, we may write this as

$$\begin{bmatrix} r_1 & r_2 \end{bmatrix} \begin{bmatrix} r_1 \\ r_2 \end{bmatrix} = 1,$$

or

$$\mathbf{r}^{\mathrm{T}}\mathbf{r} = 1. \tag{7.11}$$

If we apply a linear map to the vector $\mathbf{r}$, we have $\mathbf{r}' = A\mathbf{r}$. Inserting this into (7.11), we get

$$\mathbf{r}'^{\mathrm{T}}\mathbf{r}' = [A\mathbf{r}]^{\mathrm{T}}[A\mathbf{r}] = 1.$$

Using the rules for transpose matrices (Section 4.3), this becomes

$$\mathbf{r}^{\mathrm{T}}A^{\mathrm{T}}A\mathbf{r} = 1. \tag{7.12}$$

This is therefore the equation of an ellipse!

Figure 7.9.
Magnitude of eigenvalues: $\lambda_1'/\lambda_2' = 900$.

The matrix $A^{\mathrm{T}}A$ is *symmetric*, and thus has real eigenvalues λ_1' and λ_2'. Order them so that $\lambda_1' \geq \lambda_2'$. If λ_1' is very large and λ_2' is very small, then the ellipse (7.12) will be very elongated (see Figure 7.9). The ratio λ_1'/λ_2' of is called the *condition number* c_A of the original matrix A. The larger c_A, the "worse" A distorts. The eigenvalues λ_1' and λ_2' of $A^{\mathrm{T}}A$ are called A's *singular values*.

If you solve a linear system $A\mathbf{x} = \mathbf{b}$, you will be in trouble if c_A becomes large. In that case, the columns of A are close to being collinear, i.e., being linearly dependent.

7.8 Higher Dimensional Eigen Things

Eigenvalues and eigenvectors have nice geometric interpretations for the case of 2×2 matrices. For $n \times n$ matrices, the intuitive geometry is lost, yet the basic principles stay the same.

If you have a square matrix A, does it have fixed directions, i.e., are there vectors $\mathbf{r}$ which are mapped to multiples λ_i of themselves by A? Such vectors are characterized by

$$A\mathbf{r} = \lambda\mathbf{r}$$

or

$$[A - \lambda I]\mathbf{r} = \mathbf{0}. \tag{7.13}$$

Since $\mathbf{r} = \mathbf{0}$ trivially performs this way, we will not consider it (the zero vector) from now on. In (7.13), we see that the matrix $[A - \lambda I]$ maps a nonzero vector $\mathbf{r}$ to the zero vector $\mathbf{0}$. Thus its determinant must vanish (see Section 4.9):

$$\det[A - \lambda I] = 0. \tag{7.14}$$

Equation (7.14) is a polynomial equation of degree n in λ, as the following example shows.

Example 7.4 Let

$$A = \begin{bmatrix} 1 & 1 & 0 & 0 \\ 0 & 3 & 1 & 0 \\ 0 & 0 & 4 & 1 \\ 0 & 0 & 0 & 2 \end{bmatrix}.$$

We find

$$\det[A - \lambda I] = \begin{vmatrix} 1-\lambda & 1 & 0 & 0 \\ 0 & 3-\lambda & 1 & 0 \\ 0 & 0 & 4-\lambda & 1 \\ 0 & 0 & 0 & 2-\lambda \end{vmatrix}$$

This is a degree four polynomial $p(\lambda)$:

$$p(\lambda) = (1-\lambda)(3-\lambda)(4-\lambda)(2-\lambda).$$

The zeroes of this polynomial are found by solving $p(\lambda) = 0$. In our slightly contrived example, we find $\lambda_1 = 1, \lambda_2 = 3, \lambda_3 = 4, \lambda_4 = 3$.

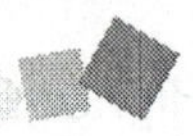

The bad news is that one does not always have trivial matrices like the above to deal with. A general $n \times n$ matrix has a characteristic polynomial $p(\lambda) = \det[A - \lambda I]$ of degree n, and the eigenvalues are the zeroes of this polynomial. Finding the zeroes of an n–th degree polynomial is a nontrivial numerical task — exact algorithms do not exist for $n > 4$.

Needless to say, not all eigenvalues of a matrix are real in general. But the important class of *symmetric* matrices always does have real eigenvalues.

Having found the λ_i, we can now solve linear systems

$$[A - \lambda_i I]\mathbf{r}_i = \mathbf{0}$$

in order to find the eigenvectors $\mathbf{r}_i$. These are homogeneous systems, and thus have no unique solutions. This is rarely of practical relevance, however: while finding eigenvalues is of great importance in engineering, eigenvectors of matrices larger than 2×2 are almost never needed.

7.9 Exercises

1. Find the eigenvalues and eigenvectors of the matrix

$$A = \begin{bmatrix} 1 & -2 \\ -2 & 1 \end{bmatrix}.$$

2. Find the eigenvalues and eigenvectors of

$$A = \begin{bmatrix} 1 & -2 \\ -2 & 0 \end{bmatrix}.$$

3. What is the condition number of the matrix

$$A = \begin{bmatrix} 0.7 & 0.3 \\ -1 & 1 \end{bmatrix}$$

which generated Figure 7.8?

4. What can you say about the condition number of a rotation matrix?

5. If all eigenvalues of a matrix have absolute value less than one, what will happen as you keep repeating the map?

6. Let

$$A = \begin{bmatrix} 1 & -1.5 \\ -1.5 & 0 \end{bmatrix}.$$

Modify file `Matdecomp.ps` to show how A's action can be broken down into two rotations and a scaling. Don't try to understand the details of the PostScript program — it's fairly involved. Just replace the entries of `mat1` through `mat4` as described in the comments.

Breaking It Up: Triangles

8

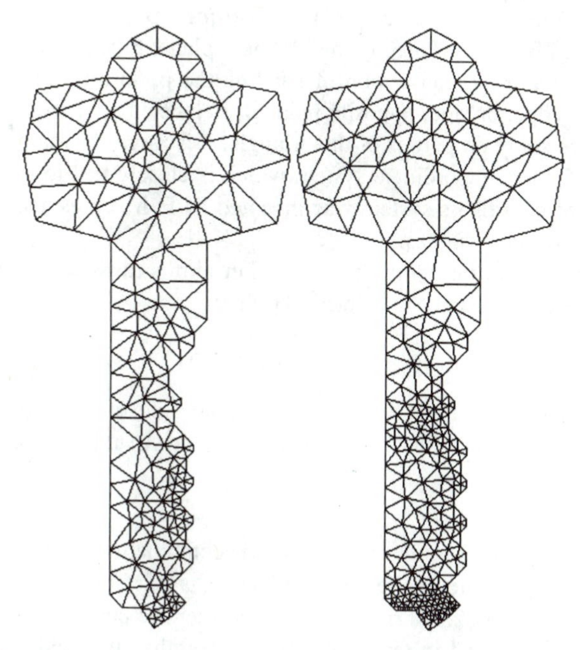

Figure 8.1.
2D FEM: Refinement of a triangulation based on stress and strain calculations. (Source: J. Shewchuk, http://www.cs.cmu.edu/~quake/triangle.html)

Triangles are as old as geometry. They were of interest to the ancient Greeks, and in fact the roots of trigonometry can be found in their study. Triangles also become an indispensable tool in computer graphics and advanced disciplines such as finite element analysis (FEM). In graphics, objects are broken down into triangular facets for display purposes; in FEM, 2D shapes are broken down into triangles in order to facilitate complicated algorithms. Figure 8.1 illustrates a refinement procedure based on stress and strain calculations.

A few pointers to triangulations and related topics can be found at the following web sites:

http://netlib.bell-labs.com/netlib/voronoi/hull.html,
http://www.cs.sunysb.edu/˜algorith/implement/geompack/implement.shtml
http://www.geom.umn.edu/software/download/qhull.html
http://www.cs.cmu.edu/˜quake/triangle.html
http://ltssg3.epfl.ch/˜bossen/triangulator.html.

8.1 Barycentric Coordinates

A *triangle* T is given by three points, its *vertices*, $\mathbf{p}_1, \mathbf{p}_2, \mathbf{p}_3$. The vertices may live in 2D or 3D. Three points define a plane, thus a triangle is a 2D element. We use the convention of labeling the $\mathbf{p}_i$ in a counterclockwise sense. The edge, or side, opposite point $\mathbf{p}_i$ is labeled $\mathbf{s}_i$. (See Sketch 70.)

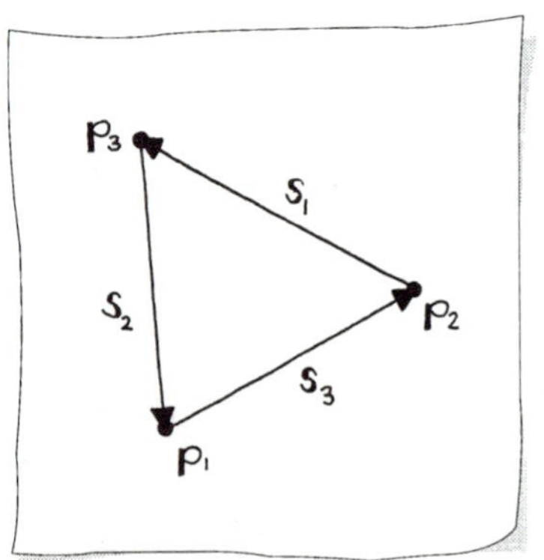

Sketch 70
Vertices and edges of a triangle.

When we study properties of this triangle, it is more convenient to work in terms of a local coordinate system which is closely tied to the triangle. This type of coordinate system was invented by F. Moebius and is known as *barycentric coordinates*.

Let $\mathbf{p}$ be an arbitrary point inside T. Our aim is to write it as a combination of the vertices $\mathbf{p}_i$, in a form like this:

$$\mathbf{p} = u\mathbf{p}_1 + v\mathbf{p}_2 + w\mathbf{p}_3. \tag{8.1}$$

We know one thing already: the right hand side of this equation is a combination of points, and so the coefficients must sum to one:

$$u + v + w = 1.$$

Otherwise, we would not have a barycentric combination! (See Sketch 71.)

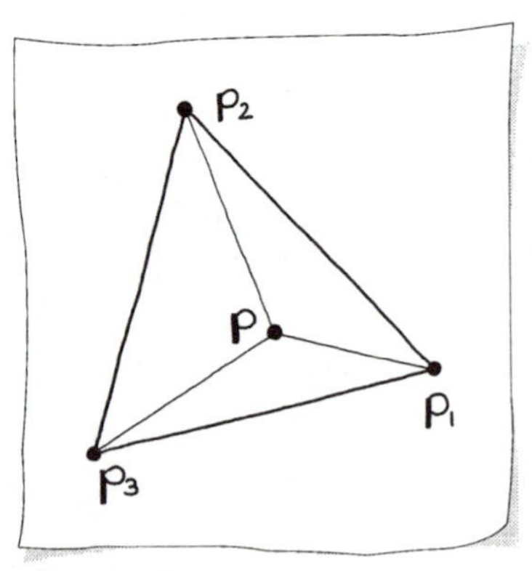

Sketch 71
Barycentric coordinates.

Before we give the explicit form of (u, v, w), let us revisit linear interpolation (Section 2.1) briefly. There, barycentric coordinates on a line segment were defined in terms of ratios of lengths. It sounds reasonable to try the analogous ratios of areas in the triangle case, and so we get:

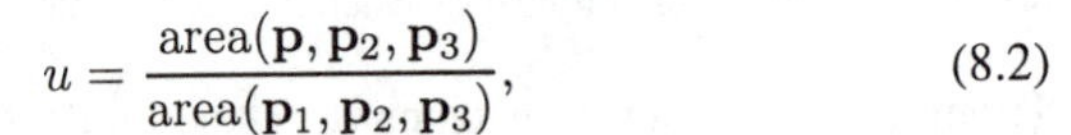

$$u = \frac{\text{area}(\mathbf{p}, \mathbf{p}_2, \mathbf{p}_3)}{\text{area}(\mathbf{p}_1, \mathbf{p}_2, \mathbf{p}_3)}, \tag{8.2}$$

$$v = \frac{\text{area}(\mathbf{p}, \mathbf{p}_3, \mathbf{p}_1)}{\text{area}(\mathbf{p}_1, \mathbf{p}_2, \mathbf{p}_3)}, \tag{8.3}$$

$$w = \frac{\text{area}(\mathbf{p}, \mathbf{p}_1, \mathbf{p}_2)}{\text{area}(\mathbf{p}_1, \mathbf{p}_2, \mathbf{p}_3)}, \tag{8.4}$$

Recall that areas are easily computed using determinants (Section 4.9) or as we will learn later (Section10.2) using cross products.

Let's see why this works. First, we observe that (u, v, w) do indeed sum to one. Next, let $\mathbf{p} = \mathbf{p}_2$. Now (8.2)–(8.4) tell us that $v = 1$ and

$u = w = 0$, just as expected. One more check: if $\mathbf{p}$ is on the edge $\mathbf{s}_1$, say, then $u = 0$, again as expected. Try for yourself that the remaining vertices and edges work the same way!

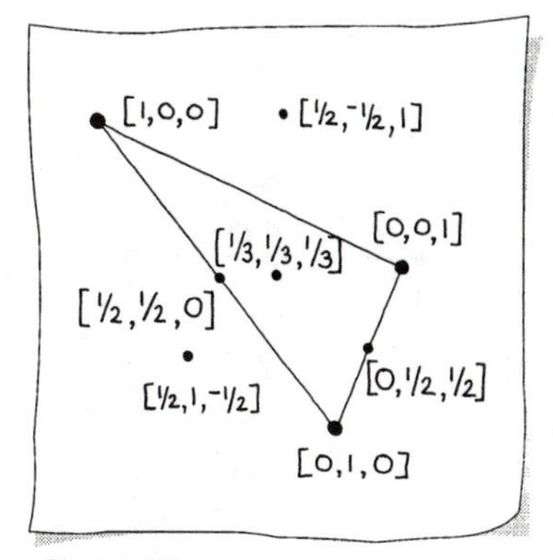

Sketch 72
Examples of barycentric coordinates.

We call (u, v, w) *barycentric coordinates* and denote them by boldface: $\mathbf{u} = (u, v, w)$. Although they are not independent of each other (we may set $w = 1 - u - v$), they behave much like "normal" coordinates: if $\mathbf{p}$ is given, then we can find $\mathbf{u}$ from (8.2)–(8.4). If $\mathbf{u}$ is given, then we can find $\mathbf{p}$ from (8.1).

The three vertices of the triangle have barycentric coordinates

$$\mathbf{p}_1 \cong (1, 0, 0),$$
$$\mathbf{p}_2 \cong (0, 1, 0),$$
$$\mathbf{p}_3 \cong (0, 0, 1).$$

The $\cong$ symbol will be used to indicate the barycentric coordinates of a point. These and several other examples are shown in Sketch 72.

As you see, even points outside of T can be given barycentric coordinates! This works since the areas involved in (8.2)–(8.4) are *signed.* So points inside T have positive barycentric coordinates, and those outside have mixed signs.[1]

This observation is the basis for one of the most frequent uses of barycentric coordinates: the *triangle inclusion test* If a triangle T and a point $\mathbf{p}$ are given, how do we determine if $\mathbf{p}$ is inside T or not? We simply compute $\mathbf{p}$'s barycentric coordinates and check their signs! If they are all of the same sign, inside — else, outside. Theoretically, one or two of the barycentric coordinates could be zero, indicating that $\mathbf{p}$ is on one of the edges. In "real" situations, you are not likely to encounter values which are *exactly* equal to zero; be sure not to test for a barycentric coordinate to be *equal* to zero! Instead, use a tolerance ϵ, and flag a point as being on an edge if one of its barycentric coordinates is less than ϵ in absolute value. A good value for ϵ? Obviously this is application dependent, but something like $1.0E - 6$ should work for most cases.

Finally, Sketch 73 shows how we may think of the whole plane as being covered by a grid of coordinate lines. Note that the plane is divided into seven regions by the (extended) edges of T!

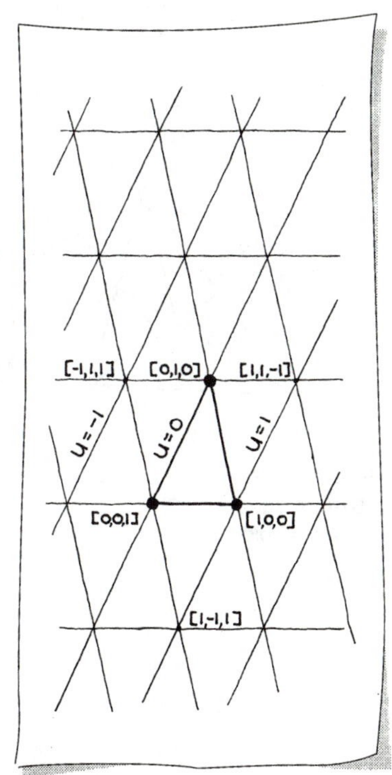

Sketch 73
Barycentric coordinates: coordinate lines.

[1]This assumes the triangle to be oriented counterclockwise. If it is oriented clockwise, then the points inside have all negative barycentric coordinates, and the outside ones still have mixed signs.

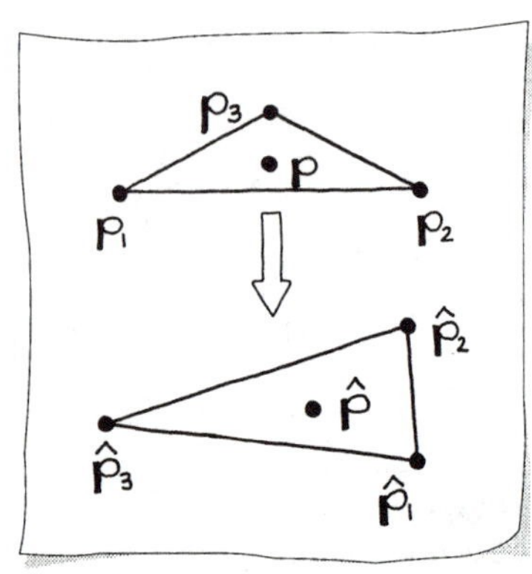

Sketch 74
Affine invariance of barycentric coordinates: an example.

8.2 Affine Invariance

In this short section, we will discuss the statement: *barycentric coordinates are affinely invariant.*

Let $\hat{T}$ be an affine image of T, having vertices $\hat{\mathbf{p}_1}, \hat{\mathbf{p}_2}, \hat{\mathbf{p}_3}$. Let $\mathbf{p}$ be a point with barycentric coordinates $\mathbf{u}$ relative to T. We may apply the affine map to $\mathbf{p}$ also, and then we ask: What are the barycentric coordinates of $\hat{\mathbf{p}}$ with respect to $\hat{T}$?

While at first sight this looks like a daunting task, simple geometry yields the answer quickly. Note that in (8.2)–(8.4), we employ *ratios of areas*. These are, as per Section 6.1, unchanged by affine maps! So while the individual areas in (8.2)–(8.4) do change, their quotients do not. Thus $\hat{\mathbf{p}}$ also has barycentric coordinates $\mathbf{u}$ with respect to $\hat{T}$.

This fact, namely that affine maps do not change barycentric coordinates, is what is meant by the statement at the beginning of this section. See Sketch 74 for an illustration.

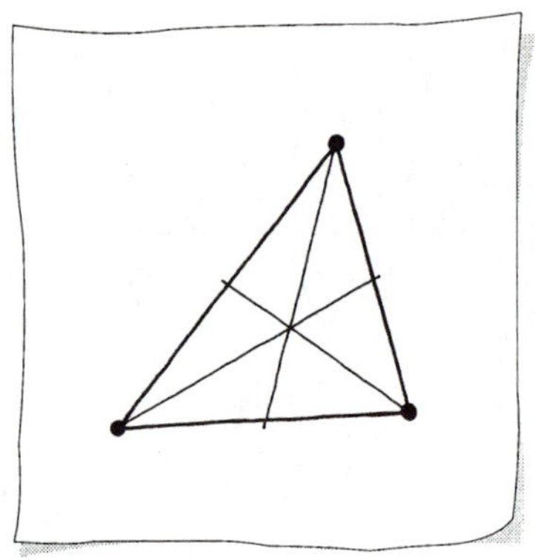

Sketch 75
The centroid.

8.3 Some Special Points

In classical geometry, many special points relative to a triangle have been discovered, but for our purposes, just three will do: the centroid, the incenter, and the circumcenter. They are used for a multitude of geometric computations.

The *centroid* $\mathbf{c}$ of a triangle is given by the intersection of the three medians. (A median is the connection of an edge midpoint to the opposite vertex.) Its barycentric coordinates (see Sketch 75) are given by

$$\mathbf{c} \cong (\frac{1}{3}, \frac{1}{3}, \frac{1}{3}). \tag{8.5}$$

We verify this by writing

$$(\frac{1}{3}, \frac{1}{3}, \frac{1}{3}) = \frac{1}{3}(0,1,0) + \frac{2}{3}(\frac{1}{2}, 0, \frac{1}{2}),$$

thus asserting that $(\frac{1}{3}, \frac{1}{3}, \frac{1}{3})$ lies on the median associated with $\mathbf{p}_2$. In the same way, we show that it is also on the remaining two medians.

The centroid may be given a physical interpretation: if all three triangle vertices were planets with equal mass, then the centroid is their center of gravity. If they had unequal masses, (u, v, w), say, then (8.1) would compute the center of gravity. Incidentally, "barycenter" means "center of gravity!"

We also observe that a triangle and its centroid are related in an affinely invariant way.

The *incenter* $\mathbf{i}$ of a triangle is the intersection of the three angle bisectors (see Sketch 76).

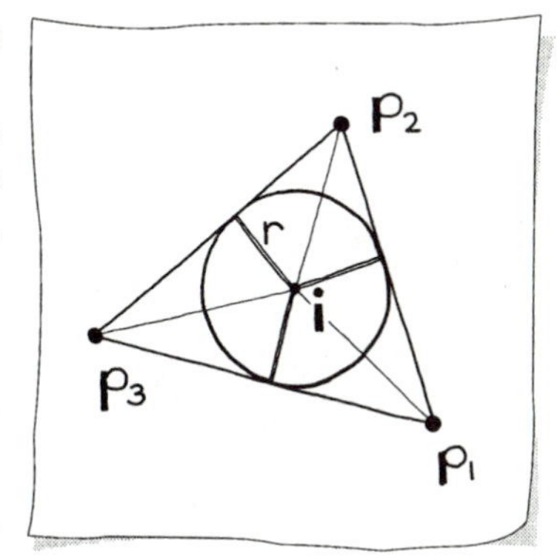

Sketch 76
The incenter.

There is a circle, called the *incircle*, that has $\mathbf{i}$ as its center and it touches all three triangle edges. Let s_i be the length of the triangle edge opposite vertex $\mathbf{p}_i$. Let r be the radius of the incircle — there is a formula for it, but we won't need it here.

If the barycentric coordinates of $\mathbf{i}$ are (i_1, i_2, i_3), then we see that

$$i_1 = \frac{\text{area}(\mathbf{i}, \mathbf{p}_2, \mathbf{p}_3)}{\text{area}(\mathbf{p}_1, \mathbf{p}_2, \mathbf{p}_3)}.$$

This may be rewritten as

$$i_1 = \frac{rs_1}{rs_1 + rs_2 + rs_3},$$

using the "1/2 base times height" rule for triangle areas.

Simplifying, we obtain

$$i_1 = s_1/c,$$
$$i_2 = s_2/c,$$
$$i_3 = s_3/c,$$

where $c = s_1 + s_2 + s_3$ is the circumference of T.

A triangle is not affinely related to its incenter — affine maps change the barycentric coordinates of $\mathbf{i}$.

The *circumcenter* $\mathbf{cc}$ of a triangle is the center of the circle through its vertices. It is obtained as the intersection of the edge bisectors, see Sketch 77. Notice that the circumcenter might not be inside the triangle. This circle is called the *circumcircle* and we will refer to its radius as R.

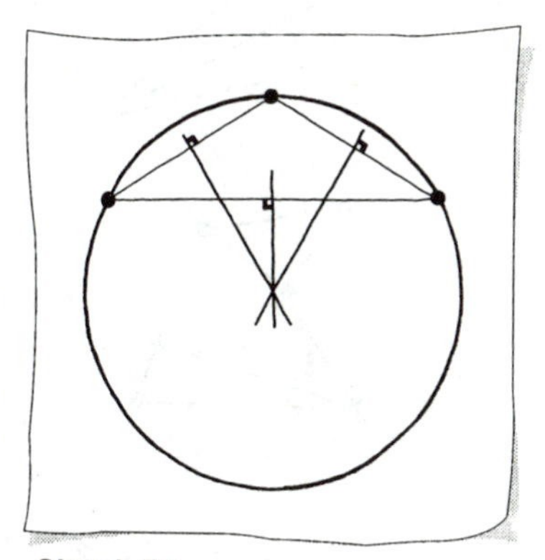

Sketch 77
The circumcenter.

The barycentric coordinates (cc_1, cc_2, cc_3) of the circumcenter are

$$cc_1 = d_1(d_2 + d_3)/D$$
$$cc_2 = d_2(d_1 + d_3)/D$$
$$cc_3 = d_3(d_1 + d_2)/D$$

where

$$d_1 = (\mathbf{p}_2 - \mathbf{p}_1) \cdot (\mathbf{p}_3 - \mathbf{p}_1)$$
$$d_2 = (\mathbf{p}_1 - \mathbf{p}_2) \cdot (\mathbf{p}_3 - \mathbf{p}_2)$$
$$d_3 = (\mathbf{p}_1 - \mathbf{p}_3) \cdot (\mathbf{p}_2 - \mathbf{p}_3)$$
$$D = 2(d_1 d_2 + d_2 d_3 + d_3 d_1).$$

Furthermore,

$$R = \frac{1}{2}\sqrt{\frac{(d_1 + d_2)(d_2 + d_3)(d_3 + d_1)}{D/2}}.$$

These formulas are due to [8].

Again, no affine invariance. Note also that some of the cc_i may be negative. If T has an angle close to 180 degrees, then the corresponding cc_i will be *very* negative, leading to serious numerical problems! As a result, the circumcenter will be far away from the vertices, and thus not be of practical use.

8.4 2D Triangulations

The study of *one* triangle is the realm of classical geometry; in computer applications, one often encounters millions of triangles. Typically, they are connected in some well-defined way; the most basic one being the 2D *triangulation*. Triangulations have also been used in surveying for centuries.

Here is the formal definition: A triangulation of a set of 2D points $\{\mathbf{p}_i\}_{i=1}^N$ is a connected set of triangles such that

1. The vertices of the triangles consist of the given points.

2. The interiors of any two triangles do not intersect.

3. If two triangles are not disjoint, then they share a vertex or have coinciding edges.

4. The union of all triangles equals the convex hull of the $\mathbf{p}_i$.

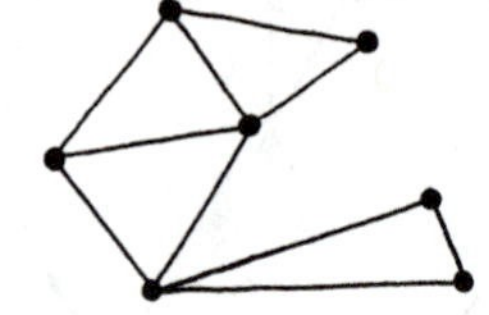

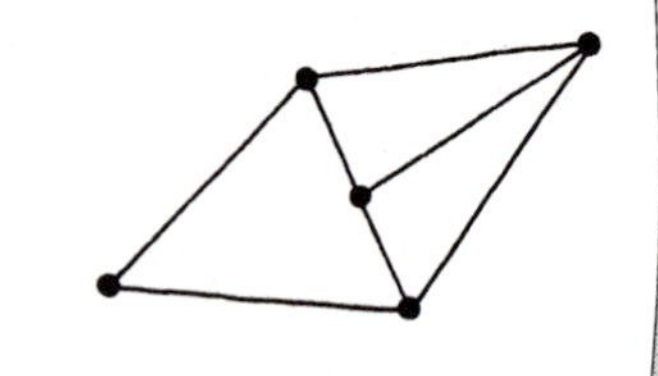

Sketch 78
Triangulations: invalid examples.

These rules sound abstract, but some examples will shed light on them. In Figure 8.2 you see a triangulation which satisfies the above rules. On the other hand, in Sketch 78 you see three illegal triangulations, violating the above rules.

If we are given a point set, is there a unique triangulation? Certainly not, as Sketch 79 shows.

Among the many possible triangulations, there is one that is most commonly agreed to be the "best": this is the *Delaunay triangulation*. Describing the details of this method is beyond the scope of this text, however a wealth of information can be found on the Web. Search or visit the sites listed in the introduction of this chapter or see [3].

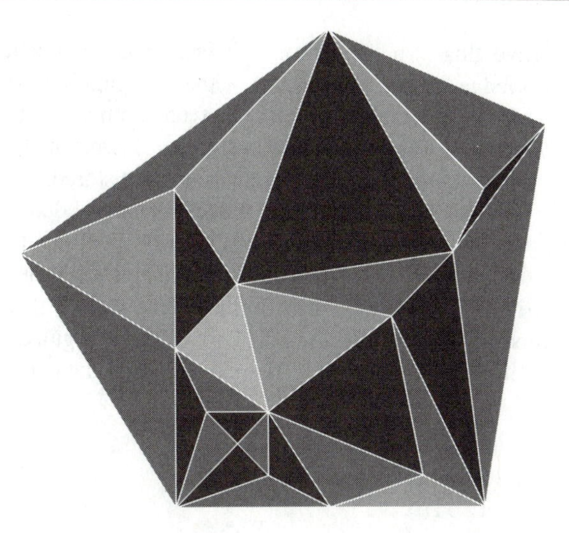

Figure 8.2.
Triangulation: a valid triangulation of the convex hull.

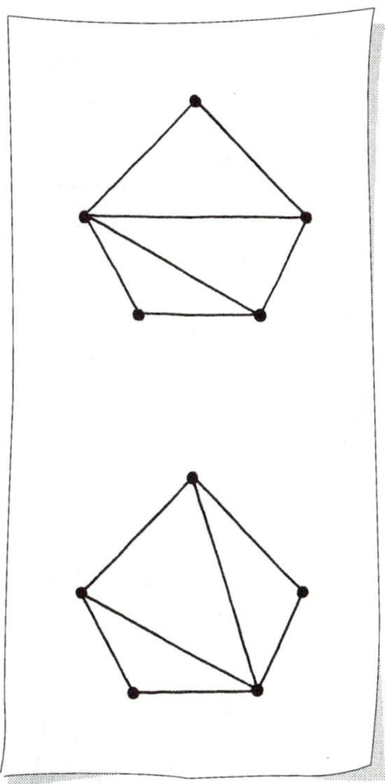

Sketch 79
Triangulations: non-uniqueness.

8.5 A Data Structure

What is the best data structure for storing a triangulation? The factors which determine the best structure include storage requirements and accessibility. Let's build the "best" structure based on the point set and triangulation illustrated in Sketch 80.

In order to minimize storage, it is an accepted practice to store each point once only. Since these are floating point values, they take up the most space. Thus a basic triangulation structure would be a listing of the point set followed by the triangulation information. This constitutes pointers into the point set, indicating which points are joined to form a triangle. Store the triangles in a counterclockwise orientation! This is the data structure for the triangulation in Sketch 80:

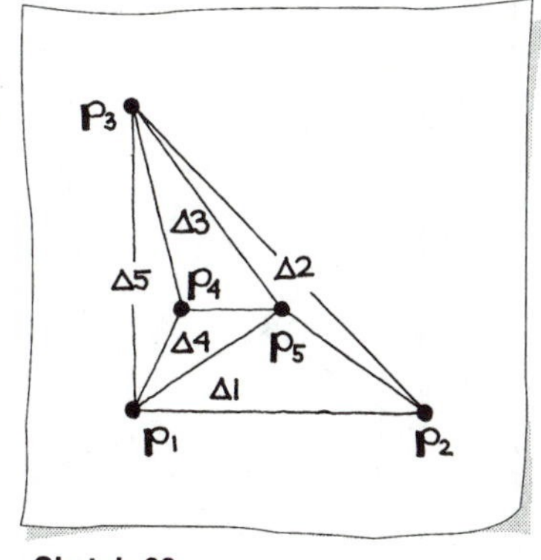

Sketch 80
A sample triangulation.

```
5          (number of points)
0.0 0.0      (point #1)
1.0 0.0
0.0 1.0
0.25 0.3
0.5 0.3
5          (number of triangles)
1 2 5       (first triangle - connects points #1,2,5)
2 3 5
4 5 3
1 5 4
1 4 3
```

We can improve this structure. We will encounter applications which require a knowledge of the connectivity of the triangulation, as described in Section 8.6. To facilitate this, it is not uncommon to also see the *neighbor information* of the triangulation stored. This means that for each triangle, the indices of the triangles surrounding it are stored. For example, in Sketch 80, triangle 1 defined by points 1,2,5 is surrounded by triangles 2,4,-1. The neighboring triangles are listed corresponding to the point across from the shared edge. Triangle -1 indicates that there is not a neighboring triangle across this edge. Immediately, we see that this gives us a fast method for determining the boundary of the triangulation! Listing the neighbor information after each triangle, the final data structure is as follows.

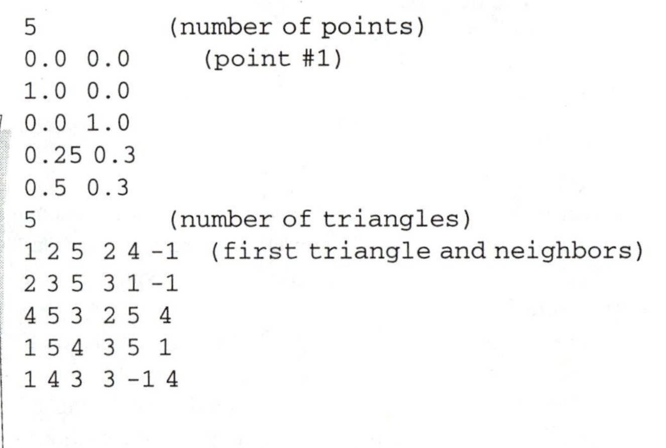

```
5           (number of points)
0.0 0.0       (point #1)
1.0 0.0
0.0 1.0
0.25 0.3
0.5 0.3
5           (number of triangles)
1 2 5  2 4 -1  (first triangle and neighbors)
2 3 5  3 1 -1
4 5 3  2 5  4
1 5 4  3 5  1
1 4 3  3 -1 4
```

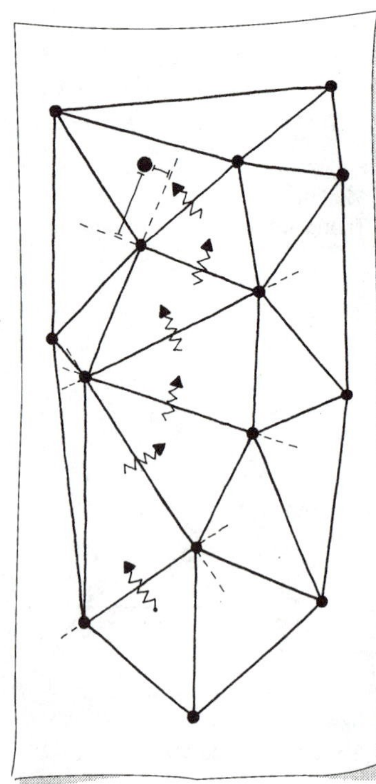

Sketch 81
Neighbor check.

8.6 Point Location

Given a triangulation of points $\mathbf{p}_i$, assume we are given a point $\mathbf{p}$ which has not been used in building the triangulation. Question: Which triangle is $\mathbf{p}$ in, if any? The easiest way is to compute $\mathbf{p}$'s barycentric coordinates with respect to all triangles; if all of them are positive with respect to some triangle, then that is the desired one, else, $\mathbf{p}$ is in none of the triangles.

While simple, this algorithm is expensive. In the worst case, every triangle has to be considered; on average, half of all triangles have to be considered. A much more efficient algorithm may be based upon the following observation. Suppose $\mathbf{p}$ is not in a particular triangle T. Then at least one of its barycentric coordinates with respect to T must be negative; let's assume it is u. We then know that $\mathbf{p}$ has no chance of being inside T's two neighbors along edges $\mathbf{s}_2$ or $\mathbf{s}_3$ (see Sketch 81).

So a likely candidate to check is the neighbor along $\mathbf{s}_1$ — recall that we have stored the neighboring information in a data structure. In this way — always searching in the direction of the currently most negative barycentric coordinate — we create a path from a starting triangle to the one that actually contains $\mathbf{p}$.

8.6.1 Point Location Algorithm. Input: triangulation and neighbor information, plus one point $\mathbf{p}$. Output: triangle that $\mathbf{p}$ is in.

Step 0: Set the "current triangle" to be the first triangle in the triangulation.
Step 1: Perform the *triangle inclusion test* (Section 8.1) for $\mathbf{p}$ and the current triangle. If all barycentric coordinates are positive, output the current triangle. If the barycentric coordinates are mixed in sign, then determine the barycentric coordinate of $\mathbf{p}$ with respect to the current triangle that has the most negative value. Set the current triangle to be the corresponding neighbor and repeat Step 1.
Notes:

- Try improving the speed of this algorithm by not completing the division for determining the barycentric coordinates in (8.2)–(8.4). This division does not change the sign. Keep in mind the test for which triangle to move to changes.

- Suppose the algorithm is to be executed for more than one point. Consider using the triangle that was output from the previous run as input, rather than always using the first triangle. Many times a data set has some *coherence*, and the output for the next run might be the same triangle, or one very near to the triangle from the previous run.

8.7 3D Triangulations

In computer applications, one often encounters millions of triangles, connected in some well-defined way, describing a geometric object. In particular, shading algorithms require this type of structure. The rules for 3D triangulations are the same as for 2D. Additionally, the data structure is the same, except that now each point has three instead of two coordinates.

Figure 8.3 shows a 3D object that is composed of triangles, and Figure 8.4 is the same object shaded. Another example is provided by Figure 12.1: all objects in that image were broken down into triangles before display.

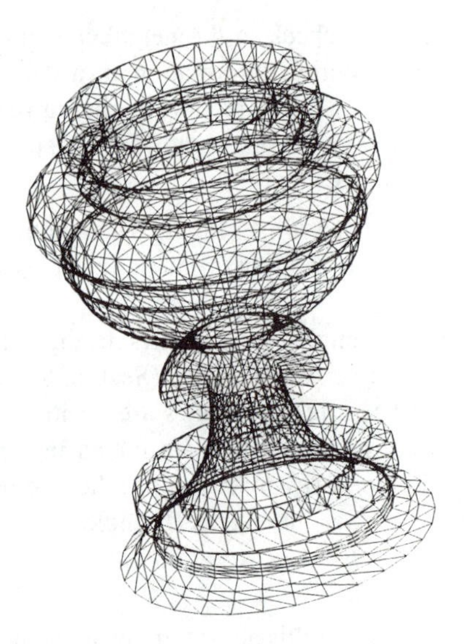

Figure 8.3.
3D Triangulated chalice.

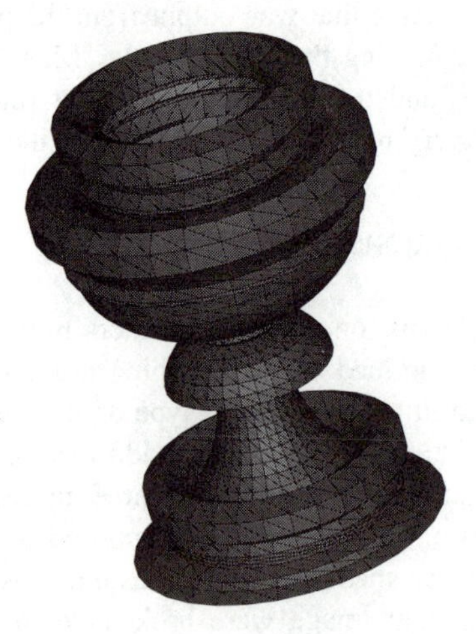

Figure 8.4.
3D Triangulated chalice shaded.

8.8 Exercises

Let a triangle T_1 be given by the vertices

$$\mathbf{p}_1 = \begin{bmatrix} 1 \\ 1 \end{bmatrix}, \quad \mathbf{p}_2 = \begin{bmatrix} 2 \\ 2 \end{bmatrix}, \quad \mathbf{p}_3 = \begin{bmatrix} -1 \\ 2 \end{bmatrix}.$$

Let a triangle T_2 be given by the vertices

$$\mathbf{q}_1 = \begin{bmatrix} 0 \\ 0 \end{bmatrix}, \quad \mathbf{q}_2 = \begin{bmatrix} 0 \\ -1 \end{bmatrix}, \quad \mathbf{q}_3 = \begin{bmatrix} -1 \\ 0 \end{bmatrix}.$$

1. Using T_1,

   a) What are the barycentric coordinates of $\mathbf{p} = \begin{bmatrix} 0 \\ 1.5 \end{bmatrix}$?

   b) What are the barycentric coordinates of $\mathbf{p} = \begin{bmatrix} 0 \\ 0 \end{bmatrix}$?

   c) Find the triangle's incenter.

   d) Find the triangle's circumcenter?

2. Repeat the above for T_2.
3. What are the areas of T_1 and T_2?
4. Let an affine map be given by

$$\mathbf{x}' = \begin{bmatrix} 1 & 2 \\ -1 & 2 \end{bmatrix} \mathbf{x} + \begin{bmatrix} -1 \\ 0 \end{bmatrix}.$$

   What are the areas of mapped triangles T_1' and T_2'? Compare the ratios

$$\frac{T_1}{T_2} \quad \text{and} \quad \frac{T_1'}{T_2'}.$$

5. Using the file sample.tri in our ftp-site, write a program that determines the triangle that an input point $\mathbf{p}$ is in. Test with $\mathbf{p}$ being the barycenter of the data. Starting with triangle 1 in the data structure, print the numbers of all triangles encountered in your search. (Note: this file gives the first point the index 0 and the first triangle the index 0.) This triangulation is illustrated in Figure 8.2.

Conics 9

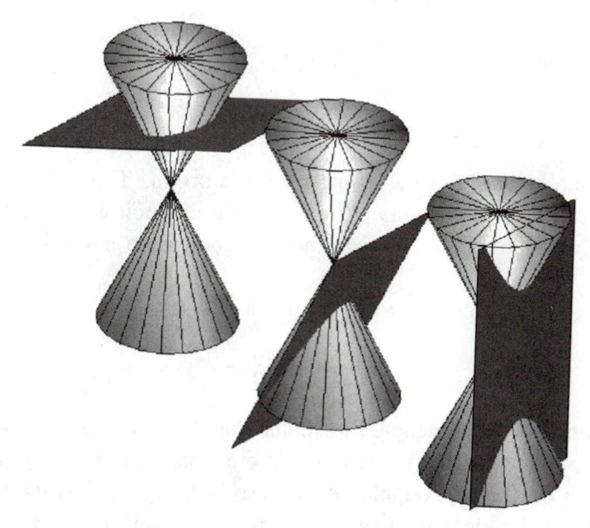

Figure 9.1.
Conic sections: three types of curves formed by the intersection of a plane and a cone. From left to right: ellipse, parabola, and hyperbola.

Take a flashlight and shine it straight onto a wall. You will see a circle. Tilt the light, and the circle will turn into an ellipse. Tilt further, and the ellipse will become more and more elongated, and will become a parabola eventually. Tilt a little more, and you will have a hyperbola — actually one branch of it. The beam of your flashlight is a *cone*, and the image it generates on the wall is the intersection of that cone with a *plane* (i.e., the wall). Thus we have the name *conic section* for curves that are the intersections of cones and planes. See Figure 9.1.

The three curves, ellipses, parabolas, and hyperbolas, arise in many situations and are the subject of this chapter. The basic tools for handling them are nothing but the matrix theory developed earlier.

Before we delve into the theory of conic sections, we list some "real life" occurances.

- The paths of the planets around the sun are ellipses.
- If you sharpen a pencil, you generate hyperbolas (see Sketch 82).
- If you water your lawn, the water leaving the hose traces a parabolic arc.

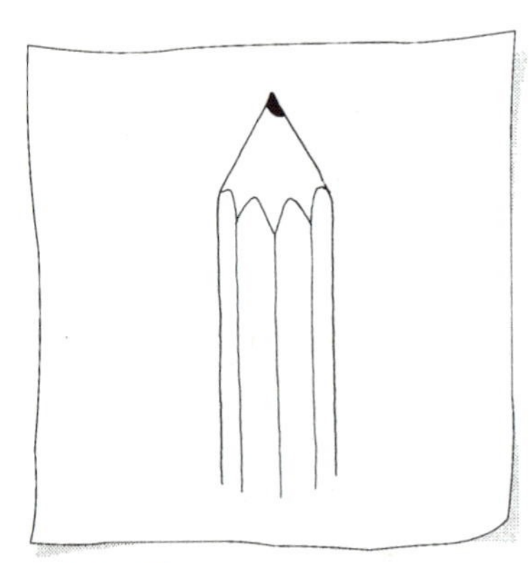

Sketch 82
A pencil with hyperbolic arcs.

9.1 The General Conic

We know that all points $\mathbf{x}$ satisfying

$$x_1^2 + x_2^2 = r^2 \tag{9.1}$$

are on a circle of radius r, centered at the origin. This type of equation is called an *implicit equation*. Similar to the implicit equation for a line, this type of equation is satisfied only for coordinate pairs that lie on the circle.

A little more generality will give us an ellipse:

$$\lambda_1 x_1^2 + \lambda_2 x_2^2 = c. \tag{9.2}$$

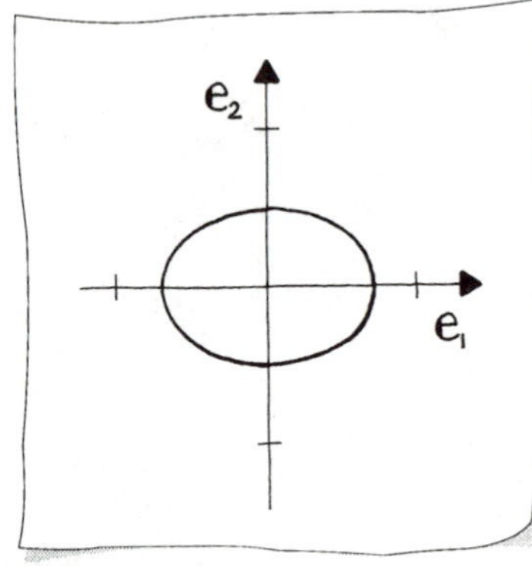

Sketch 83
An ellipse with $\lambda_1 = 2, \lambda_2 = 4$, and $c = 1$.

The factors λ_1 and λ_2 denote how much the ellipse deviates from a circle. For example, if $\lambda_1 > \lambda_2$, the ellipse is more elongated in the x_2−direction. See Sketch 83 for an example. An ellipse in this form has its *minor and major axes* coincident with the coordinate axes, and the *center* is at the origin.

We will now rewrite this simple equation in a much more complicated form:

$$\begin{bmatrix} x_1 & x_2 \end{bmatrix} \begin{bmatrix} \lambda_1 & 0 \\ 0 & \lambda_2 \end{bmatrix} \begin{bmatrix} x_1 \\ x_2 \end{bmatrix} - c = 0. \tag{9.3}$$

You will see the wisdom of this in a short while. The above equation allows a significant compactification:

$$\mathbf{x}^{\mathrm{T}} D \mathbf{x} - c = 0. \tag{9.4}$$

Suppose that we now wish to rotate the ellipse around the origin. This is achieved by replacing $\mathbf{x}$ by $R\mathbf{x}$, where R is a rotation matrix, as covered in Section 4.6. Inserting $R\mathbf{x}$ into (9.4), we now get

$$\mathbf{x}^{\mathrm{T}} R^{\mathrm{T}} D R \mathbf{x} - c = 0. \tag{9.5}$$

Just to make things more complicated, we now *translate* the ellipse to a new position using a translation vector $\mathbf{v}$. Then $\mathbf{x}$ is moved to $\mathbf{x}+\mathbf{v}$, and (9.5) becomes (with the abbreviation $R^{\mathrm{T}}DR = A$):

$$[\mathbf{x}^{\mathrm{T}} + \mathbf{v}^{\mathrm{T}}]A[\mathbf{x}+\mathbf{v}] - c = 0.$$

If we expand this,[1] we obtain

$$\mathbf{x}^{\mathrm{T}}A\mathbf{x} + 2\mathbf{x}^{\mathrm{T}}A\mathbf{v} + \mathbf{v}^{\mathrm{T}}A\mathbf{v} - c = 0. \tag{9.6}$$

This denotes an ellipse in general position and may be slightly abbreviated as

$$\mathbf{x}^{\mathrm{T}}A\mathbf{x} + 2\mathbf{x}^{\mathrm{T}}\mathbf{b} + d = 0, \tag{9.7}$$

with $\mathbf{b} = A\mathbf{v}$ and $d = \mathbf{v}^{\mathrm{T}}A\mathbf{v} - c$.

Example 9.1 Let's start with the ellipse $2x_1^2 + 4x_2^2 - 1 = 0$. In matrix form, corresponding to (9.3), we have

$$\begin{bmatrix} x_1 & x_2 \end{bmatrix} \begin{bmatrix} 2 & 0 \\ 0 & 4 \end{bmatrix} \begin{bmatrix} x_1 \\ x_2 \end{bmatrix} - 1 = 0.$$

This ellipse is shown in Sketch 83.

Now we rotate by 45°, using the rotation matrix

$$R = \begin{bmatrix} s & -s \\ s & s \end{bmatrix}$$

with $s = \sin 45° = \cos 45° = \sqrt{2}/2$. The matrix $A = R^{\mathrm{T}}DR$ becomes

$$A = \begin{bmatrix} 3 & 1 \\ 1 & 3 \end{bmatrix}.$$

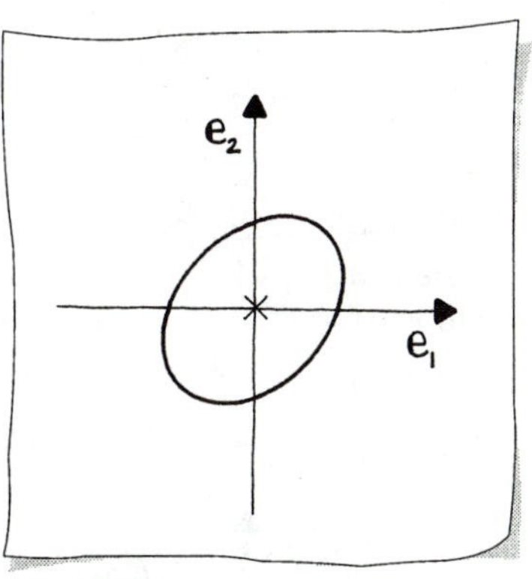

Sketch 84
A rotated ellipse.

This ellipse is illustrated in Sketch 84.

If we now translate by a vector $\mathbf{v} = \begin{bmatrix} 2 \\ -1 \end{bmatrix}$, then (9.7) becomes

$$\mathbf{x}^{\mathrm{T}} \begin{bmatrix} 3 & 1 \\ 1 & 3 \end{bmatrix} \mathbf{x} + 2\mathbf{x}^{\mathrm{T}} \begin{bmatrix} 5 \\ -1 \end{bmatrix} + 10 = 0.$$

Expanding the previous equation, the conic is

$$3x_1^2 + 2x_1x_2 + 3x_2^2 + 10x_1 - 2x_2 + 10 = 0.$$

[1] Note that A is symmetric so $\mathbf{x}^{\mathrm{T}}A\mathbf{v} = \mathbf{v}^{\mathrm{T}}A\mathbf{x}$.

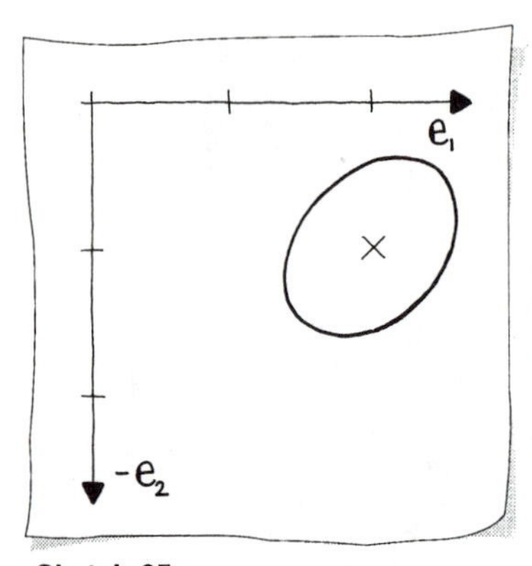

Sketch 85
A rotated and translated ellipse.

This ellipse is illustrated in Sketch 85.

This was a lot of work, just to find the general form of an ellipse! However, as we shall see, a lot more has been achieved here: the form (9.7) does not just represent ellipses, but *any* conic. However, only ellipses and hyperbolas can be written in the form (9.6).

To see that (9.7) represents any conic, let's examine two remaining conics, a hyperbola and a parabola.

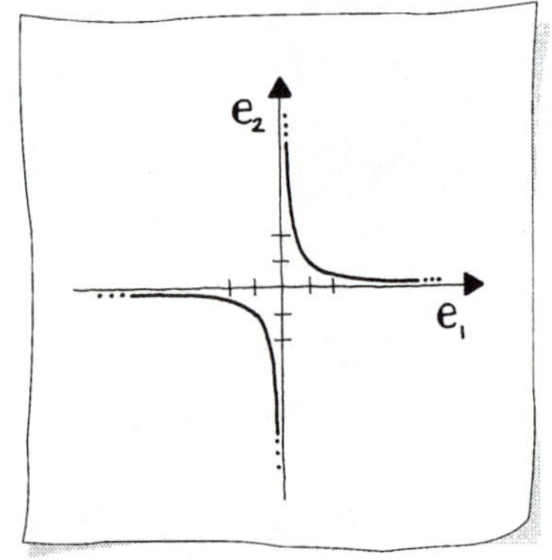

Sketch 86
A hyperbola.

Example 9.2 Sketch 86 illustrates the conic

$$x_2 = \frac{1}{x_1}$$

which is a hyperbola.

This may be written as

$$\begin{bmatrix} x_1 & x_2 \end{bmatrix} \begin{bmatrix} 0 & \frac{1}{2} \\ \frac{1}{2} & 0 \end{bmatrix} \begin{bmatrix} x_1 \\ x_2 \end{bmatrix} - 1 = 0,$$

which is clearly of the form (9.7).

Example 9.3 A parabola is illustrated in Sketch 87,

$$x_2 = x_1^2.$$

This may be written as

$$\begin{bmatrix} x_1 & x_2 \end{bmatrix} \begin{bmatrix} -1 & 0 \\ 0 & 0 \end{bmatrix} \begin{bmatrix} x_1 \\ x_2 \end{bmatrix} + \begin{bmatrix} x_1 & x_2 \end{bmatrix} \begin{bmatrix} 0 \\ 1 \end{bmatrix} = 0,$$

and thus is also of the form (9.7)!

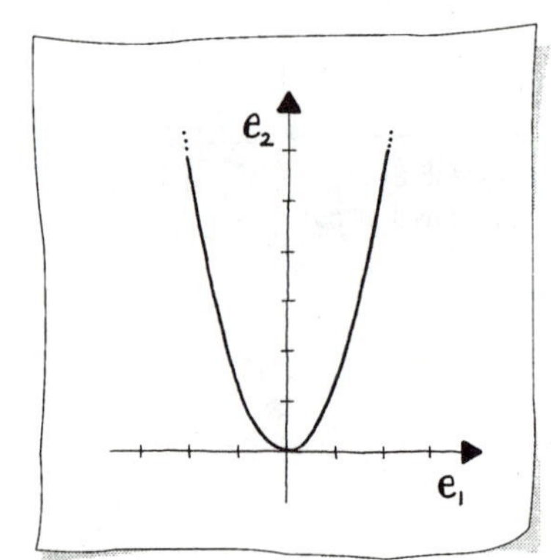

Sketch 87
A parabola.

The above examples should have built enough confidence so that we may now state that (9.7) describes all possible conics. In fact, many texts simply start out by using (9.7) as the initial *definition* of a conic.

9.2 Analyzing Conics

If you are given the equation of a conic, how can you tell which of the three basic types it is? A general conic equation might look like this:

$$c_1x_1^2 + c_2x_2^2 + c_3x_1x_2 + c_4x_1 + c_5x_2 + c_6 = 0.$$

This does not resemble (9.7) much! Yet it is the same, for we can rewrite it as

$$\begin{bmatrix} x_1 & x_2 \end{bmatrix} \begin{bmatrix} c_1 & \frac{1}{2}c_3 \\ \frac{1}{2}c_3 & c_2 \end{bmatrix} \begin{bmatrix} x_1 \\ x_2 \end{bmatrix} + 2 \begin{bmatrix} x_1 & x_2 \end{bmatrix} \begin{bmatrix} \frac{1}{2}c_4 \\ \frac{1}{2}c_5 \end{bmatrix} + c_6 = 0. \tag{9.8}$$

If we could write the symmetric matrix A as a product $R^{\mathrm{T}}DR$, with a diagonal matrix D and a rotation matrix R, then we could tell what kind of conic we have.

- If D is the zero matrix, then the conic is degenerate and simply consists of a straight line.
- If D has only one nonzero entry, then the conic is a parabola.
- If D has two nonzero entries of the same sign, then the conic is an ellipse.
- If D has two nonzero entries with opposite sign, then the conic is a hyperbola.

The problem of finding the diagonal matrix D has already been solved earlier in this book, namely in Section 7.5. All we have to do is find the eigenvalues of A; they determine the diagonal matrix D, and it in turn determines what kind of conic we have.

Example 9.4 Let a conic be given by

$$3x_1^2 + 3x_2^2 + 2x_1x_2 + 10x_1 - 2x_2 + 10 = 0. \tag{9.9}$$

We first rewrite this as

$$\mathbf{x}^{\mathrm{T}} \begin{bmatrix} 3 & 1 \\ 1 & 3 \end{bmatrix} \mathbf{x} + 2\mathbf{x}^{\mathrm{T}} \begin{bmatrix} 5 \\ -1 \end{bmatrix} + 10 = 0.$$

The eigenvalues of the 2×2 matrix are the solution of the quadratic equation $(3 - \lambda)^2 - 1 = 0$, and thus are $\lambda_1 = 2$ and $\lambda_2 = 4$.

Our desired diagonal matrix is:

$$D = \begin{bmatrix} 2 & 0 \\ 0 & 4 \end{bmatrix},$$

and thus this conic is an *ellipse*. It is illustrated in Sketch 85.

Summarizing: In this section we found that affine maps take a particular type of conic to another one of the same type. The conic type is determined by D: it is unchanged by affine maps.

9.3 The Position of a Conic

Finding the eigenvalues of A is our way to determine the conic type. If we are interested in its position in the plane, then we have to do more.

First, let's consider the linear terms

$$2 \begin{bmatrix} x_1 & x_2 \end{bmatrix} \begin{bmatrix} \frac{1}{2}c_4 & \frac{1}{2}c_5 \end{bmatrix}$$

in (9.8). They appear in slightly different forms in (9.7) and (9.6). If we are given an arbitrary conic equation, then we have the terms as they appear in (9.7), and we have to find the terms as given in (9.6). Equating the coefficients of the linear terms in those two equations gives

$$A\mathbf{v} = \mathbf{b}. \tag{9.10}$$

Since both A and $\mathbf{b}$ are given from (9.8), we see that this is a 2×2 linear system for the unknown translation vector $\mathbf{v}$.

Let's assume that A has full rank. This is equivalent to A having two nonzero eigenvalues, and so the given conic is either an ellipse or a hyperbola. In this case, (9.10) always has a unique solution $\mathbf{v}$. The coordinate transformation

$$\mathbf{x} = \mathbf{x} - \mathbf{v}$$

will then move the center of the conic to the origin, such that the conic is now of the form

$$\mathbf{x}^{\mathrm{T}} A \mathbf{x} + d = 0,$$

where

$$d = -c.$$

Example 9.5 Let's do the above for the conic in (9.9). The linear system is

$$\begin{bmatrix} 3 & 1 \\ 1 & 3 \end{bmatrix} \begin{bmatrix} v_1 \\ v_2 \end{bmatrix} = \begin{bmatrix} 5 \\ -1 \end{bmatrix}$$

and has the solution $v_1 = 2$, $v_2 = -1$. Also,

$$d = -1.$$

Thus the translated conic now is

$$\mathbf{x}^{\mathrm{T}} \begin{bmatrix} 3 & 1 \\ 1 & 3 \end{bmatrix} \mathbf{x} - 1 = 0.$$

Recall this conic is illustrated in Sketch 84.

Unless A was a diagonal matrix to begin with, the translated conic still does not have its axes lined up with the coordinate axes. If we want to achieve this, we have to *rotate* the conic. For that, we recall from Section 7.5 that the symmetric matrix A can be written in terms of a rotation R and a scaling D:

$$A = R^{\mathrm{T}} D R.$$

If λ_1 and λ_2 are the eigenvalues and $\mathbf{r}_1$ and $\mathbf{r}_2$ are the eigenvectors of A, then the orthogonal matrix

$$R = \begin{bmatrix} \mathbf{r}_1 & \mathbf{r}_2 \end{bmatrix}$$

and

$$D = \begin{bmatrix} \lambda_1 & 0 \\ 0 & \lambda_2 \end{bmatrix}.$$

To undo the rotation, apply the transformation

$$\mathbf{x} = R^{\mathrm{T}} \mathbf{x}.$$

This will rotate the conic so that now its axes agree with the coordinate axes. Thus (9.5) is modified to be

$$(R^{\mathrm{T}}\mathbf{x})^{\mathrm{T}} R^{\mathrm{T}} D R (R^{\mathrm{T}}\mathbf{x}) - c = 0,$$

giving the conic the form

$$\mathbf{x}^{\mathrm{T}} D \mathbf{x} - c = 0.$$

Example 9.6 Returning to our example, we need to find the eigenvector $\mathbf{r}_1$ for the eigenvalue $\lambda_1 = 2$ and then the eigenvector $\mathbf{r}_2$ for the eigenvalue $\lambda_2 = 4$.

For the first case, we get the linear system

$$\begin{bmatrix} 1 & 1 \\ 1 & 1 \end{bmatrix} \begin{bmatrix} r_{1,1} \\ r_{2,1} \end{bmatrix} = \begin{bmatrix} 0 \\ 0 \end{bmatrix}$$

with solution

$$\mathbf{r}_1 = \begin{bmatrix} r_{1,1} \\ r_{2,1} \end{bmatrix} = \begin{bmatrix} 1 \\ -1 \end{bmatrix}.$$

For the second case, we get the linear system

$$\begin{bmatrix} -1 & 1 \\ 1 & -1 \end{bmatrix} \begin{bmatrix} r_{1,2} \\ r_{2,2} \end{bmatrix} = \begin{bmatrix} 0 \\ 0 \end{bmatrix}$$

with solution

$$\mathbf{r}_2 = \begin{bmatrix} r_{1,2} \\ r_{2,2} \end{bmatrix} = \begin{bmatrix} 1 \\ 1 \end{bmatrix}.$$

After normalizing both eigenvectors, our rotation matrix $R = [\mathbf{r}_1, \mathbf{r}_2]$ is given by

$$R = \frac{1}{\sqrt{2}} \begin{bmatrix} 1 & 1 \\ -1 & 1 \end{bmatrix}.$$

The matrix R^{T} would be applied to each point $\mathbf{x}$ to undo the rotation. Applying this to the conic

$$\mathbf{x}^{\mathrm{T}} \begin{bmatrix} 3 & 1 \\ 1 & 3 \end{bmatrix} \mathbf{x} - 1 = 0,$$

produces the conic

$$2x_1 + 4x_2 - 1 = 0$$

centered at the origin and rotated to align with the coordinate axes. This is the conic we began with! See Sketch 83.

9.4 Exercises

1. Let $x_1^2 - 2x_1x_2 - 4 = 0$ be the equation of a conic section. What type is it?

2. What affine map takes the circle

$$(x_1 - 3)^2 + (x_2 + 1)^2 - 4 = 0$$

to the ellipse

$$2x_1^2 + 4x_2^2 - 1 = 0?$$

3. How many intersections does a straight line have with a conic? Given a conic in the form (9.2) and a parametric form of a line $\mathbf{l}(t)$, what are the t-values of the intersection points? Explain any singularities.

3D Geometry 10

Figure 10.1.
3D objects: Planar facets joined to form 3D objects.

This chapter introduces the essential building blocks of 3D geometry by first extending the 2D tools from Chapters 2 and 3 to 3D. But beyond that, we will also encounter some concepts that are "truly" 3D, i.e., those that do not have 2D counterparts. With the geometry presented in this chapter, we will be ready to create and analyze simple 3D objects, such as those illustrated in Figure 10.1.

10.1 From 2D to 3D

Moving from 2D to 3D geometry requires a coordinate system with one more dimension. Sketch 88 illustrates the $[\mathbf{e}_1, \mathbf{e}_2, \mathbf{e}_3]$-system which consists of the vectors

$$\mathbf{e}_1 = \begin{bmatrix} 1 \\ 0 \\ 0 \end{bmatrix}, \quad \mathbf{e}_2 = \begin{bmatrix} 0 \\ 1 \\ 0 \end{bmatrix}, \quad \text{and} \quad \mathbf{e}_3 = \begin{bmatrix} 0 \\ 0 \\ 1 \end{bmatrix}.$$

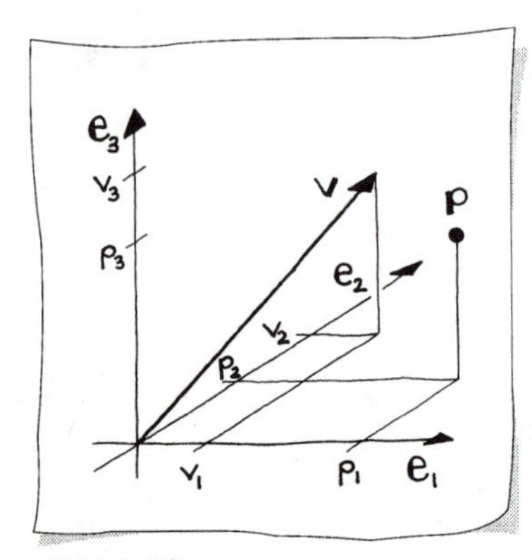

Sketch 88
The $[\mathbf{e}_1, \mathbf{e}_2, \mathbf{e}_3]$-axes, a point, and a vector.

Thus a *vector* in 3D is given as

$$\mathbf{v} = \begin{bmatrix} v_1 \\ v_2 \\ v_3 \end{bmatrix}. \tag{10.1}$$

The three *components* of $\mathbf{v}$ indicate the displacement along each axis in the $[\mathbf{e}_1, \mathbf{e}_2, \mathbf{e}_3]$-system. This is illustrated in Sketch 88. A 3D vector $\mathbf{v}$ is said to live in real 3D space, or $\mathbb{R}^3$, that is $\mathbf{v} \in \mathbb{R}^3$.

A *point* is a reference to a *location*. Points in 3D are given as

$$\mathbf{p} = \begin{bmatrix} p_1 \\ p_2 \\ p_3 \end{bmatrix}. \tag{10.2}$$

The *coordinates* indicate the point's location in the $[\mathbf{e}_1, \mathbf{e}_2, \mathbf{e}_3]$-system, as illustrated in Sketch 88. A point $\mathbf{p}$ is said to live in Euclidean 3D-space, or $\mathbb{E}^3$, that is $\mathbf{p} \in \mathbb{E}^3$.

Let's look briefly at some basic 3D vector properties, as we did for 2D vectors. First of all, the 3D *zero vector*:

$$\mathbf{0} = \begin{bmatrix} 0 \\ 0 \\ 0 \end{bmatrix}.$$

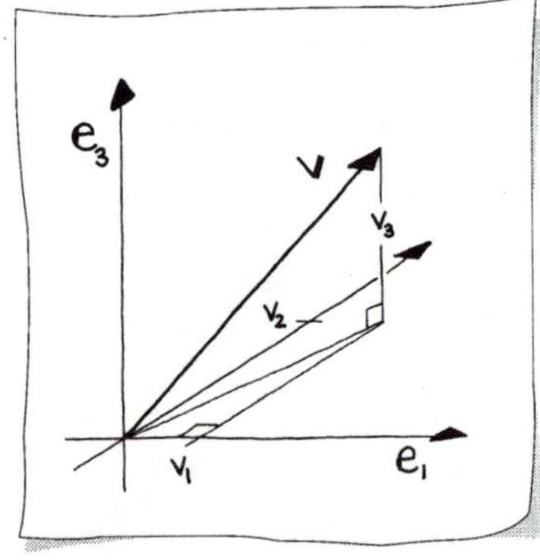

Sketch 89
Length of a 3D vector.

Sketch 89 illustrates a 3D vector $\mathbf{v}$ along with its components. Notice the two right triangles. Applying the *Pythagorean Theorem* twice, the *length* of $\mathbf{v}$, denoted as $\|\mathbf{v}\|$, is

$$\|\mathbf{v}\| = \sqrt{v_1^2 + v_2^2 + v_3^2}. \tag{10.3}$$

The length or magnitude of a 3D vector can be interpreted as distance, speed, or force.

Scaling a vector by an amount k yields $\|k\mathbf{v}\| = k\|\mathbf{v}\|$. Also, a *normalized vector* has unit length, $\|\mathbf{v}\| = 1$.

Example 10.1 We will get some practice working with 3D vectors. The first task is to normalize the vector

$$\mathbf{v} = \begin{bmatrix} 1 \\ 2 \\ 3 \end{bmatrix}.$$

First calculate the length of $\mathbf{v}$ as

$$\|\mathbf{v}\| = \sqrt{1^2 + 2^2 + 3^2} = \sqrt{14},$$

then the normalized vector $\mathbf{w}$ is

$$\mathbf{w} = \frac{\mathbf{v}}{\|\mathbf{v}\|} = \begin{bmatrix} 1/\sqrt{14} \\ 2/\sqrt{14} \\ 3/\sqrt{14} \end{bmatrix} \approx \begin{bmatrix} 0.27 \\ 0.53 \\ 0.80 \end{bmatrix}.$$

Check for yourself that $\|\mathbf{w}\| = 1$.

Scale $\mathbf{v}$ by $k = 2$:

$$2\mathbf{v} = \begin{bmatrix} 2 \\ 4 \\ 6 \end{bmatrix}.$$

Now calculate

$$\|2\mathbf{v}\| = \sqrt{2^2 + 4^2 + 6^2} = 2\sqrt{14}.$$

Thus we verified that $\|2\mathbf{v}\| = 2\|\mathbf{v}\|$.

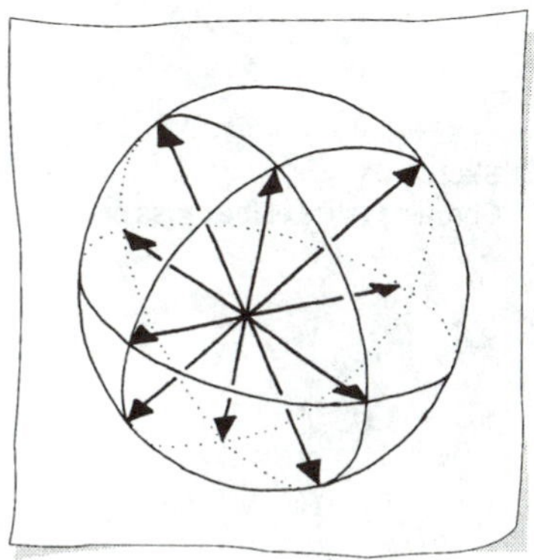

Sketch 90
All 3D unit vectors define a sphere.

There are infinitely many 3D unit vectors. In Sketch 90 a few of these are drawn emanating from the origin. The sketch is a sphere of radius one.

All the rules for combining points and vectors in 2D from Section 2.2 carry over to 3D. The *dot product* of two 3D vectors, $\mathbf{v}$ and $\mathbf{w}$, becomes

$$\mathbf{v} \cdot \mathbf{w} = v_1 w_1 + v_2 w_2 + v_3 w_3.$$

The cosine of the angle θ between the two vectors can be determined as

$$\cos\theta = \frac{\mathbf{v} \cdot \mathbf{w}}{\|\mathbf{v}\|\|\mathbf{w}\|}. \tag{10.4}$$

10.2 Cross Product

The dot product is a type of multiplication for two vectors which reveals geometric information, namely the angle between them. However, this does not reveal information about their orientation in relation to $\mathbb{R}^3$. Two vectors define a plane — which is a *subspace* of $\mathbb{R}^3$. Thus it would be useful to have yet another vector in order to create a 3D coordinate system which is *embedded* in the $[\mathbf{e}_1, \mathbf{e}_2, \mathbf{e}_3]$-system. This is the purpose of another form of vector multiplication called the *cross product*.

In other words, the cross product of $\mathbf{v}$ and $\mathbf{w}$, written as

$$\mathbf{u} = \mathbf{v} \wedge \mathbf{w},$$

produces the vector $\mathbf{u}$ which satisfies the following.

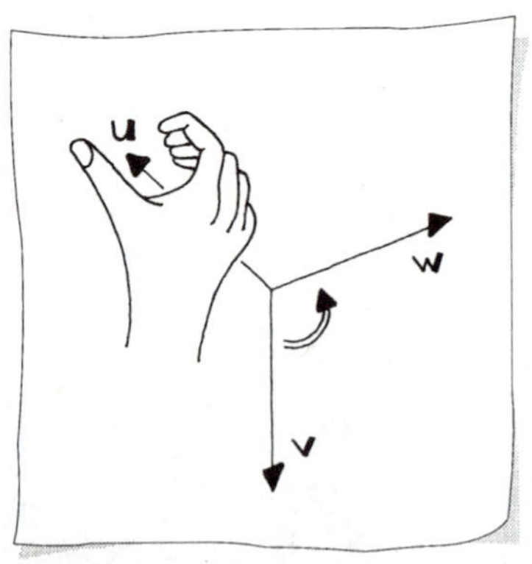

Sketch 91
Characteristics of the cross product.

1. The vector $\mathbf{u}$ is perpendicular to $\mathbf{v}$ and $\mathbf{w}$, that is

$$\mathbf{u} \cdot \mathbf{v} = 0 \quad \text{and} \quad \mathbf{u} \cdot \mathbf{w} = 0$$

2. The orientation of the vector $\mathbf{u}$ follows the *right-hand rule*. This means that if you curl the fingers of your right hand from $\mathbf{v}$ to $\mathbf{w}$, your thumb will point in the direction of $\mathbf{u}$.

3. The magnitude of $\mathbf{u}$ is the area of the parallelogram defined by $\mathbf{v}$ and $\mathbf{w}$.

These items are illustrated in Sketch 91. Because the cross product produces a vector, it is also called a *vector product.*

Items 1 and 2 determine the direction of $\mathbf{u}$ and item 3 determines the length of $\mathbf{u}$. The cross product is defined as

$$\mathbf{v} \wedge \mathbf{w} = \begin{bmatrix} v_2 w_3 - w_2 v_3 \\ v_3 w_1 - w_3 v_1 \\ v_1 w_2 - w_1 v_2 \end{bmatrix}. \tag{10.5}$$

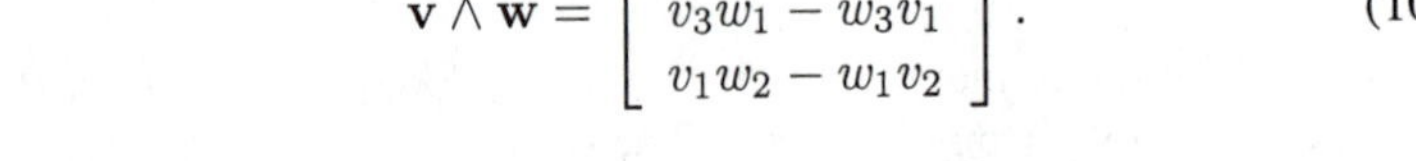

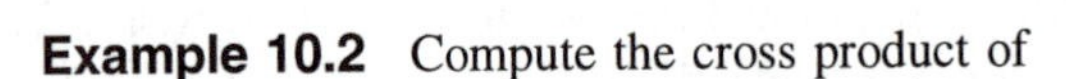

Example 10.2 Compute the cross product of

$$\mathbf{v} = \begin{bmatrix} 1 \\ 0 \\ 2 \end{bmatrix} \quad \text{and} \quad \mathbf{w} = \begin{bmatrix} 0 \\ 3 \\ 4 \end{bmatrix}.$$

The cross product is

$$\mathbf{u} = \mathbf{v} \wedge \mathbf{w} = \begin{bmatrix} 0 \times 4 - 3 \times 2 \\ 2 \times 0 - 4 \times 1 \\ 1 \times 3 - 0 \times 0 \end{bmatrix} = \begin{bmatrix} -6 \\ -4 \\ 3 \end{bmatrix}. \tag{10.6}$$

Section 4.9 described why the 2×2 determinant, formed from two 2D vectors, is equal to the area P of the parallelogram defined by these two vectors. The analogous result for two vectors in 3D is

$$P = \|\mathbf{v} \wedge \mathbf{w}\|. \tag{10.7}$$

Recall that P is also defined by measuring a height and side length of the parallelogram, as illustrated in Sketch 92. The height h is

$$h = \|\mathbf{w}\| \sin\theta,$$

and the side length is $\|\mathbf{v}\|$, which makes

$$P = \|\mathbf{v}\|\|\mathbf{w}\| \sin\theta. \tag{10.8}$$

Equating (10.7) and (10.8) results in

$$\|\mathbf{v} \wedge \mathbf{w}\| = \|\mathbf{v}\|\|\mathbf{w}\| \sin\theta. \tag{10.9}$$

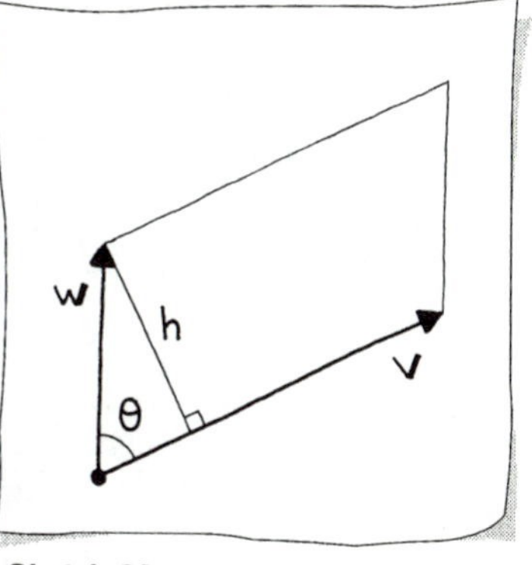

Sketch 92
Area of a parallelogram.

Example 10.3 Compute the area of the parallelogram formed by

$$\mathbf{v} = \begin{bmatrix} 2 \\ 2 \\ 0 \end{bmatrix} \quad \text{and} \quad \mathbf{w} = \begin{bmatrix} 0 \\ 0 \\ 1 \end{bmatrix}.$$

Set up the cross product:

$$\mathbf{v} \wedge \mathbf{w} = \begin{bmatrix} 2 \\ -2 \\ 0 \end{bmatrix}.$$

Then the area is

$$P = \|\mathbf{v} \wedge \mathbf{w}\| = 2\sqrt{2}.$$

Since the parallelogram is a rectangle, the area is the product of the edge lengths, so this is the correct result.

Verifying (10.9),

$$P = 2\sqrt{2} \sin 90° = 2\sqrt{2}.$$

In order to derive another useful expression in terms of the cross product, square both sides of (10.9). Thus we have

$$\begin{aligned} \|\mathbf{v} \wedge \mathbf{w}\|^2 &= \|\mathbf{v}\|^2\|\mathbf{w}\|^2 \sin^2\theta \\ &= \|\mathbf{v}\|^2\|\mathbf{w}\|^2(1 - \cos^2\theta) \\ &= \|\mathbf{v}\|^2\|\mathbf{w}\|^2 - \|\mathbf{v}\|^2\|\mathbf{w}\|^2 \cos^2\theta \\ &= \|\mathbf{v}\|^2\|\mathbf{w}\|^2 - (\mathbf{v} \cdot \mathbf{w})^2 \end{aligned} \tag{10.10}$$

The last line is refered to as *Lagrange's identity.*

To get a better feeling for the behaviour of the cross product, let's look at some of its properties.

- Parallel vectors result in the zero vector: $\mathbf{v} \wedge c\mathbf{v} = \mathbf{0}$.
- Homogeneous: $c\mathbf{v} \wedge \mathbf{w} = c(\mathbf{v} \wedge \mathbf{w})$.
- Anti-symmetric: $\mathbf{v} \wedge \mathbf{w} = -(\mathbf{w} \wedge \mathbf{v})$.
- Non-associative: $\mathbf{u} \wedge (\mathbf{v} \wedge \mathbf{w}) \neq (\mathbf{u} \wedge \mathbf{v}) \wedge \mathbf{w}$, in general.
- Distributive: $\mathbf{u} \wedge (\mathbf{v} + \mathbf{w}) = \mathbf{u} \wedge \mathbf{v} + \mathbf{u} \wedge \mathbf{w}$.
- Right-hand rule:

$$\mathbf{e}_1 \wedge \mathbf{e}_2 = \mathbf{e}_3$$
$$\mathbf{e}_2 \wedge \mathbf{e}_3 = \mathbf{e}_1$$
$$\mathbf{e}_3 \wedge \mathbf{e}_1 = \mathbf{e}_2$$

Example 10.4 Let's test these properties of the cross product with

$$\mathbf{u} = \begin{bmatrix} 1 \\ 1 \\ 1 \end{bmatrix} \quad \mathbf{v} = \begin{bmatrix} 2 \\ 0 \\ 0 \end{bmatrix} \quad \mathbf{w} = \begin{bmatrix} 0 \\ 3 \\ 0 \end{bmatrix}.$$

Make your own sketches and don't forget the right-hand rule to guess the resulting vector direction.

Parallel vectors:

$$\mathbf{v} \wedge 3\mathbf{v} = \begin{bmatrix} 0 \times 0 - 0 \times 0 \\ 0 \times 6 - 0 \times 2 \\ 2 \times 0 - 6 \times 0 \end{bmatrix} = \mathbf{0}.$$

Homogeneous:

$$4\mathbf{v} \wedge \mathbf{w} = \begin{bmatrix} 0 \times 0 - 3 \times 0 \\ 0 \times 0 - 0 \times 8 \\ 8 \times 3 - 0 \times 0 \end{bmatrix} = \begin{bmatrix} 0 \\ 0 \\ 24 \end{bmatrix},$$

and

$$4(\mathbf{v} \wedge \mathbf{w}) = 4 \begin{bmatrix} 0 \times 0 - 3 \times 0 \\ 0 \times 0 - 0 \times 2 \\ 2 \times 3 - 0 \times 0 \end{bmatrix} = 4 \begin{bmatrix} 0 \\ 0 \\ 6 \end{bmatrix} = \begin{bmatrix} 0 \\ 0 \\ 24 \end{bmatrix}.$$

Anti-symmetric:

$$\mathbf{v} \wedge \mathbf{w} = \begin{bmatrix} 0 \\ 0 \\ 6 \end{bmatrix} \quad \text{and} \quad -(\mathbf{w} \wedge \mathbf{v}) = -\left(\begin{bmatrix} 0 \\ 0 \\ -6 \end{bmatrix} \right).$$

Non-associative:

$$\mathbf{u} \wedge (\mathbf{v} \wedge \mathbf{w}) = \begin{bmatrix} 1 \times 6 - 0 \times 1 \\ 1 \times 0 - 6 \times 1 \\ 1 \times 0 - 0 \times 1 \end{bmatrix} = \begin{bmatrix} 6 \\ -6 \\ 0 \end{bmatrix}$$

which is not the same as

$$(\mathbf{u} \wedge \mathbf{v}) \wedge \mathbf{w} = \begin{bmatrix} 0 \\ 2 \\ -2 \end{bmatrix} \wedge \begin{bmatrix} 0 \\ 3 \\ 0 \end{bmatrix} = \begin{bmatrix} 6 \\ 0 \\ 0 \end{bmatrix}.$$

Distributive:

$$\mathbf{u} \wedge (\mathbf{v} + \mathbf{w}) = \begin{bmatrix} 1 \\ 1 \\ 1 \end{bmatrix} \wedge \begin{bmatrix} 2 \\ 3 \\ 0 \end{bmatrix} = \begin{bmatrix} -3 \\ 2 \\ 1 \end{bmatrix}$$

which is equal to

$$(\mathbf{u} \wedge \mathbf{v}) + (\mathbf{u} \wedge \mathbf{w}) = \begin{bmatrix} 0 \\ 2 \\ -2 \end{bmatrix} + \begin{bmatrix} -3 \\ 0 \\ 3 \end{bmatrix} = \begin{bmatrix} -3 \\ 2 \\ 1 \end{bmatrix}$$

The cross product is an invaluable tool for engineering. One reason: it facilitates the construction of a coordinate independent frame of reference.

10.3 Lines

Specifying a line with 3D geometry differs a bit from 2D. In terms of points and vectors, two pieces of information define a line, however we are restricted to specifying

- two points or
- a point and a vector parallel to the line.

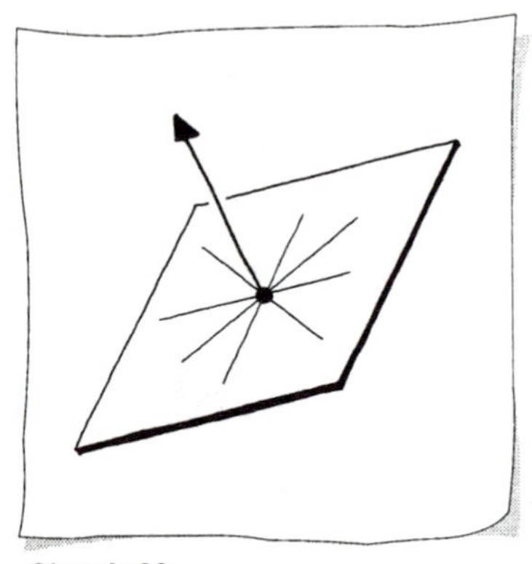

Sketch 93
Point and perpendicular don't define a line.

The 2D geometry item

- a point and a vector perpendicular to the line,

no longer works. It isn't specific enough. See Sketch 93. In other words, an entire family of lines satisfies this specification; this family lies in a plane. (More on planes in Section 10.4.) As a consequence, the concept of a *normal* to a 3D line does not exist.

Let's look at the mathematical representations of a 3D line. Clearly, from the discussion above, there cannot be an *implicit form*.

The *parametric form* of a 3D line does not differ from the 2D line except for the fact that the given information lives in 3D. A line $\mathbf{l}(t)$ has the form

$$\mathbf{l}(t) = \mathbf{p} + t\mathbf{v}, \tag{10.11}$$

where $\mathbf{p} \in \mathbb{E}^3$ and $\mathbf{v} \in \mathbb{R}^3$. Points are generated on the line as the parameter t varies.

In 2D, two lines either intersect or they are parallel. In 3D this is not the case; a third possibility is that the lines are *skew*. Sketch 94 illustrates skew lines using a cube as a reference frame.

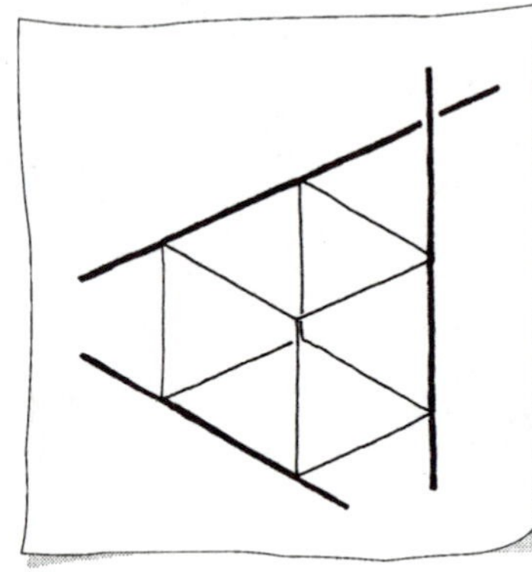

Sketch 94
Skew lines.

Because lines in 3D can be skew, the intersection of two lines might not have a solution. Revisiting the problem of the intersection of two lines given in parametric form from Section 3.8, we can see the algebraic truth in this statement. Now the two lines are

$$\begin{aligned} \mathbf{l}_1 : \quad \mathbf{l}_1(t) &= \mathbf{p} + t\mathbf{v} \\ \mathbf{l}_2 : \quad \mathbf{l}_2(s) &= \mathbf{q} + s\mathbf{w} \end{aligned}$$

where $\mathbf{p}, \mathbf{q} \in \mathbb{E}^3$ and $\mathbf{v}, \mathbf{w} \in \mathbb{R}^3$. To find the intersection point, we solve for t or s. Repeating (3.15), we have the linear system

$$\hat{t}\mathbf{v} - \hat{s}\mathbf{w} = \mathbf{q} - \mathbf{p}.$$

However, now there are three equations and still only two unknowns. Thus the system is *overdetermined*; more information on this type of system is given in Section 14.5. No solution exists when the lines are skew. In many applications it is important to know the closest point on a line to another line. This problem is solved in Section 11.2.

We still have the concepts of perpendicular and parallel lines in 3D.

10.4 Planes

While exploring the possibility of a 3D implicit line, we encountered a plane. We'll essentially repeat that here, however with a little change in notation. Suppose we are given a point $\mathbf{p}$ and a vector $\mathbf{n}$ bound to $\mathbf{p}$. The locus of all points $\mathbf{x}$ which satisfy the equation

$$\mathbf{n} \cdot (\mathbf{x} - \mathbf{p}) = 0 \tag{10.12}$$

defines the *implicit form* of a plane. This is illustrated in Sketch 95.The vector $\mathbf{n}$ is called the *normal* to the plane if $\|\mathbf{n}\| = 1$. If this is the case, then (10.12) is called the *point normal plane equation.*

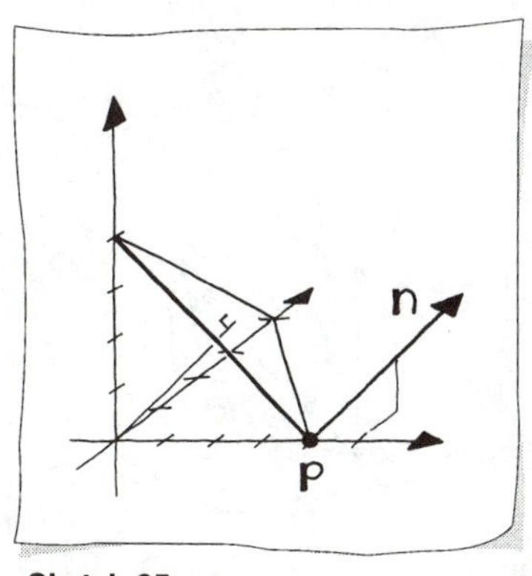

Sketch 95
Point normal plane equation.

Expanding (10.12), we have

$$n_1x_1 + n_2x_2 + n_3x_3 - (n_1p_1 + n_2p_2 + n_3p_3) = 0.$$

Typically this is written as

$$Ax_1 + Bx_2 + Cx_3 + D = 0, \tag{10.13}$$

where

$$\begin{aligned} A &= n_1 \\ B &= n_2 \\ C &= n_3 \\ D &= -(n_1p_1 + n_2p_2 + n_3p_3). \end{aligned}$$

Example 10.5 Compute the implicit form of the plane through the point

$$\mathbf{p} = \begin{bmatrix} 4 \\ 0 \\ 0 \end{bmatrix}$$

which is perpendicular to the vector

$$\mathbf{n} = \begin{bmatrix} 1 \\ 1 \\ 1 \end{bmatrix}.$$

All we need to compute is D:

$$D = -(1 \times 4 + 1 \times 0 + 1 \times 0) = -4.$$

Thus the plane equation is

$$x_1 + x_2 + x_3 - 4 = 0.$$

Similar to a 2D implicit line, if the coefficients A, B, C correspond to the normal to the plane, then $|D|$ describes the distance of the plane to the origin (Sketch 95). The point normal form reflects the distance of a point from a plane. This situation is illustrated in Sketch 96. The distance d of an arbitrary point $\hat{\mathbf{x}}$ from the point normal form of the plane is

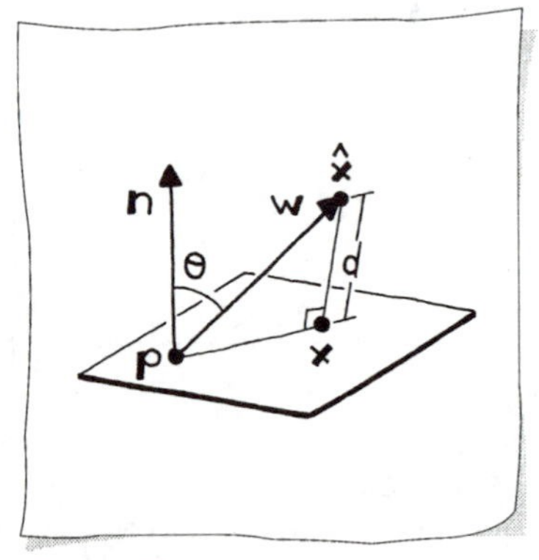

Sketch 96
Point to plane distance.

$$d = A\hat{x}_1 + B\hat{x}_2 + C\hat{x}_3 + D.$$

The reason for this follows precisely as it did for the implicit line in Section 3.3. See Section 11.1 for more on this topic.

Suppose we would like to find the distance of many points to a given plane. Then it is computationally more efficient to have the plane in (10.13) corresponding to the point normal form: the new coefficients will be A', B', C', D'. In order to do this we need to know one point $\mathbf{p}$ in the plane. This can be found by setting two x_i coordinates to zero and solving for the third, e.g., $x_1 = x_2 = 0$ and solving for x_3. With this point, we normalize the vector of coefficients and define:

$$\begin{bmatrix} A' \\ B' \\ C' \end{bmatrix} = \frac{\mathbf{n}}{\|\mathbf{n}\|}.$$

Now solve for D':

$$D' = -(A'p_1 + B'p_2 + C'p_3).$$

Example 10.6 Let's continue with the plane from the previous example,

$$x_1 + x_2 + x_3 - 4 = 0.$$

Clearly it is not in point normal form because the length of the vector $\|\mathbf{n}\| \neq 1$.

In order to convert it to point normal form, we need one point $\mathbf{p}$ in the plane. Set $x_2 = x_3 = 0$ and solve for $x_1 = 4$, thus

$$\mathbf{p} = \begin{bmatrix} 4 \\ 0 \\ 0 \end{bmatrix}.$$

Normalize $\mathbf{n}$ above, thus forming

$$\begin{bmatrix} A' \\ B' \\ C' \end{bmatrix} = \frac{\mathbf{n}}{\|\mathbf{n}\|} = \begin{bmatrix} \frac{1}{\sqrt{3}} \\ \frac{1}{\sqrt{3}} \\ \frac{1}{\sqrt{3}} \end{bmatrix}.$$

Now solve for the plane coefficient

$$D' = -(\frac{1}{\sqrt{3}} \times 4 + \frac{1}{\sqrt{3}} \times 0 + \frac{1}{\sqrt{3}} \times 0) = \frac{-4}{\sqrt{3}},$$

making the point normal plane equation

$$\frac{1}{\sqrt{3}}x_1 + \frac{1}{\sqrt{3}}x_2 + \frac{1}{\sqrt{3}}x_3 - \frac{4}{\sqrt{3}} = 0.$$

Determine the distance d of the point

$$\mathbf{q} = \begin{bmatrix} 4 \\ 4 \\ 4 \end{bmatrix}$$

from the plane:

$$d = \frac{1}{\sqrt{3}} \times 4 + \frac{1}{\sqrt{3}} \times 4 + \frac{1}{\sqrt{3}} \times 4 - \frac{4}{\sqrt{3}} = \frac{8}{\sqrt{3}} \approx 4.6.$$

Notice that $d > 0$; this is because the point $\mathbf{q}$ is on the same side of the plane as the normal direction. The distance of the origin to the plane is $d = -4/\sqrt{3}$, which is negative because it is on the opposite side of the plane. This is analogous to the 2D implicit line.

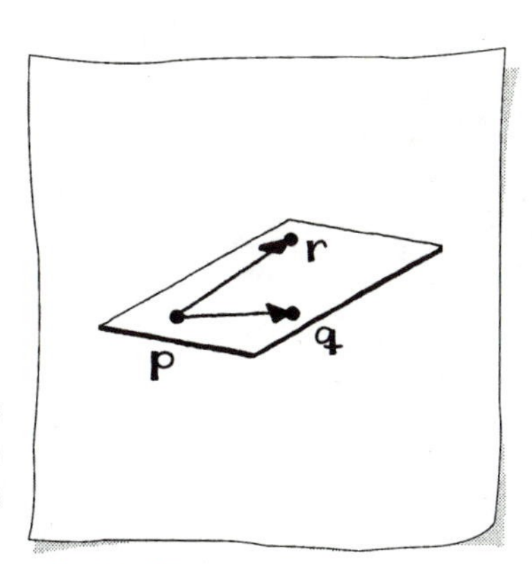

Sketch 97
Parametric plane.

The implicit plane equation is wonderful for determining if a point is in a plane, however it is not so useful for creating points in a plane. For this we have the *parametric form* of a plane.

The *given* information for defining a parametric representation of a plane usually comes in one of two ways:

- three points, or
- a point and two vectors.

If we start with the first scenario, we choose three points $\mathbf{p}, \mathbf{q}, \mathbf{r}$, then choose one of these points and form two vectors $\mathbf{v}$ and $\mathbf{w}$ bound to that point as shown in Sketch 97:

$$\mathbf{v} = \mathbf{q} - \mathbf{p} \quad \text{and} \quad \mathbf{w} = \mathbf{r} - \mathbf{p}.$$

Why not just specify one point and a vector in the plane, analogous to the implicit form of a plane? Sketch 98 illustrates that this is not enough information to uniquely define a plane. Many planes fit that data.

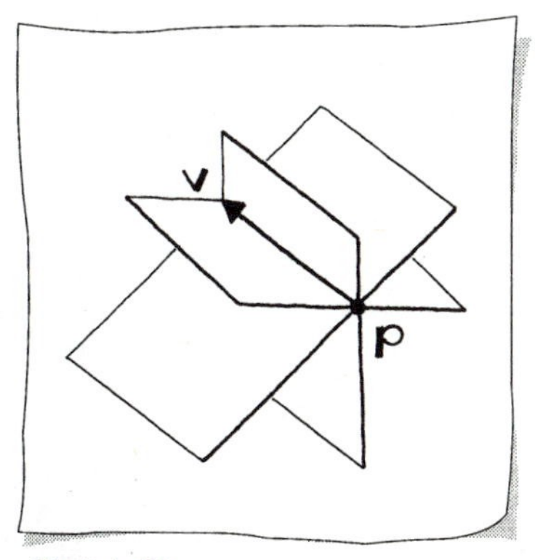

Sketch 98
Family of planes through a point and vector.

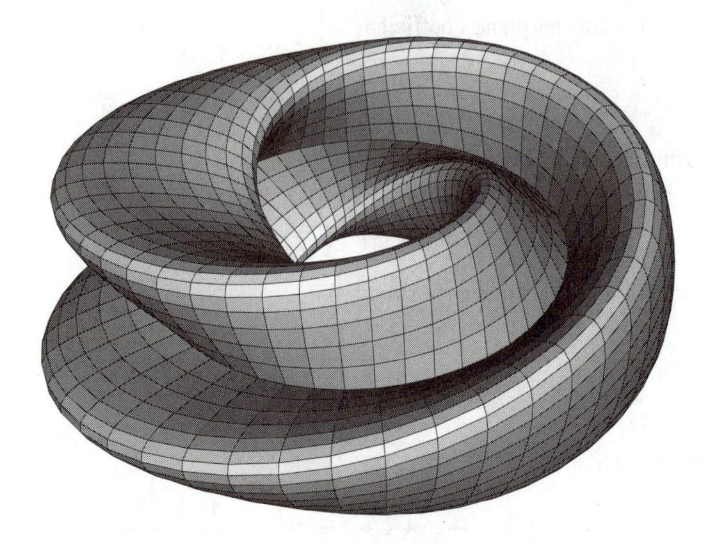

Figure 10.2.
Shading uses the normal of the planar facets.

Two vectors bound to a point is the data we'll use to define a plane $\mathbf{P}$ in parametric form as

$$\mathbf{P}(s,t) = \mathbf{p} + s\mathbf{v} + t\mathbf{w}. \tag{10.14}$$

The two independent parameters, s and t, determine a point $\mathbf{P}(s,t)$ in the plane.[1] Notice that (10.14) can be rewritten as

$$\begin{aligned} \mathbf{P}(s,t) &= \mathbf{p} + s(\mathbf{q} - \mathbf{p}) + t(\mathbf{r} - \mathbf{p}) \\ &= (1 - s - t)\mathbf{p} + s\mathbf{q} + t\mathbf{r}. \end{aligned} \tag{10.15}$$

As described in Section 8.1, $(1-s-t, s, t)$ are the *barycentric coordinates* of a point $\mathbf{P}(s,t)$ with respect to the triangle with vertices $\mathbf{p}, \mathbf{q}$, and $\mathbf{r}$.

In graphics, for instance, it is often necessary to know the normal to a plane in order to calculate shading for a lighting model. A nice example is illustrated in Figure 10.2. The shade of each rectangular piece of a plane, or facet, is calculated based on the normal to the plane, the light source location, and our eye location. We calculate the normal by using the *cross product* from Section 10.2. The normal $\mathbf{n}$ is

$$\mathbf{n} = \frac{\mathbf{v} \wedge \mathbf{w}}{\|\mathbf{v} \wedge \mathbf{w}\|}.$$

[1]This is a slight deviation in notation: an uppercase boldface letter rather than a lowercase one denoting a point.

The normal is by convention considered to be of unit length. Why use $\mathbf{v} \wedge \mathbf{w}$ instead of $\mathbf{w} \wedge \mathbf{v}$? Consider the point $\mathbf{p}$ in (10.14) as a base point. Since we have ordered the parameters — (s, t) — this implies the triangle has vertices with the orientation $\mathbf{p}, \mathbf{q}, \mathbf{r}$. Following the right-hand rule, we would like the normal to the plane to be in the direction $\mathbf{v} \wedge \mathbf{w}$. It is important to have a rule, as just described, to determine the side of the plane the normal points, so that all facets are treated in the same manner.

Another method for specifying a plane is as the bisector of two points. This is how a plane is defined in Euclidean geometry — the locus of points equidistant from two points. The line between two given points defines the normal to the plane, and the midpoint of this line segment defines a point in the plane. With this information it is most natural to express the plane in implicit form.

10.5 Scalar Triple Product

In Section 10.2 we encountered the area P of the parallelogram formed by vectors $\mathbf{v}$ and $\mathbf{w}$ measured as

$$P = \|\mathbf{v} \wedge \mathbf{w}\|.$$

The next natural question is how do we measure the *volume* of the *parallelepiped*, or skew box, formed by three vectors. See Sketch 99. The volume is a product of a face area and the corresponding height of the skew box. As illustrated in the sketch, after choosing $\mathbf{v}$ and $\mathbf{w}$ to form the face, the height is $\|\mathbf{u}\| \cos\theta$. Thus the volume V is

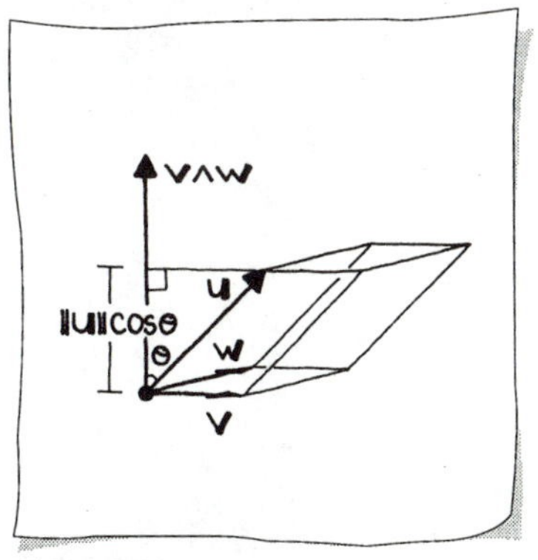

Sketch 99
Scalar triple product for the volume.

$$V = \|\mathbf{u}\|\|\mathbf{v} \wedge \mathbf{w}\| \cos\theta.$$

Recall the definition of the dot product, then

$$V = \mathbf{u} \cdot (\mathbf{v} \wedge \mathbf{w}). \tag{10.16}$$

This is called the *scalar triple product*, and it is a number representing a signed volume.

The sign reveals something about the orientation of the three vectors. If $\cos\theta > 0$ resulting in a positive volume, then $\mathbf{u}$ is on the same side of the plane formed by $\mathbf{v}$ and $\mathbf{w}$ as $\mathbf{v} \wedge \mathbf{w}$. If $\cos\theta < 0$ resulting in a negative volume, then $\mathbf{u}$ is on the opposite side of the plane as $\mathbf{v} \wedge \mathbf{w}$. If $\cos\theta = 0$ resulting in zero volume, then $\mathbf{u}$ lies in this plane — the vectors are *coplanar*.

The scalar triple product is really just a fancy name for a 3×3 determinant. Another way to express (10.16) is as

$$\begin{aligned} V &= \begin{vmatrix} u_1 & v_1 & w_1 \\ u_2 & v_2 & w_2 \\ u_3 & v_3 & w_3 \end{vmatrix} \\ &= u_1 \begin{vmatrix} v_2 & w_2 \\ v_3 & w_3 \end{vmatrix} - u_2 \begin{vmatrix} v_1 & w_1 \\ v_3 & w_3 \end{vmatrix} + u_3 \begin{vmatrix} v_1 & w_1 \\ v_2 & w_2 \end{vmatrix}. \end{aligned} \tag{10.17}$$

Interchanging columns (or rows) in a determinant causes a sign change. Therefore, two subsequent interchanges cause no change. As a result, the scalar triple product is invariant under *cyclic permutations*. This means that the we get the same volume for the following:

$$\begin{aligned} V &= \mathbf{u} \cdot (\mathbf{v} \wedge \mathbf{w}) \\ &= \mathbf{w} \cdot (\mathbf{u} \wedge \mathbf{v}) \\ &= \mathbf{v} \cdot (\mathbf{w} \wedge \mathbf{u}). \end{aligned} \tag{10.18}$$

10.6 Exercises

For the following exercises, use the following points and vectors.

$$\mathbf{p} = \begin{bmatrix} 0 \\ 0 \\ 1 \end{bmatrix} \quad \mathbf{q} = \begin{bmatrix} 1 \\ 1 \\ 1 \end{bmatrix} \quad \mathbf{r} = \begin{bmatrix} 4 \\ 2 \\ 4 \end{bmatrix} \quad \mathbf{v} = \begin{bmatrix} 1 \\ 0 \\ 0 \end{bmatrix} \quad \mathbf{w} = \begin{bmatrix} 1 \\ 1 \\ 1 \end{bmatrix} \quad \mathbf{u} = \begin{bmatrix} 0 \\ 0 \\ 1 \end{bmatrix}$$

1. Normalize the vector $\mathbf{r}$. What is the length of the vector $2\mathbf{r}$?
2. Find the angle between the vectors $\mathbf{v}$ and $\mathbf{w}$.
3. Compute $\mathbf{v} \wedge \mathbf{w}$.
4. Compute the area of the parallelogram formed by $\mathbf{v}$ and $\mathbf{w}$.
5. What is the sine of the angle between $\mathbf{v}$ and $\mathbf{w}$?
6. Find three vectors so that their cross product is associative.
7. Form the point normal plane equation for a plane through $\mathbf{p}$ and with normal direction $\mathbf{r}$.
8. Form the point normal plane equation for the plane defined by $\mathbf{p}$, $\mathbf{q}$, and $\mathbf{r}$.
9. Form a parametric plane equation for the plane defined by $\mathbf{p}$, $\mathbf{q}$, and $\mathbf{r}$.
10. Form an equation of the plane that bisects the points $\mathbf{p}$ and $\mathbf{q}$.
11. Find the volume of the parallelepiped defined by $\mathbf{v}$, $\mathbf{w}$, and $\mathbf{u}$.

11 Interactions in 3D

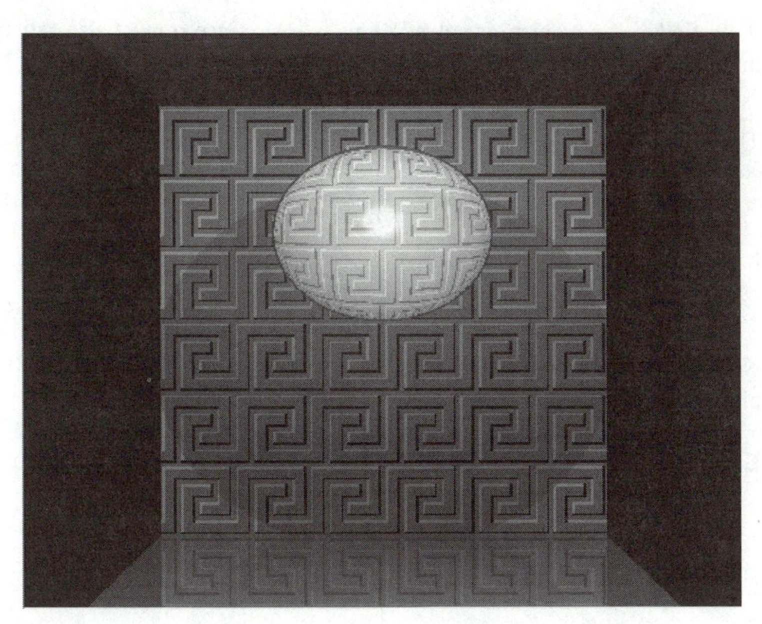

Figure 11.1.
A ray traced image. (Courtesy of Ben Steinberg, Arizona State University.)

The tools of points, lines, and planes are our most basic 3D geometry building blocks. But in order to build real objects, we must be able to compute with these building blocks. For example, a cube is defined by its six bounding planes; but in most cases, one also needs to know its eight vertices. These are found by intersecting appropriate planes. A similar problem is to determine whether a given point is inside or outside the cube? This chapter outlines the basic algorithms for these types of problems. Figure 11.1 was generated by using the tools developed in this chapter.[1]

[1] See Section 11.3 for a description of the technique used to generate this image.

11.1 Distance of Point and Plane

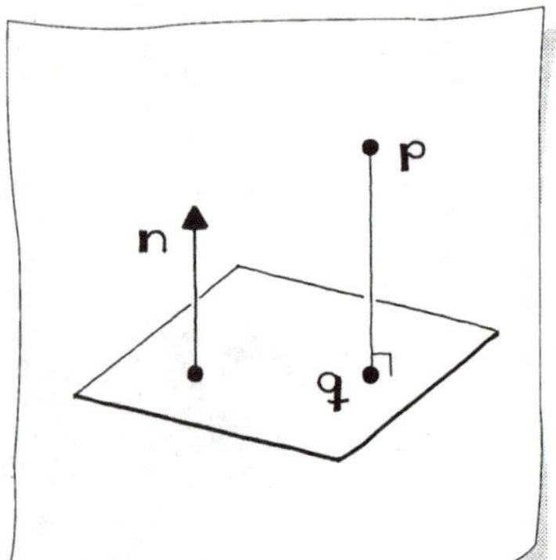

Sketch 100
A point and a plane.

Let a plane be given by its implicit form $\mathbf{n} \cdot \mathbf{x} + c = 0$. If we also have a point $\mathbf{p}$, how far is it from the plane, and what is the closest point $\mathbf{q}$ on the plane? See Sketch 100 for the geometry.

Clearly the vector $\mathbf{p} - \mathbf{q}$ must be perpendicular to the plane, i.e., parallel to the plane's normal $\mathbf{n}$. Thus $\mathbf{p}$ can be written as

$$\mathbf{p} = \mathbf{q} + t\mathbf{n};$$

if we find t, our problem is solved. This is easy, since $\mathbf{q}$ must also satisfy the plane equation:

$$\mathbf{n} \cdot [\mathbf{p} - t\mathbf{n}] + c = 0.$$

Thus

$$t = \frac{c + \mathbf{n} \cdot \mathbf{p}}{\mathbf{n} \cdot \mathbf{n}}. \tag{11.1}$$

It is good practice to assure that $\mathbf{n}$ is normalized, i.e, $\mathbf{n} \cdot \mathbf{n} = 1$, and then

$$t = c + \mathbf{n} \cdot \mathbf{p}. \tag{11.2}$$

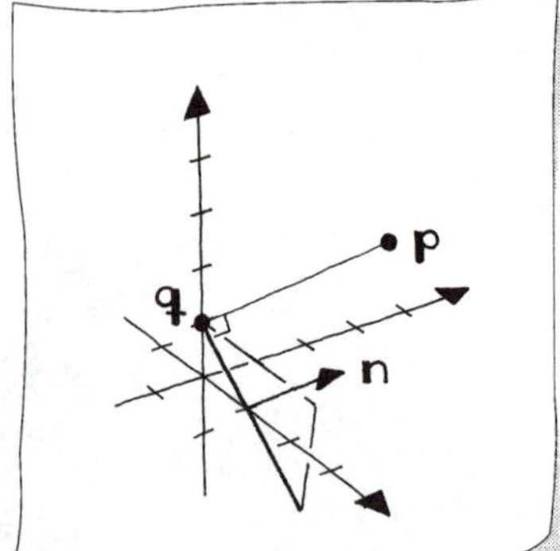

Sketch 101
Example point and a plane.

Note that $t = 0$ is equivalent to $\mathbf{n} \cdot \mathbf{p} + c = 0$; in that case, $\mathbf{p}$ is on the plane to begin with!

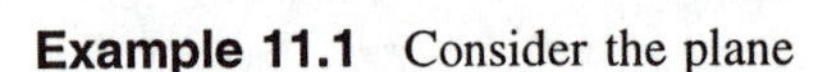

Example 11.1 Consider the plane

$$x_1 + x_2 + x_3 - 1 = 0$$

and the point

$$\mathbf{p} = \begin{bmatrix} 2 \\ 2 \\ 3 \end{bmatrix},$$

as shown in Sketch 101.

According to (11.1), we find $t = 2$. Thus

$$\mathbf{q} = \begin{bmatrix} 2 \\ 2 \\ 3 \end{bmatrix} - 2 \times \begin{bmatrix} 1 \\ 1 \\ 1 \end{bmatrix} = \begin{bmatrix} 0 \\ 0 \\ 1 \end{bmatrix}.$$

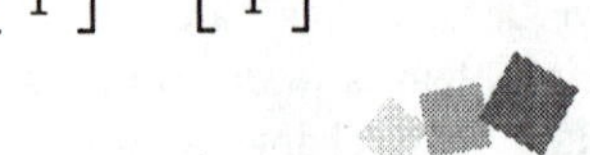

The vector $\mathbf{p} - \mathbf{q}$ is given by

$$\mathbf{p} - \mathbf{q} = t\mathbf{n}.$$

Thus the length of $\mathbf{p} - \mathbf{q}$, or the *distance* of $\mathbf{p}$ to the plane, is given by $t\|\mathbf{n}\|$. If $\mathbf{n}$ is normalized, then $\|\mathbf{p} - \mathbf{q}\| = t$; this means that we simply insert $\mathbf{p}$ into the plane equation!

It is also clear that if $t > 0$, then $\mathbf{n}$ points towards $\mathbf{p}$, and away from it if $t < 0$ (see Sketch 102) where the plane is drawn "edge on." Compare with the almost identical Sketch 33!

Again, a numerical caveat: If a point is very close to a plane, it becomes very hard numerically to decide which side it is on!

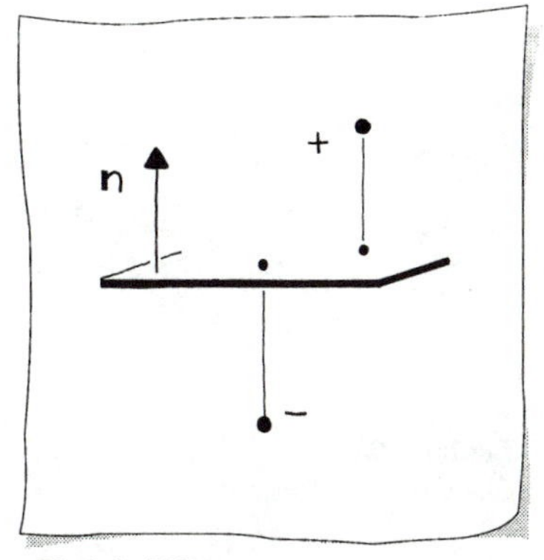

Sketch 102
Points around a plane.

11.2 The Distance between Two Lines

Two 3D lines typically do not meet — then they are called *skew*. It might be of interest to know how close they are to meeting; in other words, what is the *distance* between the lines? See Sketch 103 for an illustration.

Let the two lines $\mathbf{l}_1$ and $\mathbf{l}_2$ be given by

$$\mathbf{x}_1(s_1) = \mathbf{p}_1 + s_1\mathbf{v}_1, \quad \text{and}$$
$$\mathbf{x}_2(s_2) = \mathbf{p}_2 + s_2\mathbf{v}_2,$$

respectively. Let $\mathbf{x}_1$ be the point on $\mathbf{l}_1$ closest to $\mathbf{l}_2$, also let $\mathbf{x}_2$ be the point on $\mathbf{l}_2$ closest to $\mathbf{l}_1$. It should be clear that the vector $\mathbf{x}_2 - \mathbf{x}_1$ is perpendicular to both $\mathbf{l}_1$ and $\mathbf{l}_2$. Thus

$$[\mathbf{x}_2 - \mathbf{x}_1]\mathbf{v}_1 = 0,$$
$$[\mathbf{x}_2 - \mathbf{x}_1]\mathbf{v}_2 = 0,$$

or

$$[\mathbf{p}_2 - \mathbf{p}_1]\mathbf{v}_1 = s_1\mathbf{v}_1 \cdot \mathbf{v}_1 - s_2\mathbf{v}_1 \cdot \mathbf{v}_2,$$
$$[\mathbf{p}_2 - \mathbf{p}_1]\mathbf{v}_2 = s_1\mathbf{v}_1 \cdot \mathbf{v}_2 - s_2\mathbf{v}_2 \cdot \mathbf{v}_2.$$

These are two equations in the two unknowns s_1 and s_2, and are thus readily solved using the methods from Chapter 5.

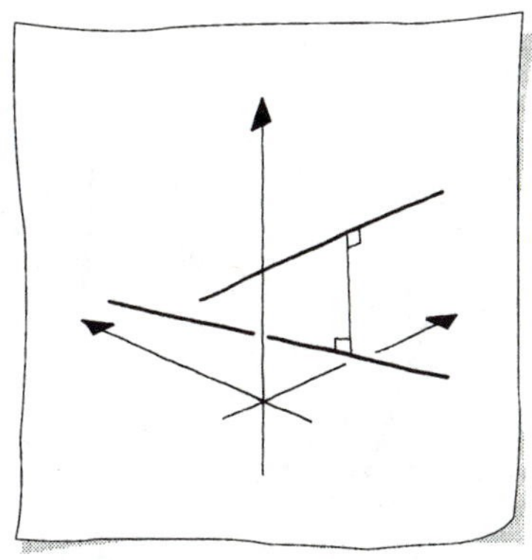
Sketch 103
Skew lines in 3D.

Example 11.2 Let $\mathbf{l}_1$ be given by

$$\mathbf{x}_1(s_1) = \begin{bmatrix} 0 \\ 0 \\ 0 \end{bmatrix} + s_1 \begin{bmatrix} 1 \\ 0 \\ 0 \end{bmatrix}.$$

This means, of course, that $\mathbf{l}_1$ is the $\mathbf{e}_1$-axis. For $\mathbf{l}_2$, we assume

$$\mathbf{x}_2(s_2) = \begin{bmatrix} 0 \\ 1 \\ 1 \end{bmatrix} + s_2 \begin{bmatrix} 0 \\ 1 \\ 0 \end{bmatrix}.$$

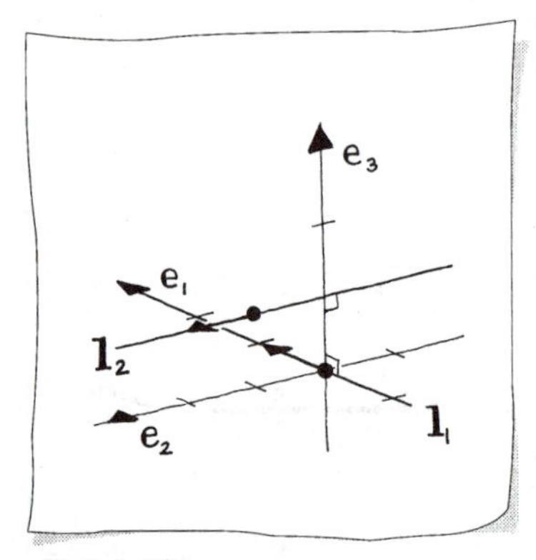

Sketch 104
Skew lines distance example.

This line is parallel to the $\mathbf{e}_2$-axis; both lines are shown in Sketch 104.

Our linear system becomes

$$0 = s_1,$$
$$1 = -s_2.$$

Inserting these values, we have

$$\mathbf{x}_1(0) = \begin{bmatrix} 0 \\ 0 \\ 0 \end{bmatrix} \quad \text{and} \quad \mathbf{x}_2(-1) = \begin{bmatrix} 0 \\ 0 \\ 1 \end{bmatrix}.$$

These are the two points of closest proximity.

Two 3D lines intersect if the two points $\mathbf{x}_1$ and $\mathbf{x}_2$ are identical.[2]

A condition for two 3D lines to intersect is found from the observation that the three vectors $\mathbf{v}_1, \mathbf{v}_2, \mathbf{p}_2 - \mathbf{p}_1$ must be coplanar, or linearly dependent. This would lead to the condition

$$\det[\mathbf{v}_1, \mathbf{v}_2, \mathbf{p}_2 - \mathbf{p}_1] = 0.$$

From a numerical viewpoint, it is safer to compare the distance between the points $\mathbf{x}_1$ and $\mathbf{x}_2$; in the field of computer-aided design (CAD), one usually has known tolerances (e.g., 0.001") for distances. It is much harder to come up with a meaningful tolerance for a determinant.

11.3 Lines and Planes: Intersections

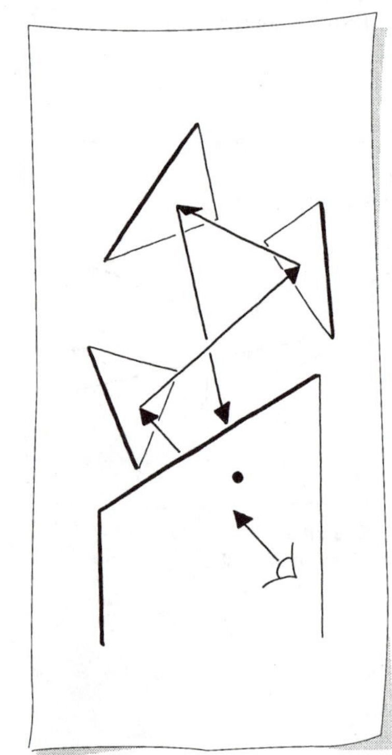
Sketch 105
A ray is traced through a scene.

One of the basic techniques in computer graphics is called *ray tracing*. Figure 11.1 illustrates this technique. A scene is given as an assembly of planes (usually restricted to triangles). A computer-generated image needs to compute proper lighting, and this is done by tracing light rays through the scene. The ray intersects a plane, it is reflected, then it intersects the next plane, etc. Sketch 105 gives an example.

The basic problem to be solved is this: Given a plane $\mathbf{P}$ and a line $\mathbf{l}$, what is their *intersection point* $\mathbf{x}$? It is most convenient to represent the plane by assuming that we know a point $\mathbf{q}$ on it as well as its normal vector $\mathbf{n}$ (see Sketch 106). Then the unknown point $\mathbf{x}$, being on $\mathbf{P}$, must satisfy

$$[\mathbf{x} - \mathbf{q}] \cdot \mathbf{n} = 0. \tag{11.3}$$

[2] In "real life," that means within a tolerance!

By definition, the intersection point is also on the line (the ray, in computer graphics jargon), given by a point $\mathbf{p}$ and a vector $\mathbf{v}$:

$$\mathbf{x} = \mathbf{p} + t\mathbf{v}. \tag{11.4}$$

At this point, we do not know the correct value for t; once we have it, our problem is solved.

The solution is obtained by substituting the expression for $\mathbf{x}$ from (11.4) into (11.3):

$$[\mathbf{p} + t\mathbf{v} - \mathbf{q}] \cdot \mathbf{n} = 0.$$

Thus

$$[\mathbf{p} - \mathbf{q}] \cdot \mathbf{n} + t\mathbf{v} \cdot \mathbf{n} = 0$$

and

$$t = \frac{[\mathbf{q} - \mathbf{p}] \cdot \mathbf{n}}{\mathbf{v} \cdot \mathbf{n}}. \tag{11.5}$$

The intersection point $\mathbf{x}$ is now computed as

$$\mathbf{x} = \mathbf{p} + \frac{[\mathbf{q} - \mathbf{p}] \cdot \mathbf{n}}{\mathbf{v} \cdot \mathbf{n}}\mathbf{v}. \tag{11.6}$$

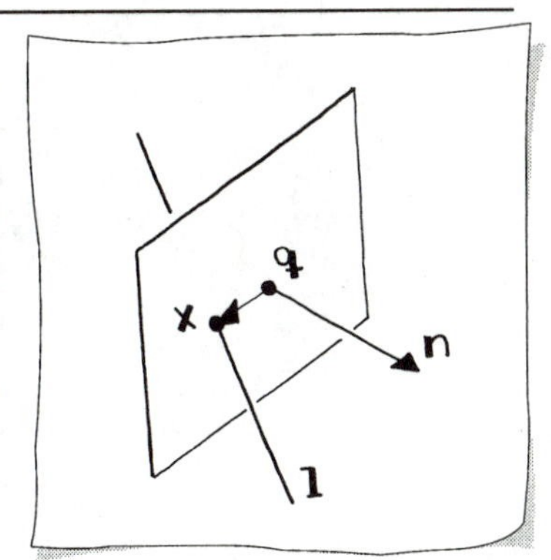

Sketch 106
A line and a plane.

Example 11.3 Take the plane

$$x_1 + x_2 + x_3 - 1 = 0$$

and the line

$$\mathbf{p}(t) = \begin{bmatrix} 1 \\ 1 \\ 2 \end{bmatrix} + t \begin{bmatrix} 0 \\ 0 \\ 1 \end{bmatrix}$$

as shown in Sketch 107.

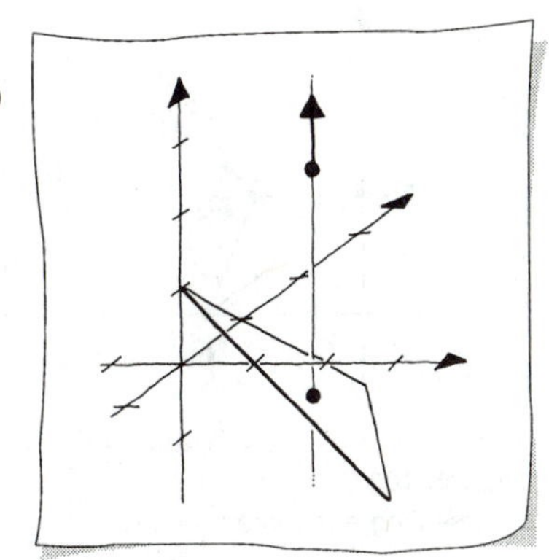
Sketch 107
Intersecting a line and a plane.

We need a point $\mathbf{q}$ on the plane; set $x_1 = x_2 = 0$ and solve for x_3, resulting in $x_3 = 1$. This amounts to intersecting the plane with the $\mathbf{e}_3$−axis. From (11.5), we find $t = -3$ and then (11.6) gives the intersection point as

$$\mathbf{x} = \begin{bmatrix} 1 \\ 1 \\ 2 \end{bmatrix} - 3 \begin{bmatrix} 0 \\ 0 \\ 1 \end{bmatrix} = \begin{bmatrix} 1 \\ 1 \\ -1 \end{bmatrix}$$

It never hurts to carry out a sanity check: Verify that this $\mathbf{x}$ does indeed satisfy the plane equation!

A word of caution: In (11.5), we happily divide by the dot product $\mathbf{v} \cdot \mathbf{n}$ — but that better not be zero! If it is,[3] then the ray "grazes" the plane, i.e., it is parallel to it. Then no intersection exists.

The same problem — intersecting a line with a plane — may be solved if the plane is given in *parametric form.* Then the unknown intersection point $\mathbf{x}$ must satisfy

$$\mathbf{x} = \mathbf{q} + u_1\mathbf{r}_1 + u_2\mathbf{r}_2.$$

Since we know that $\mathbf{x}$ is also on the line $\mathbf{l}$, we may set

$$\mathbf{p} + t\mathbf{v} = \mathbf{q} + u_1\mathbf{r}_1 + u_2\mathbf{r}_2.$$

This equation is short for three individual equations, one for each coordinate. We thus have three equations in three unknowns t, u_1, u_2, and solve them according to methods of Chapter 14.

11.4 Intersecting a Triangle and a Line

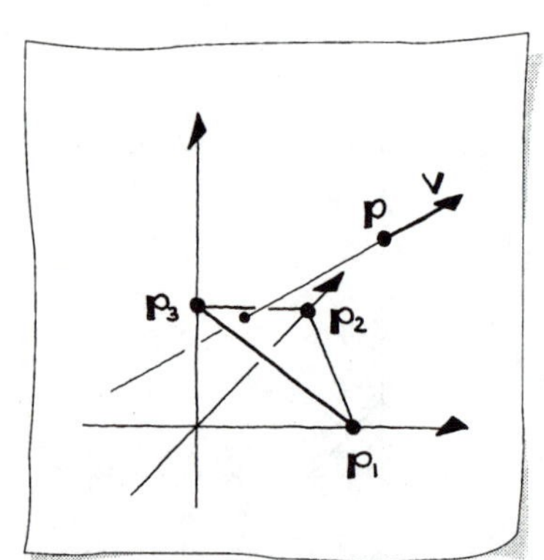

Sketch 108
Intersecting a triangle and a line.

A plane is, by definition, an unbounded object. In many applications, planes are parts of objects; one is only interested in a small part of a plane. For example, the six faces of a cube are bounded planes, so are the four faces of a tetrahedron.

We will now examine the case of a 3D *triangle* as an example of a bounded plane. If we intersect a 3D triangle with a line (a ray), then we are not interested in an intersection point *outside* the triangle — only an interior one will count.

Let the triangle be given by three points $\mathbf{p}_1, \mathbf{p}_2, \mathbf{p}_3$ and the line by a point $\mathbf{p}$ and a direction $\mathbf{v}$ (see Sketch 108).

The plane may be written in parametric form as

$$\mathbf{x}(u_1, u_2) = \mathbf{p}_1 + u_1(\mathbf{p}_2 - \mathbf{p}_1) + u_2(\mathbf{p}_3 - \mathbf{p}_1).$$

We thus arrive at

$$\mathbf{p} + t\mathbf{v} = \mathbf{p}_1 + u_1(\mathbf{p}_2 - \mathbf{p}_1) + u_2(\mathbf{p}_3 - \mathbf{p}_1),$$

a linear system in the unknowns t, u_1, u_2. The solution is inside the triangle if both u_1 and u_2 are between zero and one, and their sum is less than or equal to one. This is so since we may view $(u_1, u_2, 1 - u_1 - u_2)$ as *barycentric coordinates* of the triangle. These are positive exactly for points inside the triangle. See Section 8.1 for a review of barycentric coordinates in a triangle.

[3] Keep in mind that real numbers are rarely *equal* to zero. A tolerance needs to be used; 0.001 should work if both $\mathbf{n}$ and $\mathbf{v}$ are normalized.

11.5 Lines and Planes: Reflections

The next problem is that of line or plane *reflection*. Given a point $\mathbf{x}$ on a plane $\mathbf{P}$ and an "incoming" direction $\mathbf{v}$, what is the reflected, or "outgoing" direction $\mathbf{v}'$? See Sketch 109, where we look at the plane $\mathbf{P}$ "edge on." We assume that both $\mathbf{v}$ and $\mathbf{v}'$ are of unit length.

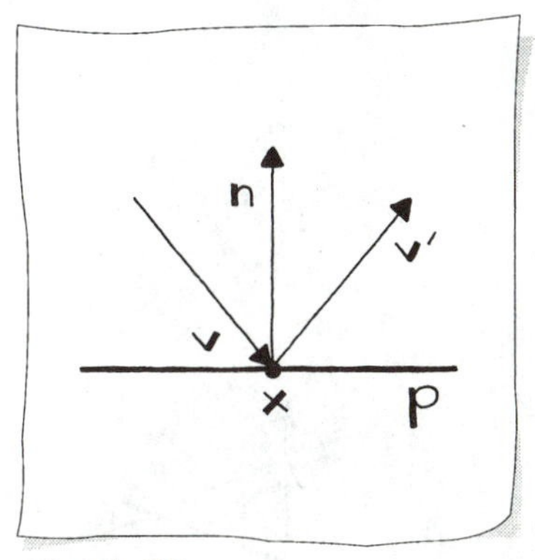

Sketch 109
A reflection.

From physics, you might recall that the angle between $\mathbf{v}$ and the plane normal $\mathbf{n}$ must equal that of $\mathbf{v}'$ and $\mathbf{n}$, except for a sign change. We conveniently record this fact using a dot product:

$$-\mathbf{v} \cdot \mathbf{n} = \mathbf{v}' \cdot \mathbf{n}. \tag{11.7}$$

The normal vector $\mathbf{n}$ is thus the *angle bisector* of $\mathbf{v}$ and $\mathbf{v}'$. From inspection of Sketch 109, we also infer the symmetry property

$$c\mathbf{n} = \mathbf{v}' - \mathbf{v} \tag{11.8}$$

for some real number c. This means that some multiple of the normal vector may be written as the sum $\mathbf{v}' + (-\mathbf{v})$.

We now solve (11.8) for $\mathbf{v}'$ and insert into (11.7):

$$-\mathbf{v} \cdot \mathbf{n} = [c\mathbf{n} + \mathbf{v}] \cdot \mathbf{n},$$

and solve for c:

$$c = -2\mathbf{v} \cdot \mathbf{n}. \tag{11.9}$$

Here, we made use of the fact that $\mathbf{n}$ is a unit vector and thus $\mathbf{n} \cdot \mathbf{n} = 1$.

The reflected vector $\mathbf{v}'$ is now given by using our value for c in (11.9):

$$\mathbf{v}' = \mathbf{v} - [2\mathbf{v} \cdot \mathbf{n}]\mathbf{n}. \tag{11.10}$$

In the special case of $\mathbf{v}$ being perpendicular to the plane, i.e., $\mathbf{v} = -\mathbf{n}$, we obtain $\mathbf{v}' = -\mathbf{v}$ as expected. Also note that the point of reflection does not enter the equations at all.

11.6 Intersecting Three Planes

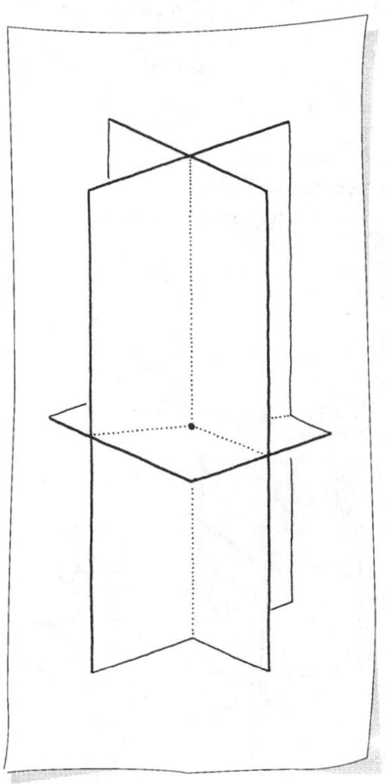
Sketch 110
Intersecting three planes.

Suppose we are given three planes with implicit equations

$$\mathbf{n}_1 \cdot \mathbf{x} + c_1 = 0,$$
$$\mathbf{n}_2 \cdot \mathbf{x} + c_2 = 0,$$
$$\mathbf{n}_3 \cdot \mathbf{x} + c_3 = 0.$$

Where do they intersect? The answer is some point $\mathbf{x}$, which lies on each of the planes. See Sketch 110 for an illustration.

The solution is surprisingly simple; just condense the three plane equations into matrix form:

$$\begin{bmatrix} \mathbf{n}_1^{\mathrm{T}} \\ \mathbf{n}_2^{\mathrm{T}} \\ \mathbf{n}_3^{\mathrm{T}} \end{bmatrix} \begin{bmatrix} x_1 \\ x_2 \\ x_3 \end{bmatrix} = \begin{bmatrix} -c_1 \\ -c_2 \\ -c_3 \end{bmatrix}. \tag{11.11}$$

We have three equations in the three unknowns x_1, x_2, x_3!

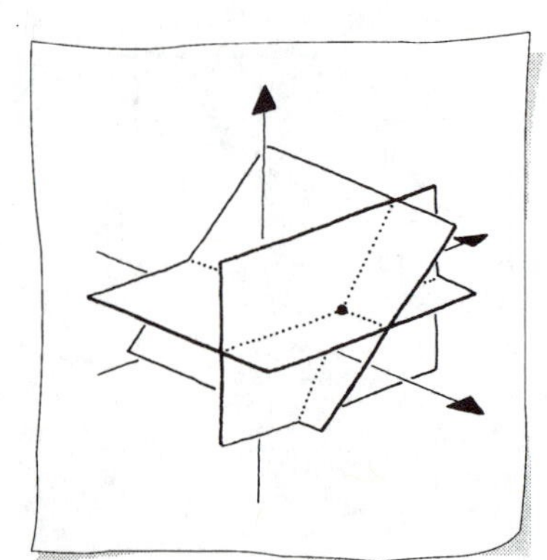

Sketch 111
Intersecting three planes example.

Example 11.4 The following example is shown in Sketch 111. The equations of the planes in that sketch are

$$x_1 + x_3 = 1, \quad x_3 = 1, \quad x_2 = 2.$$

The linear system is

$$\begin{bmatrix} 1 & 0 & 1 \\ 0 & 0 & 1 \\ 0 & 1 & 0 \end{bmatrix} \begin{bmatrix} x_1 \\ x_2 \\ x_3 \end{bmatrix} = \begin{bmatrix} 1 \\ 1 \\ 2 \end{bmatrix}.$$

Solving it by Gauss elimination (Chapter 14), we obtain

$$\begin{bmatrix} x_1 \\ x_2 \\ x_3 \end{bmatrix} = \begin{bmatrix} 0 \\ 2 \\ 1 \end{bmatrix}.$$

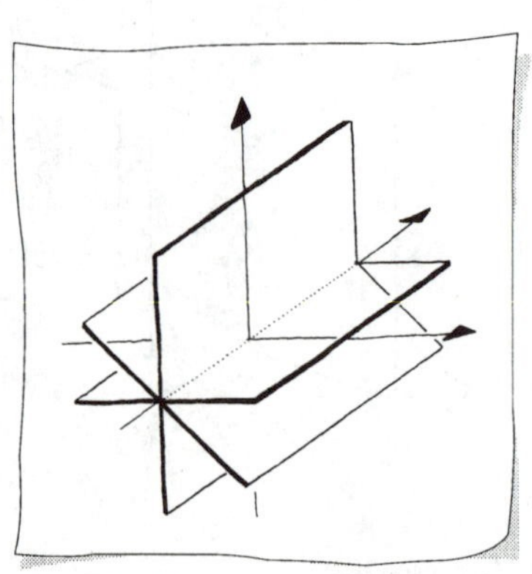

Sketch 112
Three nonintersecting planes.

While simple to solve, the three-planes problem does not always have a solution. Two lines in 2D do not intersect if they are parallel; in this case, their normal vectors are also parallel, or linearly dependent. The situation is analogous in 3D. If the normal vectors $\mathbf{n}_1, \mathbf{n}_2, \mathbf{n}_3$ are linearly dependent, then there is no solution to the intersection problem.

Example 11.5 The normal vectors are

$$\mathbf{n}_1 = \begin{bmatrix} 1 \\ 0 \\ 0 \end{bmatrix}, \quad \mathbf{n}_2 = \begin{bmatrix} 1 \\ 0 \\ 1 \end{bmatrix}, \quad \mathbf{n}_3 = \begin{bmatrix} 0 \\ 0 \\ 1 \end{bmatrix}.$$

Since $\mathbf{n}_2 = \mathbf{n}_1 + \mathbf{n}_3$, they are indeed linearly dependent, and thus the planes defined by them do not intersect in one point (see Sketch 112).

11.7 Intersecting Two Planes

Odd as it may seem, intersecting two planes is harder than intersecting three of them. The problem is this: Two planes are given in their implicit form

$$\mathbf{n} \cdot \mathbf{x} + c = 0, \tag{11.12}$$

$$\mathbf{m} \cdot \mathbf{x} + d = 0. \tag{11.13}$$

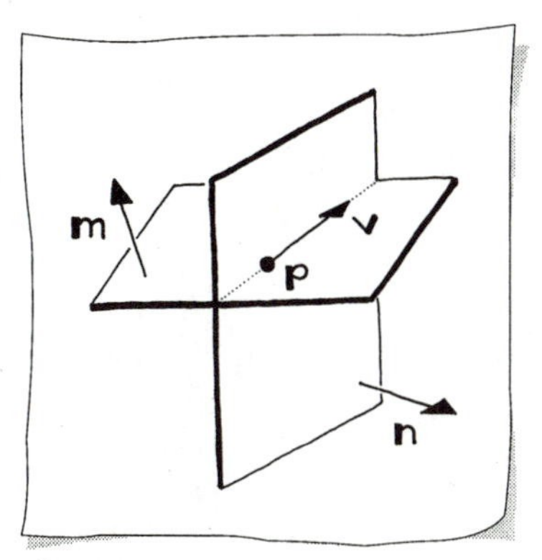

Sketch 113
Intersecting two planes.

Find their intersection, which is a line. We would like the solution to be of the form

$$\mathbf{x}(t) = \mathbf{p} + t\mathbf{v}. \tag{11.14}$$

This situation is depicted in Sketch 113.

The direction vector $\mathbf{v}$ of this line is easily found; since it lies in each of the planes, it must be perpendicular to both their normal vectors:

$$\mathbf{v} = \mathbf{n} \wedge \mathbf{m}.$$

We still need a point $\mathbf{p}$ on the line.

To this end, we come up with an auxiliary plane that intersects both given planes. The intersection point is clearly on the desired line. Let us assume for now that not both c and d are zero. Define the third plane by

$$\mathbf{v} \cdot \mathbf{x} = 0.$$

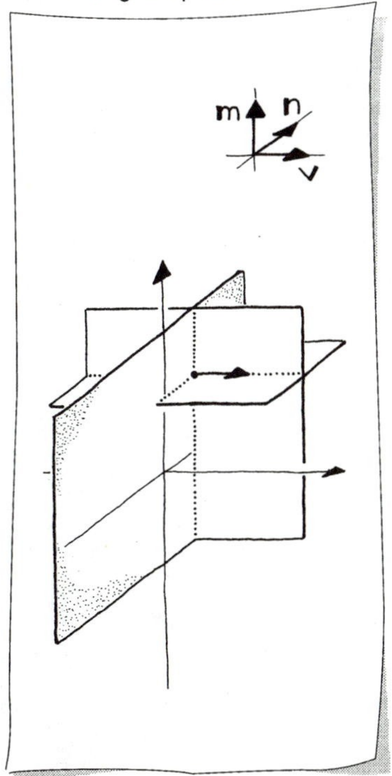

Sketch 114
The auxiliary plane (shaded).

This plane passes through the origin and has normal vector $\mathbf{v}$, i.e., it is perpendicular to the desired line (see Sketch 114).

We now solve the three-plane intersection problem for the two given planes and the auxiliary plane for the missing point $\mathbf{p}$, and our line is determined.

In the case $c = d = 0$, both given planes pass through the origin, and it can serve as the point $\mathbf{p}$.

11.8 Exercises

For exercises 1 and 2, we will deal with two planes. $\mathbf{P}_1$ goes through a point $\mathbf{p}$ and has normal vector $\mathbf{n}$:

$$\mathbf{p} = \begin{bmatrix} 1 \\ 2 \\ 0 \end{bmatrix}, \quad \mathbf{n} = \begin{bmatrix} -1 \\ 0 \\ 0 \end{bmatrix}.$$

The plane $\mathbf{P}_2$ is given by its implicit form

$$x_1 + 2x_2 - 2x_3 - 1 = 0.$$

Also, let a line $\mathbf{l}$ go through the point $\mathbf{q}$ and have direction $\mathbf{v}$:

$$\mathbf{q} = \begin{bmatrix} -1 \\ 2 \\ 0 \end{bmatrix}, \quad \mathbf{v} = \begin{bmatrix} 0 \\ 1 \\ 0 \end{bmatrix}.$$

1. Find the intersection of $\mathbf{P}_1$ with the line $\mathbf{l}$.
2. Find the intersection of $\mathbf{P}_2$ with the line $\mathbf{l}$.
3. Does the ray $\mathbf{p} + t\mathbf{v}$ with

$$\mathbf{p} = \begin{bmatrix} -1 \\ -1 \\ 0 \end{bmatrix}, \quad \mathbf{v} = \begin{bmatrix} 1 \\ 1 \\ 1 \end{bmatrix}$$

intersect the triangle with vertices

$$\begin{bmatrix} 3 \\ 0 \\ 0 \end{bmatrix}, \quad \begin{bmatrix} 0 \\ 2 \\ 1 \end{bmatrix}, \quad \begin{bmatrix} 2 \\ 2 \\ 3 \end{bmatrix}?$$

4. Revisit the example from Section 11.2, but set the point defining the line $\mathbf{l}_2$ to be

$$\mathbf{p}_2 = \begin{bmatrix} 0 \\ 0 \\ 1 \end{bmatrix}.$$

The lines have not changed; how do you obtain the (unchanged) solutions $\mathbf{x}_1$ and $\mathbf{x}_2$?

5. Let $\mathbf{a}$ be an arbitrary vector. It may be projected along a direction $\mathbf{v}$ onto the plane $\mathbf{P}$ with normal vector $\mathbf{n}$. What is its image $\mathbf{a}'$?

12 Linear Maps in 3D

Figure 12.1.
A flight simulator scene from the NASA / Langley web page
http://bigben.larc.nasa.gov/fltsim/fltsim.html.

An important part in the training of airplane pilots is the flight simulator. It has a real cockpit, but what you see outside the windows is computer imagery. As you take a right turn, the terrain below changes accordingly; as you dive downwards, it comes closer to you. When you change the (simulated) position of your plane, the simulation software must recompute a new view of the terrain, clouds, or other aircraft. This is done through the application of 3D affine and linear maps.[1] Figure 12.1 shows an image that was generated by an actual flight simulator. For each frame of the

[1] Actually, perspective maps are also needed here. They will be discussed in Section 13.5.

simulated scene, complex 3D computations are necessary, most of them consisting of the types of maps discussed in this section.

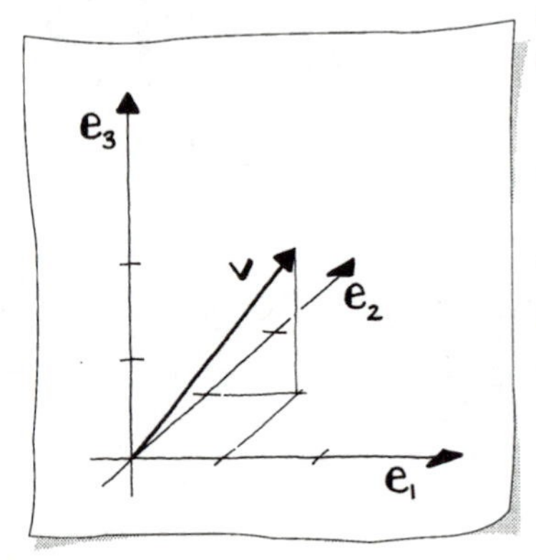

Sketch 115
A vector in the $[\mathbf{e}_1, \mathbf{e}_2, \mathbf{e}_3]$-coordinate system.

12.1 Matrices and Linear Maps

The general concept of a linear map in 3D is the same as that for a 2D map. Let $\mathbf{v}$ be a vector in the standard $[\mathbf{e}_1, \mathbf{e}_2, \mathbf{e}_3]$-coordinate system, i.e.,

$$\mathbf{v} = v_1\mathbf{e}_1 + v_2\mathbf{e}_2 + v_3\mathbf{e}_3.$$

See Sketch 115 for an illustration.

Let another coordinate system, the $[\mathbf{a}_1, \mathbf{a}_2, \mathbf{a}_3]$-coordinate system, be given by its origin $\mathbf{p}$ and three vectors $\mathbf{a}_1, \mathbf{a}_2, \mathbf{a}_3$. What vector $\mathbf{v}'$ in the $[\mathbf{a}_1, \mathbf{a}_2, \mathbf{a}_3]$-system corresponds to $\mathbf{v}$ in the $[\mathbf{e}_1, \mathbf{e}_2, \mathbf{e}_3]$-system? Simply the vector with the same coordinates relative to the $[\mathbf{a}_1, \mathbf{a}_2, \mathbf{a}_3]$-system! Thus:

$$\mathbf{v}' = v_1\mathbf{a}_1 + v_2\mathbf{a}_2 + v_3\mathbf{a}_3. \tag{12.1}$$

This is illustrated by Sketch 116 and the following example.

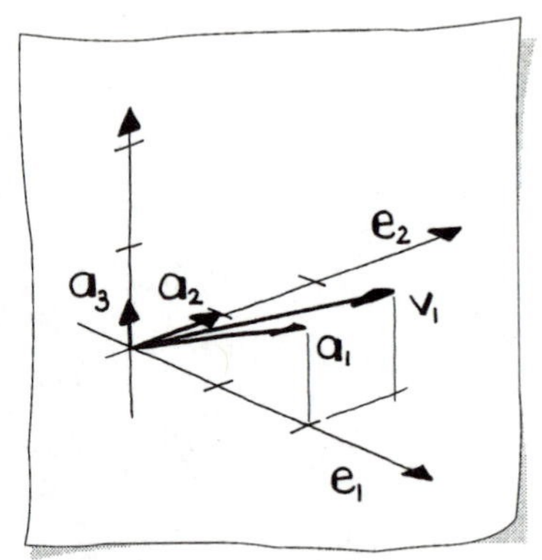

Sketch 116
The corresponding vector in the $[\mathbf{a}_1, \mathbf{a}_2, \mathbf{a}_3]$-coordinate system.

Example 12.1 Let

$$\mathbf{v} = \begin{bmatrix} 1 \\ 1 \\ 2 \end{bmatrix}, \quad \mathbf{a}_1 = \begin{bmatrix} 2 \\ 0 \\ 1 \end{bmatrix}, \quad \mathbf{a}_2 = \begin{bmatrix} 0 \\ 1 \\ 0 \end{bmatrix}, \quad \mathbf{a}_3 = \begin{bmatrix} 0 \\ 0 \\ 1/2 \end{bmatrix}.$$

Then

$$\mathbf{v}' = 1 \cdot \begin{bmatrix} 2 \\ 0 \\ 1 \end{bmatrix} + 1 \cdot \begin{bmatrix} 0 \\ 1 \\ 0 \end{bmatrix} + 2 \cdot \begin{bmatrix} 0 \\ 0 \\ 1/2 \end{bmatrix} = \begin{bmatrix} 2 \\ 1 \\ 2 \end{bmatrix}.$$

You should recall that we had the same configuration earlier for the 2D case — (12.1) corresponds directly to (4.3) of Section 4.1. In Section 4.2, we then introduced the matrix form. That is now an easy project for this chapter — nothing changes except the matrices will be 3×3 instead of 2×2. In 3D, a matrix equation looks like this:

$$\mathbf{v}' = A\mathbf{v}, \tag{12.2}$$

i.e., just the same as for the 2D case. Written out in detail, there is a difference:

$$\begin{bmatrix} v_1' \\ v_2' \\ v_3' \end{bmatrix} = \begin{bmatrix} a_{1,1} & a_{1,2} & a_{1,3} \\ a_{2,1} & a_{2,2} & a_{2,3} \\ a_{3,1} & a_{3,2} & a_{3,3} \end{bmatrix} \begin{bmatrix} v_1 \\ v_2 \\ v_3 \end{bmatrix} \tag{12.3}$$

All matrix properties from Sections 4.2 and 4.3 carry over almost verbatim.

Example 12.2 Returning to our example, it is quite easy to condense it into a matrix equation:

$$\begin{bmatrix} 2 & 0 & 0 \\ 0 & 1 & 0 \\ 1 & 0 & 1/2 \end{bmatrix} \begin{bmatrix} 1 \\ 1 \\ 2 \end{bmatrix} = \begin{bmatrix} 2 \\ 1 \\ 2 \end{bmatrix}.$$

Again, if we multiply a matrix A by a vector $\mathbf{v}$, the i–th component of the result vector is obtained as the dot product of the i–th row of A and $\mathbf{v}$.

The matrix A represents a *linear map*: Given the vector $\mathbf{v}$ in the $[\mathbf{e}_1, \mathbf{e}_2, \mathbf{e}_3]$-system, there is a vector $\mathbf{v}'$ in the $[\mathbf{a}_1, \mathbf{a}_2, \mathbf{a}_3]$-system such that $\mathbf{v}'$ has the same components in the $[\mathbf{a}_1, \mathbf{a}_2, \mathbf{a}_3]$-system as did $\mathbf{v}$ in the $[\mathbf{e}_1, \mathbf{e}_2, \mathbf{e}_3]$-system. The matrix A finds the components of $\mathbf{v}'$ relative to the $[\mathbf{e}_1, \mathbf{e}_2, \mathbf{e}_3]$-system.

12.2 Scalings

A scaling is a linear map which enlarges or reduces vectors:

$$\mathbf{v}' = \begin{bmatrix} s_{1,1} & 0 & 0 \\ 0 & s_{2,2} & 0 \\ 0 & 0 & s_{3,3} \end{bmatrix} \mathbf{v} \tag{12.4}$$

If all scale factors $s_{i,i}$ are larger than one, then all vectors are enlarged, see Figure 12.2. If all $s_{i,i}$ are positive yet less than one, all vectors are shrunk.

Example 12.3 In this example,

$$\begin{bmatrix} s_{1,1} & 0 & 0 \\ 0 & s_{2,2} & 0 \\ 0 & 0 & s_{3,3} \end{bmatrix} = \begin{bmatrix} 2 & 0 & 0 \\ 0 & 1/2 & 0 \\ 0 & 0 & 1 \end{bmatrix},$$

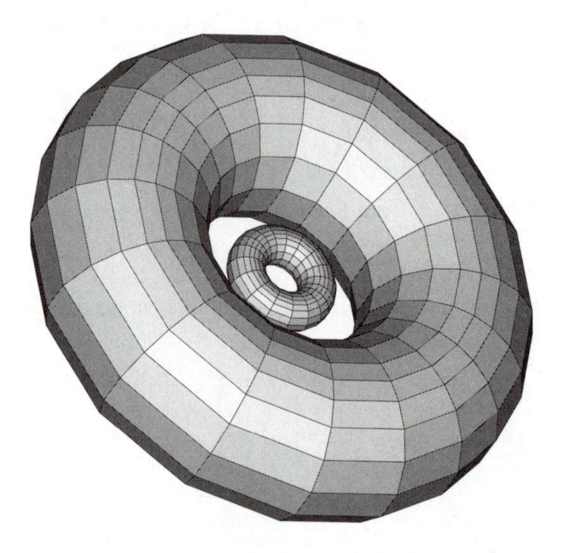

Figure 12.2.
Scalings in 3D: the small torus is scaled to form the large torus.

we stretch in the $\mathbf{e}_1$-direction, shrink in the $\mathbf{e}_2$-direction, and leave the $\mathbf{e}_3$-direction unchanged. See Figure 12.3.

Negative numbers for the $s_{i,i}$ will cause a flip in addition to a scale. So, for instance

$$\begin{bmatrix} -2 & 0 & 0 \\ 0 & 1 & 0 \\ 0 & 0 & -1 \end{bmatrix}$$

will stretch and reverse the $\mathbf{e}_1$-direction, leave the $\mathbf{e}_2$-direction unchanged, and will reverse the $\mathbf{e}_3$-direction.

How do scalings affect volumes? If we map the *unit cube*, given by the three vectors $\mathbf{e}_1, \mathbf{e}_2, \mathbf{e}_3$ by a scaling, we get a rectangular box. Its side lengths are $s_{1,1}$ in the $\mathbf{e}_1$-direction, $s_{2,2}$ in the $\mathbf{e}_2$-direction, and $s_{3,3}$ in the $\mathbf{e}_3$-direction. Hence its volume is given by $s_{1,1}s_{2,2}s_{3,3}$. A scaling thus changes the volume of an object by a factor that equals the product of its diagonal elements.[2]

[2]We have only shown this for the unit cube. But it is true for any other object as well.

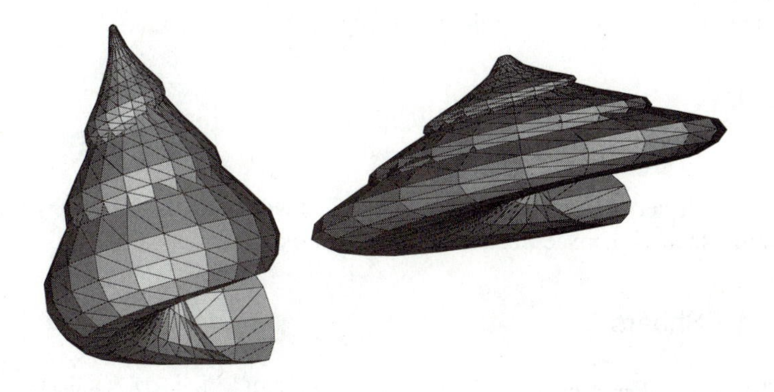

Figure 12.3.
Uneven scalings in 3D: the shell on the left is scaled by $1/2, 1, 2$ in the $\mathbf{e}_1, \mathbf{e}_2, \mathbf{e}_3$-directions, respectively, and then translated resulting in the shell on the right.

12.3 Reflections

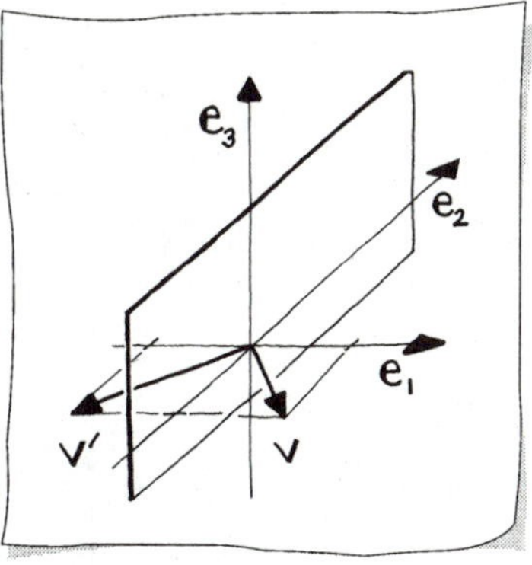

Sketch 117
Reflection of a vector about the $\mathbf{e}_2, \mathbf{e}_3$-plane.

If we reflect a vector about the $\mathbf{e}_2, \mathbf{e}_3$-plane, then its first component should change in sign:

$$\begin{bmatrix} v_1 \\ v_2 \\ v_3 \end{bmatrix} \longrightarrow \begin{bmatrix} -v_1 \\ v_2 \\ v_3 \end{bmatrix},$$

as shown in Sketch 117.

This reflection is achieved by a scaling matrix:

$$\begin{bmatrix} -v_1 \\ v_2 \\ v_3 \end{bmatrix} = \begin{bmatrix} -1 & 0 & 0 \\ 0 & 1 & 0 \\ 0 & 0 & 1 \end{bmatrix} \begin{bmatrix} v_1 \\ v_2 \\ v_3 \end{bmatrix}.$$

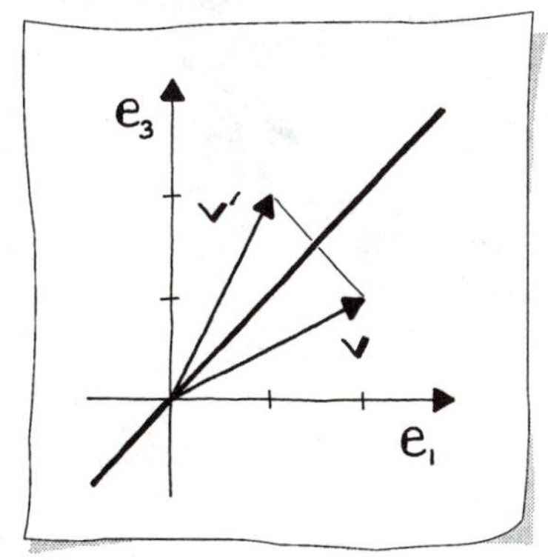

Sketch 118
Reflection of a vector about the $x_1 = x_3$ plane.

The following is also a reflection, as Sketch 118 shows:[3]

$$\begin{bmatrix} v_1 \\ v_2 \\ v_3 \end{bmatrix} \longrightarrow \begin{bmatrix} v_3 \\ v_2 \\ v_1 \end{bmatrix}.$$

[3] In that sketch, the plane $x_1 = x_3$ is shown "edge on."

It interchanges the first and third component of a vector, and is thus a reflection about the plane $x_1 = x_3$. This is an implicit plane equation, as discussed in Section 10.4.

This map is achieved by the following matrix equation:

$$\begin{bmatrix} v_1 \\ v_2 \\ v_3 \end{bmatrix} = \begin{bmatrix} 0 & 0 & 1 \\ 0 & 1 & 0 \\ 1 & 0 & 0 \end{bmatrix} \begin{bmatrix} v_1 \\ v_2 \\ v_3 \end{bmatrix}.$$

By their very nature, reflections do not change volumes — but they may change their signs; see Section 12.7 for more details.

12.4 Shears

What map takes a cube to the slanted cube of Sketch 119? That slanted cube, by the way, is called a *parallelepiped*, but *skew box* will do here. The answer: a shear. Shears in 3D are more complicated than the 2D shears from Section 4.7 because there are so many more directions to shear. Let's look at some of the shears more commonly used.

Consider the shear that maps $\mathbf{e}_1$ and $\mathbf{e}_2$ to themselves, and that also maps $\mathbf{e}_3$ to

$$\begin{bmatrix} a \\ b \\ 1 \end{bmatrix}.$$

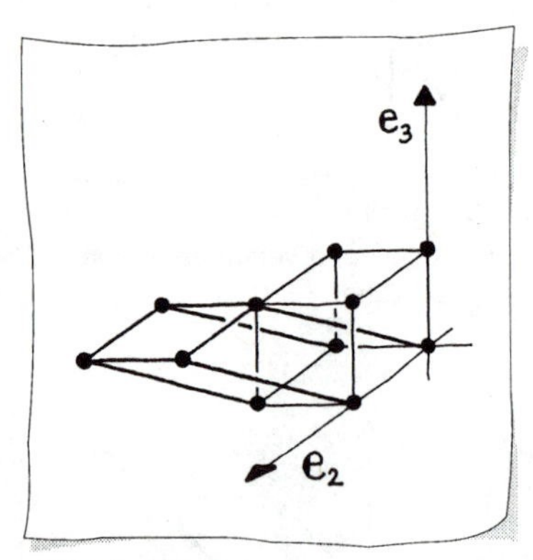

Sketch 119
A 3D shear.

The shear matrix S_1 that accomplishes the desired task is easily found:

$$S_1 = \begin{bmatrix} 1 & 0 & a \\ 0 & 1 & b \\ 0 & 0 & 1 \end{bmatrix}.$$

It is illustrated in Sketch 119 with $a = 1$ and $b = 1$, and in Figure 12.4.

Let's look at another example. What shear maps $\mathbf{e}_2$ and $\mathbf{e}_3$ to themselves, and also maps

$$\begin{bmatrix} a \\ b \\ c \end{bmatrix} \quad \text{to} \quad \begin{bmatrix} a \\ 0 \\ 0 \end{bmatrix}?$$

This shear is given by the matrix

$$S_2 = \begin{bmatrix} 1 & 0 & 0 \\ \frac{-b}{a} & 1 & 0 \\ \frac{-c}{a} & 0 & 1 \end{bmatrix}.$$

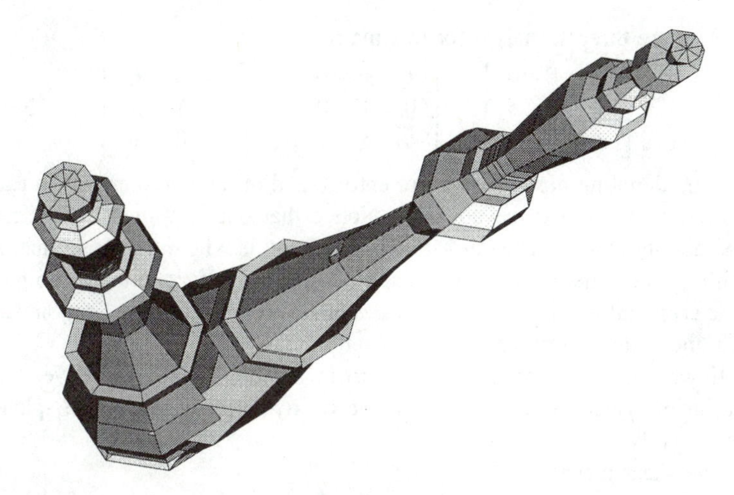

Figure 12.4.
Shears in 3D: a king chess piece sheared in the $\mathbf{e}_1$- and $\mathbf{e}_2$-directions. The $\mathbf{e}_3$-direction is the king's axis.

One quick check gives:

$$\begin{bmatrix} 1 & 0 & 0 \\ \frac{-b}{a} & 1 & 0 \\ \frac{-c}{a} & 0 & 1 \end{bmatrix} \begin{bmatrix} a \\ b \\ c \end{bmatrix} = \begin{bmatrix} a \\ 0 \\ 0 \end{bmatrix};$$

thus our map does what it was meant to do. This will be used in Section 14.2 as the shear of a Gauss elimination step.

Let's look at the matrix for the shear a little more generally. Although it is possible to shear in any direction, it is more common to shear parallel to a coordinate axis or coordinate plane. With the help of rotations, the following shear matrices should be sufficient for any need.

Write the shear matrix S as

$$S = \begin{bmatrix} 1 & s_{1,2} & s_{1,3} \\ s_{2,1} & 1 & s_{2,3} \\ s_{3,1} & s_{3,2} & 1 \end{bmatrix}. \qquad (12.5)$$

Suppose we apply this shear to a vector $\mathbf{v}$ resulting in

$$\mathbf{v}' = S\mathbf{v}.$$

An $s_{i,j}$ element is a factor by which the jth component of $\mathbf{v}$ affects the ith component of $\mathbf{v}'$. However, not all $s_{i,j}$ entries can be nonzero.

Here are three scenarios for this matrix:

$$\begin{bmatrix} 1 & 0 & 0 \\ \bigstar & 1 & \diamond \\ \bigstar & \diamond & 1 \end{bmatrix} \quad \begin{bmatrix} 1 & \bigstar & \diamond \\ 0 & 1 & 0 \\ \diamond & \bigstar & 1 \end{bmatrix} \quad \begin{bmatrix} 1 & \diamond & \bigstar \\ \diamond & 1 & \bigstar \\ 0 & 0 & 1 \end{bmatrix}$$

with $\bigstar$ denoting possible nonzero entries, and of the two $\diamond$ entries in each matrix, only one can be nonzero. Notice that one row must come from the identity matrix. The corresponding column is where the $\bigstar$ entries lie. With ones on the diagonal, these are the conditions for the matrix to have a determinant equal to one — thus volume preserving. See Section 12.7 and check the determinants for yourself.

If we take the $\diamond$ entries to be zero in the three matrices above, then we have created shears parallel to the $\mathbf{e}_2, \mathbf{e}_3$-, $\mathbf{e}_1, \mathbf{e}_3$-, and $\mathbf{e}_1, \mathbf{e}_2$-planes respectively.

The shear matrix

$$\begin{bmatrix} 1 & \bigstar & \bigstar \\ 0 & 1 & 0 \\ 0 & 0 & 1 \end{bmatrix}$$

shears parallel to the $\mathbf{e}_1$-axis. Matrices for the other axes follow similarly.

How does a shear affect volume? As stated above, the shear matrix has determinant equal to one. For more of a geometric feeling, notice the simple shear S_1 from above. It maps the unit cube to a skew box with the same base and the same height — thus it does not change volume!

12.5 Projections

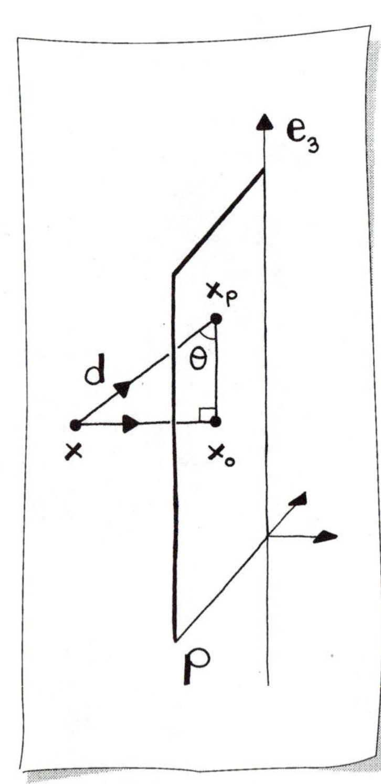

Sketch 120
Oblique and orthographic parallel projections.

Recall from 2D that a projection reduces dimensionality; it "flattens" geometry. In 3D this means that a vector is projected into a (2D) plane, and an example is illustrated in Figure 12.5. This is the technique used in computer graphics to view 3D geometry on a 2D screen. A parallel projection is a linear map, as opposed to a perspective projection which is not. A parallel projection preserves relative dimensions of an object, thus it is used in drafting to produce accurate views of a design.

As illustrated in Sketch 120, a parallel projection is defined by a *direction* of projection $\mathbf{d}$ and a *projection plane* P. A point $\mathbf{x}$ is projected into P, and is represented as $\mathbf{x}_p$ in the sketch. This information in turn defines a *projection angle* θ between $\mathbf{d}$ and the line joining the perpendicular projection point $\mathbf{x}_o$ in P. This angle is used to categorize parallel projections as *orthographic* or *oblique*. Orthographic projections are special; the direction is perpendicular to the plane. There are many special names for particular projection angles; see a computer graphics text such as [10] for more details.

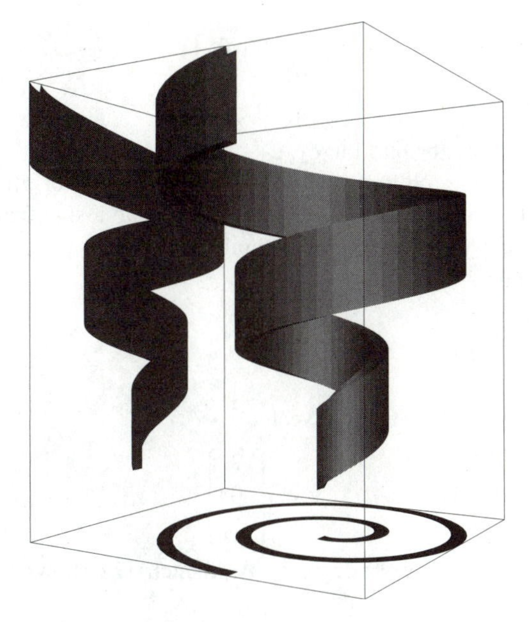

Figure 12.5.
Projections in 3D: a 3D helix is projected into two different 2D planes.

Let's construct some special orthographic projections.

Example 12.4 Take the three vectors

$$\mathbf{a}_1 = \begin{bmatrix} 2 \\ 0 \\ 1 \end{bmatrix}, \quad \mathbf{a}_2 = \begin{bmatrix} 0 \\ 2 \\ 1 \end{bmatrix}, \quad \mathbf{a}_3 = \begin{bmatrix} -1 \\ 0 \\ 1 \end{bmatrix}.$$

If we flatten them out into the $\mathbf{e}_1, \mathbf{e}_2$−plane, they become

$$\mathbf{a}'_1 = \begin{bmatrix} 2 \\ 0 \\ 0 \end{bmatrix}, \quad \mathbf{a}'_2 = \begin{bmatrix} 0 \\ 2 \\ 0 \end{bmatrix}, \quad \mathbf{a}'_3 = \begin{bmatrix} -1 \\ 0 \\ 0 \end{bmatrix},$$

(see Sketch 121). This action is achieved by the linear map

$$\begin{bmatrix} v_1 \\ v_2 \\ 0 \end{bmatrix} = \begin{bmatrix} 1 & 0 & 0 \\ 0 & 1 & 0 \\ 0 & 0 & 0 \end{bmatrix} \begin{bmatrix} v_1 \\ v_2 \\ v_3 \end{bmatrix},$$

as you should convince yourself!

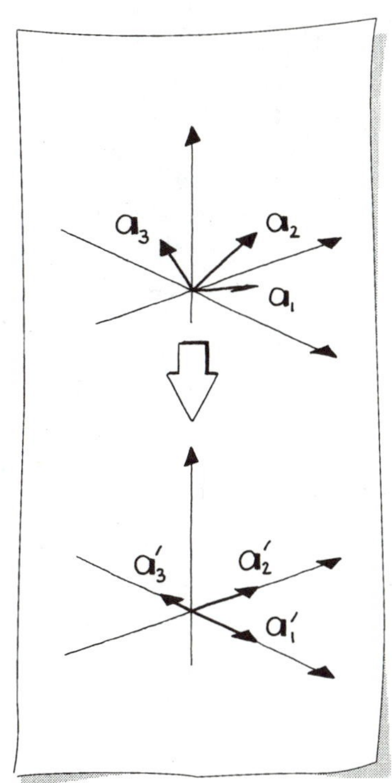

Sketch 121
Projection example.

Similarly, the matrix

$$\begin{bmatrix} 1 & 0 & 0 \\ 0 & 0 & 0 \\ 0 & 0 & 1 \end{bmatrix}$$

will flatten any vector into the $\mathbf{e}_1, \mathbf{e}_3-$plane.

We will examine oblique projections in the context of affine maps in Section 13.4. Finally we note that projections have a significant effect on the volume of objects. Since everything is flat after a projection, it has zero 3D volume.

12.6 Rotations

Suppose you want to rotate the vector

$$\mathbf{v} = \begin{bmatrix} 2 \\ 0 \\ 1 \end{bmatrix}$$

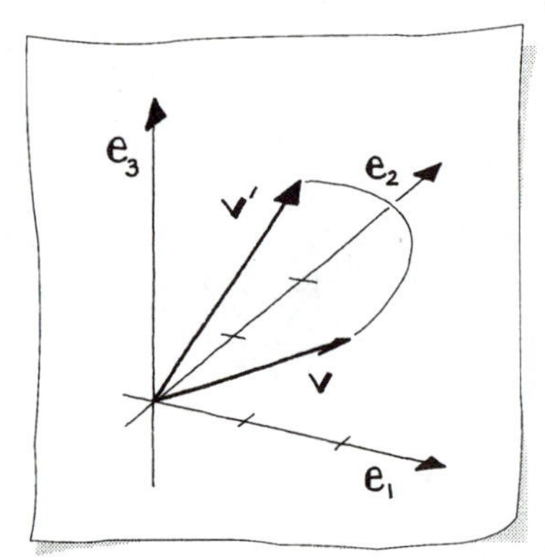

Sketch 122
Rotation example.

around the $\mathbf{e}_3-$axis by 90 degrees. As Sketch 122 shows, the resulting vector is

$$\mathbf{v}' = \begin{bmatrix} 0 \\ 2 \\ 1 \end{bmatrix}.$$

A rotation around $\mathbf{e}_3$ by different angles would result in different vectors, but they all will have one thing in common: their third components will not be changed by the rotation.

Thus if we rotate a vector around $\mathbf{e}_3$, the rotation action will only change its first and second components. This suggests another look at the 2D rotation matrices from Section 4.6! Our desired rotation matrix R_3 looks much the one from (4.16):

$$R_3 = \begin{bmatrix} \cos\alpha & -\sin\alpha & 0 \\ \sin\alpha & \cos\alpha & 0 \\ 0 & 0 & 1 \end{bmatrix}. \tag{12.6}$$

Example 12.5 Let us verify that R_3 performs as promised with $\alpha = 90°$:

$$\begin{bmatrix} 0 & -1 & 0 \\ 1 & 0 & 0 \\ 0 & 0 & 1 \end{bmatrix} \begin{bmatrix} 2 \\ 0 \\ 1 \end{bmatrix} = \begin{bmatrix} 0 \\ 2 \\ 1 \end{bmatrix},$$

so it works!

Figure 12.6.
Rotations in 3D: a barn rotated 90° about the $\mathbf{e}_1$-axis.

Similarly, we may rotate around the $\mathbf{e}_2$ axis; the corresponding matrix is

$$R_2 = \begin{bmatrix} \cos\alpha & 0 & -\sin\alpha \\ 0 & 1 & 0 \\ \sin\alpha & 0 & \cos\alpha \end{bmatrix}. \tag{12.7}$$

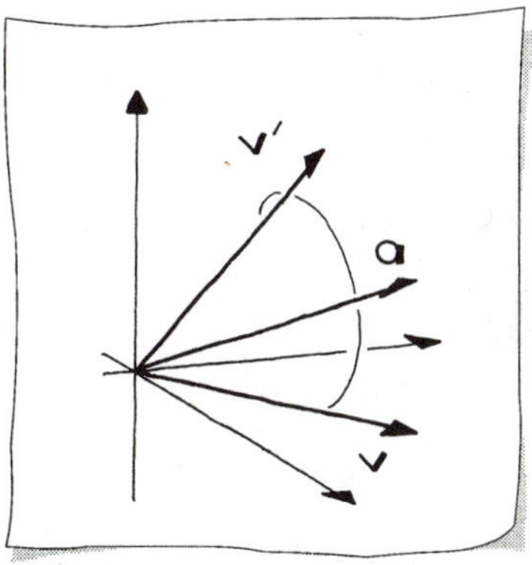

Sketch 123
Rotation about an arbitrary vector.

Notice the pattern here. The rotation matrix for a rotation about the $\mathbf{e}_i$-axis is characterized by the ith row being $\mathbf{e}_i^{\mathrm{T}}$ and the ith column being $\mathbf{e}_i$.

Figure 12.6 illustrates a barn rotated about the $\mathbf{e}_1$-axis by 90°. The barn's initial position was centered on the $\mathbf{e}_3$-axis. Notice that the direction of rotation follows the right-hand rule: curl your fingers with the rotation, and your thumb points in the direction of the rotation axis. Figure 12.7 illustrates the same rotation, however with three rotations.

How about a rotation by α degrees around an arbitrary vector $\mathbf{a}$? The principle is illustrated in Sketch 123. The derivation of the following matrix is more tedious than called for here (see [6]) — we just give the result:

$$R = \begin{bmatrix} a_1^2 + C(1-a_1^2) & a_1a_2(1-C) - a_3S & a_1a_3(1-C) + a_2S \\ a_1a_2(1-C) + a_3S & a_2^2 + C(1-a_2^2) & a_2a_3(1-C) - a_1S \\ a_1a_3(1-C) - a_2S & a_2a_3(1-C) + a_1S & a_3^2 + C(1-a_3^2) \end{bmatrix} \tag{12.8}$$

Figure 12.7.
Rotations in 3D: a barn rotated in $90°$ increments about the $\mathbf{e}_1$-axis.

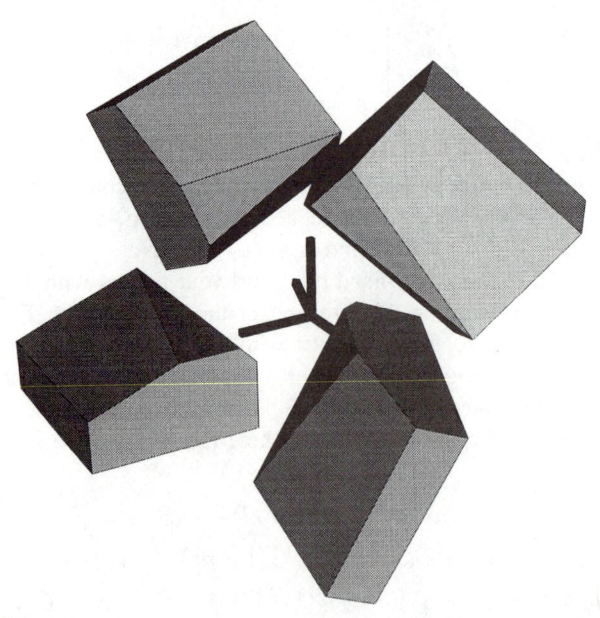

Figure 12.8.
Rotations in 3D: a barn centered on the $\mathbf{e}_1$-axis is rotated about a vector which is not an $\mathbf{e}_i$-direction.

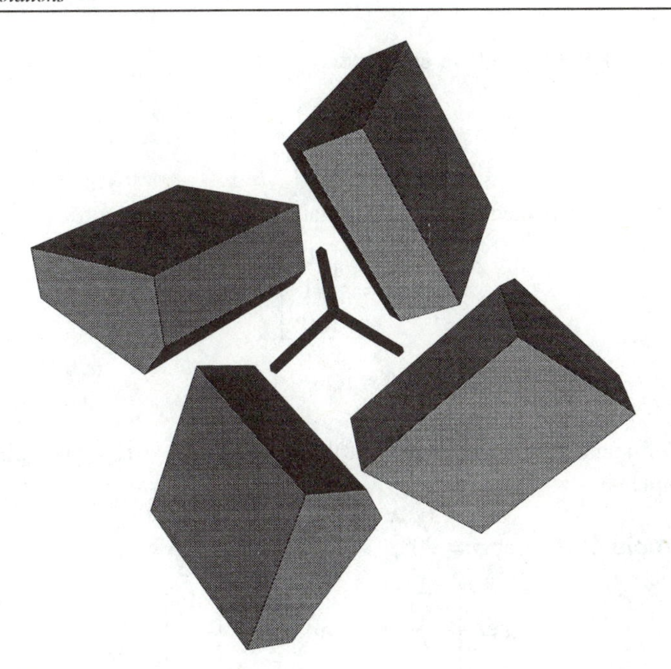

Figure 12.9.
Rotations in 3D: same as the previous figure, however from a different view.

where we have set $C = \cos\alpha$ and $S = \sin\alpha$. It is necessary that $\|\mathbf{a}\| = 1$ in order for the rotation to take place without scaling.

Figures 12.8 and 12.9 illustrate the same rotations from different views. A barn, initially positioned on the $\mathbf{e}_1$-axis is rotated in 90° increments about the vector

$$\begin{bmatrix} 1 \\ 1 \\ 1 \end{bmatrix}.$$

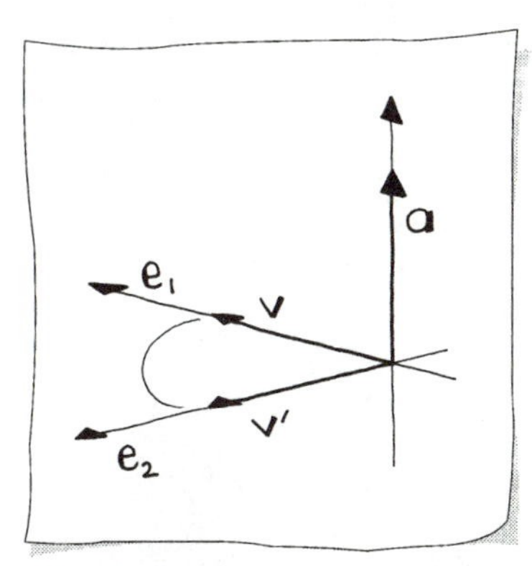

Sketch 124
A simple example of a rotation about a vector.

Example 12.6 With a complicated result as this one, a sanity check is not a bad idea. So let $\alpha = 90°$,

$$\mathbf{a} = \begin{bmatrix} 0 \\ 0 \\ 1 \end{bmatrix} \quad \text{and} \quad \mathbf{v} = \begin{bmatrix} 1 \\ 0 \\ 0 \end{bmatrix}.$$

This means that we want to rotate $\mathbf{v}$ around $\mathbf{a}$, orthe $\mathbf{e}_3$-axis, by 90° as shown in Sketch 124. In advance, we know what R should be. In (12.8),

$C = 0$ and $S = 1$, and we calculate

$$R = \begin{bmatrix} 0 & -1 & 0 \\ 1 & 0 & 0 \\ 0 & 0 & 1 \end{bmatrix},$$

which is the expected matrix! We obtain

$$\mathbf{v}' = \begin{bmatrix} 0 \\ 1 \\ 0 \end{bmatrix}.$$

With some confidence that (12.8) works, let's try a more complicated example.

Example 12.7 Let $\alpha = 90°$,

$$\mathbf{a} = \begin{bmatrix} \frac{1}{\sqrt{3}} \\ \frac{1}{\sqrt{3}} \\ \frac{1}{\sqrt{3}} \end{bmatrix} \quad \text{and} \quad \mathbf{v} = \begin{bmatrix} 1 \\ 0 \\ 0 \end{bmatrix}.$$

With $C = 0$ and $S = 1$ in (12.8), we calculate

$$R = \begin{bmatrix} \frac{1}{3} & \frac{1}{3} - \frac{1}{\sqrt{3}} & \frac{1}{3} + \frac{1}{\sqrt{3}} \\ \frac{1}{3} + \frac{1}{\sqrt{3}} & \frac{1}{3} & \frac{1}{3} - \frac{1}{\sqrt{3}} \\ \frac{1}{3} - \frac{1}{\sqrt{3}} & \frac{1}{3} + \frac{1}{\sqrt{3}} & \frac{1}{3} \end{bmatrix},$$

We obtain

$$\mathbf{v}' = \begin{bmatrix} \frac{1}{3} \\ \frac{1}{3} + \frac{1}{\sqrt{3}} \\ \frac{1}{3} - \frac{1}{\sqrt{3}} \end{bmatrix}.$$

Convince yourself that $\|\mathbf{v}'\| = \|\mathbf{v}\|$.

Continue this example with the vector

$$\mathbf{v} = \begin{bmatrix} 1 \\ 1 \\ 1 \end{bmatrix}.$$

Surprised at the result?

It should be intuitively clear that rotations do not change volumes. Recall from 2D that rotations are *rigid body motions*.

12.7 Volumes and Linear Maps: Determinants

Most linear maps change volumes; some don't. Since this is an important aspect of the action of a map, this section will study the effect of a linear map on volume. The unit cube in the $[\mathbf{e}_1, \mathbf{e}_2, \mathbf{e}_3]$-system has volume one. A linear map A will change that volume to that of the skew box spanned by the images of $\mathbf{e}_1, \mathbf{e}_2, \mathbf{e}_3$, i.e., by the volume spanned by the vectors $\mathbf{a}_1, \mathbf{a}_2, \mathbf{a}_3$ — the column vectors of A. What is the volume spanned by $\mathbf{a}_1, \mathbf{a}_2, \mathbf{a}_3$?

Recall the 2×2 determinant from Section 4.9. Through Sketch 4.9, the area of a 2D parallelogram was shown to be equivalent to a determinant. Here we want to illustrate that the volume of the parallelepiped, or skew box, is simply a 3D determinant calculation. Proceeding directly with a sketch in the 3D case would be difficult to follow. For 3D, let's augment the determinant idea with the tools from Section 5.4. There we demonstrated how shears — area preserving linear maps — can be used to transform a matrix to upper triangular. These are the Gauss elimination steps.

First, let's introduce the 3×3 determinant of a matrix A. It is easily remembered as an alternating sum of 2×2 determinants.

$$|A| = a_{1,1} \begin{vmatrix} a_{2,2} & a_{2,3} \\ a_{3,2} & a_{3,3} \end{vmatrix} - a_{2,1} \begin{vmatrix} a_{1,2} & a_{1,3} \\ a_{3,2} & a_{3,3} \end{vmatrix} + a_{3,1} \begin{vmatrix} a_{1,2} & a_{1,3} \\ a_{2,2} & a_{2,3} \end{vmatrix}. \tag{12.9}$$

Determinants and their properties are discussed in more detail in Section 10.5.

As we have seen in Section 12.4, a 3D shear preserves volume. Therefore, we can apply a series of shears to the matrix A, resulting in a new matrix

$$\tilde{A} = \begin{bmatrix} \tilde{a}_{1,1} & \tilde{a}_{1,2} & \tilde{a}_{1,3} \\ 0 & \tilde{a}_{2,2} & \tilde{a}_{2,3} \\ 0 & 0 & \tilde{a}_{3,3} \end{bmatrix}.$$

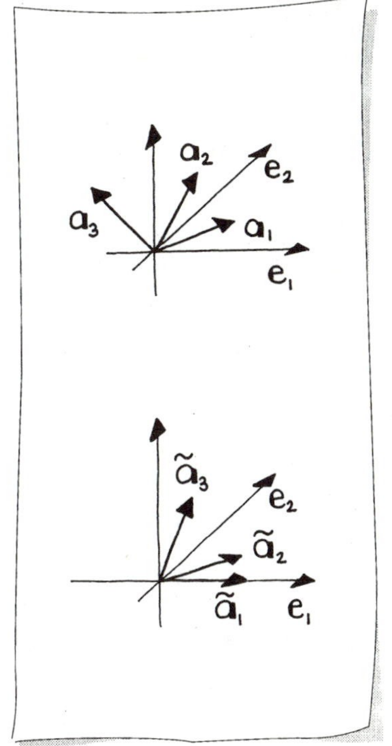

Sketch 125
Determinant and volume in 3D.

The determinant of $\tilde{A}$ is

$$|\tilde{A}| = \tilde{a}_{1,1}\tilde{a}_{2,2}\tilde{a}_{3,3},$$

with, of course, $|A| = |\tilde{A}|$. We don't actually calculate the volume of three vectors by proceeding with the Gauss elimination steps, or shears. We would just directly calculate the 3×3 determinant. What is interesting about this development is now we can illustrate, as in Sketch 125, how the determinant defines the volume of the skew box. The first two column vectors of $\tilde{A}$ lie in the $[\mathbf{e}_1, \mathbf{e}_2]$-plane. Their determinant defines the area of the parallelogram that they span; this determinant is $\tilde{a}_{1,1}\tilde{a}_{2,2}$. The height of the skew box is simply the $\mathbf{e}_3$ component of $\tilde{\mathbf{a}_3}$. Thus we have an easy to visualize interpretation of the 3×3 determinant. See Section 10.5 for another geometric interpretation of the determinant.

12.8 Combining Linear Maps

If we apply a linear map A to a vector $\mathbf{v}$ and then apply a map B to the result, we may write this as

$$\mathbf{v}' = BA\mathbf{v}.$$

Matrix multiplication is defined just as in the 2D case; the element $c_{i,j}$ of the product matrix $C = BA$ is obtained as the dot product of the $i-$th row of B with the $j-$th column of A. Instead of a complicated formula, an example should suffice:

$$\begin{bmatrix} 0 & 0 & -1 \\ 1 & -2 & 0 \\ -2 & 1 & 1 \end{bmatrix} \cdot \begin{bmatrix} 1 & 5 & -4 \\ -1 & -2 & 0 \\ 2 & 3 & -4 \end{bmatrix} = \begin{bmatrix} 2 & 3 & 4 \\ 3 & 9 & -4 \\ -1 & -9 & 4 \end{bmatrix}.$$

In this example B and A are 3×3 matrices, and thus the result is another 3×3 matrix. In the example in Section 12.1, a 3×3 matrix A is multiplied by a 3×1 matrix (vector) $\mathbf{v}$ resulting in a 3×1 matrix or vector. Thus two matrices need not be the same size in order to multiply them. There is a rule however! Suppose we are to multiply two matrices A and B together as AB. The sizes of A and B are

$$m \times n \quad \text{and} \quad n \times p, \tag{12.10}$$

respectively. The resulting matrix will be of size $m \times p$ — the "outside" dimensions in 12.10. In order to form AB, it is necessary that the "inside" dimensions be equal. The matrix multiplication scheme from Section 4.2 simplifies hand-calculations by illustrating the resulting dimensions.

As in the 2D case, matrix multiplication does not commute! That is, $AB \neq BA$ in most cases. An interesting difference between 2D and 3D is the fact that in 2D, rotations *did* commute; however in 3D they do not. For example, in 2D, rotating first by α and then by β is no different from doing it the other way around. In 3D, let's look at a rotation by 90° around the $\mathbf{e}_1$-axis with matrix R_1 and a rotation by 90° around $\mathbf{e}_2$-axis with matrix R_2.

We have

$$R_1 = \begin{bmatrix} 1 & 0 & 0 \\ 0 & 0 & -1 \\ 0 & 1 & 0 \end{bmatrix}, \quad R_2 = \begin{bmatrix} 0 & 0 & -1 \\ 0 & 1 & 0 \\ 1 & 0 & 0 \end{bmatrix}.$$

Figure 12.10.
Combining 3D rotations: the barn in initial position.

Now we simply compute

$$R_1R_2 = \begin{bmatrix} 0 & 0 & -1 \\ -1 & 0 & 0 \\ 0 & 1 & 0 \end{bmatrix}, \quad R_2R_1 = \begin{bmatrix} 0 & -1 & 0 \\ 0 & 0 & -1 \\ 1 & 0 & 0 \end{bmatrix}.$$

So it does matter which rotation we perform first! In order to really understand this phenomenon, take the vector $\mathbf{e}_1$ and apply the rotation R_1R_2 and then apply R_2R_1. Compare the results. Figure 12.10 illustrates the barn in an initial position. Figure 12.11 shows its position after the rotations R_1R_2 and Figure 12.12 shows its position after the rotations R_2R_1.

The *transpose* A^{T} of a matrix A is obtained by interchanging rows and columns, i.e.,

$$\begin{bmatrix} 2 & 3 & -4 \\ 3 & 9 & -4 \\ -1 & -9 & 4 \end{bmatrix}^T = \begin{bmatrix} 2 & 3 & -1 \\ 3 & 9 & -9 \\ -4 & -4 & 4 \end{bmatrix}.$$

Figure 12.11.
Combining 3D rotations: the barn after R_1R_2.

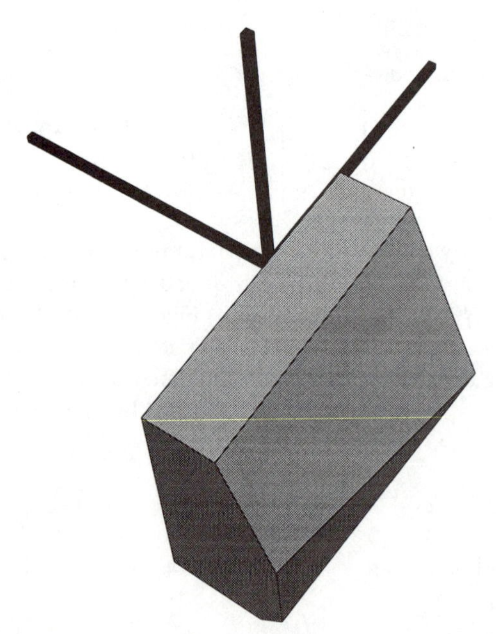

Figure 12.12.
Combining 3D rotations: the barn after R_2R_1.

As a concise formula,

$$a^{\mathrm{T}}_{i,j} = a_{j,i}.$$

We restate some results from the 2D case:

$$\begin{aligned}[A+B]^{\mathrm{T}} &= A^{\mathrm{T}} + B^{\mathrm{T}},\\ A^{\mathrm{T}^{\mathrm{T}}} &= A,\\ [cA]^{\mathrm{T}} &= cA^{\mathrm{T}}\\ [AB]^{\mathrm{T}} &= B^{\mathrm{T}}A^{\mathrm{T}}.\end{aligned}$$

12.9 Inverse Matrices

In Section 5.5, we saw how inverse matrices undo linear maps. A linear map A takes a vector $\mathbf{v}$ to its image $\mathbf{v}'$. The inverse map, A^{-1}, will take $\mathbf{v}'$ back to $\mathbf{v}$, i.e., $A^{-1}\mathbf{v}' = \mathbf{v}$ or $A^{-1}A\mathbf{v} = \mathbf{v}$. Thus the combined action of $A^{-1}A$ has no effect on any vector $\mathbf{v}$, which we can write as

$$A^{-1}A = I, \tag{12.11}$$

where I is the 3×3 identity matrix. If we applied A^{-1} to $\mathbf{v}$ first, and then applied A, there would not be any action either; in other words,

$$AA^{-1} = I \tag{12.12}$$

too.

See Section 14.6 for details on calculating A^{-1}.

12.10 Exercises

1. Describe the linear map given by the matrix

$$\begin{bmatrix} 0 & 1 & 0 \\ 0 & 0 & -1 \\ 1 & 0 & 0 \end{bmatrix}.$$

   Hint: you might want to try a few simple examples to get a feeling for what is going on.

2. What matrix scales by 2 in the $\mathbf{e}_1$-direction, scales by $1/4$ in the $\mathbf{e}_2$-direction, and reverses direction and scales by 4 in the $\mathbf{e}_3$-direction? Map the unit cube with this matrix. What is the volume of the resulting parallelepiped?

3. What is the matrix which reflects a vector about the plane $x_1 = x_2$? Map the unit cube with this matrix. What is the volume of the resulting parallelepiped?

4. What is the shear matrix which maps

$$\begin{bmatrix} a \\ b \\ c \end{bmatrix} \quad \text{to} \quad \begin{bmatrix} 0 \\ 0 \\ c \end{bmatrix}?$$

   Map the unit cube with this matrix. What is the volume of the resulting parallelepiped?

5. What matrix rotates around the $\mathbf{e}_1$−axis by α degrees?

6. What matrix rotates by 45° around the vector

$$\begin{bmatrix} -1 \\ 0 \\ -1 \end{bmatrix}?$$

7. Compute

$$\begin{bmatrix} 0 & 0 & 1 \\ 1 & -2 & 0 \\ -2 & 1 & 1 \end{bmatrix} \cdot \begin{bmatrix} 1 & 5 & -4 \\ -1 & -2 & 0 \\ 2 & 3 & -4 \end{bmatrix}.$$

8. What is the matrix for a shear parallel to an arbitrary plane in a specified direction. Assume the given information is a plane with normal $\mathbf{n}$ and a vector $\mathbf{s}$ in this plane. Hint: First write this as a vector equation, then build the matrix.

13 Affine Maps in 3D

Figure 13.1.
Affine maps in 3D: fighter jets twisting and turning through 3D space.

This chapters wraps up the basic geometry tools. Affine maps in 3D are a primary tool for modeling and computer graphics. Figure 13.1 illustrates the use of various affine maps. This chapter goes a little farther than just affine maps by introducing projective maps — the maps used to create realistic 3D images.

13.1 Affine Maps

Linear maps relate vectors to vectors. Affine maps relate points to points. A 3D affine map is written just as a 2D one, namely as

$$\mathbf{x}' = \mathbf{p} + A(\mathbf{x} - \mathbf{o}). \tag{13.1}$$

In general we will assume that the origin of $\mathbf{x}$'s coordinate system has three zero coordinates, and drop the $\mathbf{o}$ term:

$$\mathbf{x}' = \mathbf{p} + A\mathbf{x}. \tag{13.2}$$

Figure 13.2.
Parallel planes get mapped to parallel planes via an affine map.

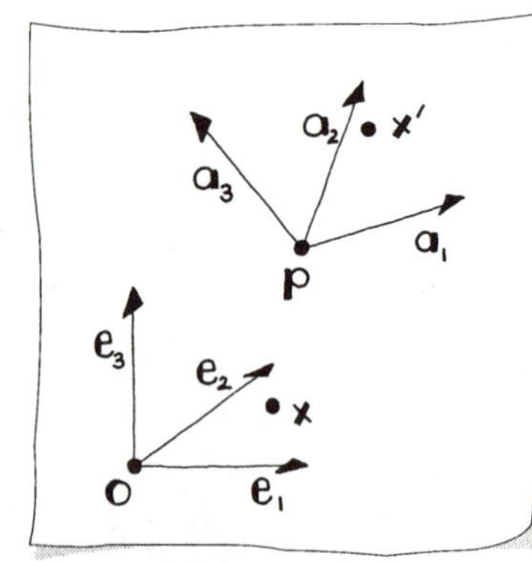

Sketch 126
An affine map in 3D.

Sketch 126 gives an example. Recall, the column vectors of A are the vectors $\mathbf{a}_1, \mathbf{a}_2, \mathbf{a}_3$. The point $\mathbf{p}$ tells us where to move the origin of the $[\mathbf{e}_1, \mathbf{e}_2, \mathbf{e}_3]$-system; again, the real action of an affine map is captured by the matrix. Thus by studying matrix actions, or linear maps, we will learn more about affine maps.

We now list some of the important properties of 3D affine maps. They are straightforward generalizations of the 2D cases, and so we just give a brief listing.

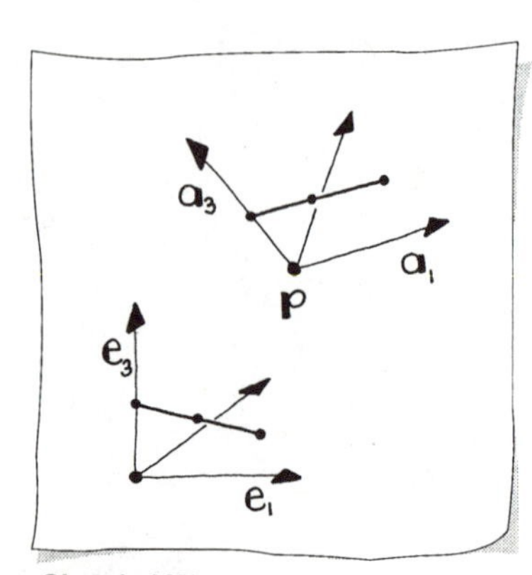

Sketch 127
Affine maps leave ratios invariant.

1. Affine maps leave *ratios* invariant (see Sketch 127).

2. Affine maps take *parallel planes* to parallel planes (see Figure 13.2).

3. Affine maps take *intersecting planes* to intersecting planes. In particular, the intersection line of the mapped planes is the map of the original intersection line.

4. Affine maps leave *barycentric combinations* invariant. If

$$\mathbf{x} = c_1\mathbf{p}_1 + c_2\mathbf{p}_2 + c_3\mathbf{p}_3 + c_4\mathbf{p}_4,$$

where $c_1 + c_2 + c_3 + c_4 = 1$, then after an affine map we have

$$\mathbf{x}' = c_1\mathbf{p}'_1 + c_2\mathbf{p}'_2 + c_3\mathbf{p}'_3 + c_4\mathbf{p}'_4.$$

For example, the *centroid* of a tetrahedron will be mapped to the centroid of the mapped tetrahedron (see Sketch 128).

Most 3D maps do not offer much over their 2D counterparts — but some do. We will go through all of them in detail now.

13.2 Translations

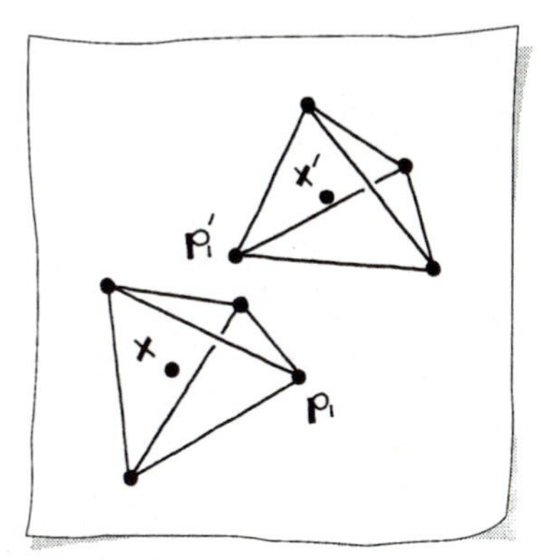

Sketch 128
The centroid is mapped to the centroid.

A translation is simply (13.2) with $A = I$, the 3×3 identity matrix:

$$I = \begin{bmatrix} 1 & 0 & 0 \\ 0 & 1 & 0 \\ 0 & 0 & 1 \end{bmatrix},$$

that is

$$\mathbf{x}' = \mathbf{p} + I\mathbf{x}.$$

Thus the new $[\mathbf{a}_1, \mathbf{a}_2, \mathbf{a}_3]$–system has its coordinate axes parallel to the $[\mathbf{e}_1, \mathbf{e}_2, \mathbf{e}_3]$-system. The term $I\mathbf{x} = \mathbf{x}$ needs to be interpreted as a *vector* in the $[\mathbf{e}_1, \mathbf{e}_2, \mathbf{e}_3]$-system for this to make sense! Figure 13.3 shows an example of repeated 3D translations.

Just as in 2D, a translation is a rigid body motion. The volume of an object is not changed.

Figure 13.3.
Translations in 3D: three translated teapots.

13.3 Mapping Tetrahedra

A 3D affine map is determined by four point pairs $\mathbf{p}_i \rightarrow \mathbf{p}'_i$ for $i = 1, 2, 3, 4$. In other words, an affine map is determined by a tetrahedron and its image. What is the image of an arbitrary point $\mathbf{x}$ under this affine map?

Affine maps leave *barycentric combinations* unchanged. This will be the key to finding $\mathbf{x}'$, the image of $\mathbf{x}$. If we can write $\mathbf{x}$ in the form

$$\mathbf{x} = u_1\mathbf{p}_1 + u_2\mathbf{p}_2 + u_3\mathbf{p}_3 + u_4\mathbf{p}_4, \tag{13.3}$$

then we know that the image has the same relationship with the $\mathbf{p}'_i$:

$$\mathbf{x}' = u_1\mathbf{p}'_1 + u_2\mathbf{p}'_2 + u_3\mathbf{p}'_3 + u_4\mathbf{p}'_4. \tag{13.4}$$

So all we need to do is find the u_i! These are called the *barycentric coordinates* of $\mathbf{x}$ with respect to the $\mathbf{p}_i$, quite in analogy to the triangle case (Section 6.4).

We observe that (13.3) is short for three individual coordinate equations. Together with the barycentric combination condition

$$u_1 + u_2 + u_3 + u_4 = 1,$$

we have four equations for the four unknowns $u_1, \ldots, u_4$, which we can solve by consulting Chapter 14.

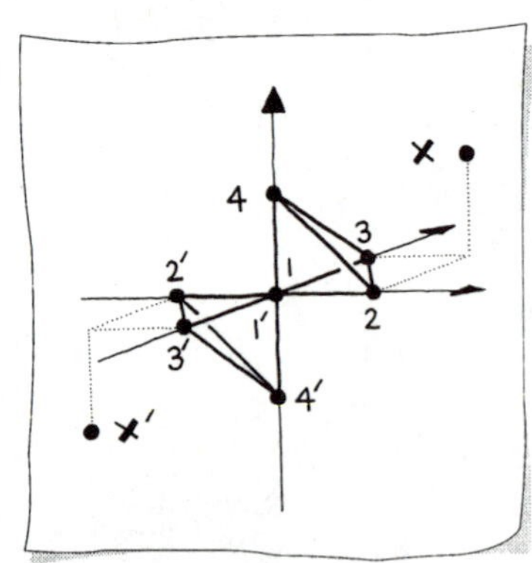

Sketch 129
An example tetrahedron map.

Example 13.1 Let the original tetrahedron be given by the four points $\mathbf{p}_i$

$$\begin{bmatrix} 0 \\ 0 \\ 0 \end{bmatrix}, \begin{bmatrix} 1 \\ 0 \\ 0 \end{bmatrix}, \begin{bmatrix} 0 \\ 1 \\ 0 \end{bmatrix}, \begin{bmatrix} 0 \\ 0 \\ 1 \end{bmatrix}.$$

Let's assume we want to map this tetrahedron to the four points $\mathbf{p}'_i$

$$\begin{bmatrix} 0 \\ 0 \\ 0 \end{bmatrix}, \begin{bmatrix} -1 \\ 0 \\ 0 \end{bmatrix}, \begin{bmatrix} 0 \\ -1 \\ 0 \end{bmatrix}, \begin{bmatrix} 0 \\ 0 \\ -1 \end{bmatrix}.$$

This is a pretty straightforward map if you consult Sketch 129.

Let's see where the point

$$\mathbf{x} = \begin{bmatrix} 1 \\ 1 \\ 1 \end{bmatrix} \quad \text{ends up!}$$

First, we find that

$$\begin{bmatrix}1\\1\\1\end{bmatrix} = -2\begin{bmatrix}0\\0\\0\end{bmatrix} + \begin{bmatrix}1\\0\\0\end{bmatrix} + \begin{bmatrix}0\\1\\0\end{bmatrix} + \begin{bmatrix}0\\0\\1\end{bmatrix},$$

i.e., the barycentric coordinates of $\mathbf{x}$ with respect to the original $\mathbf{p}_i$ are $(-2, 1, 1, 1)$. Note how they sum to one! Now it is simple to compute the image of $\mathbf{x}$; compute $\mathbf{x}'$ using the same barycentric coordinates with respect to the $\mathbf{p}'_i$:

$$\mathbf{x}' = -2\begin{bmatrix}0\\0\\0\end{bmatrix} + \begin{bmatrix}-1\\0\\0\end{bmatrix} + \begin{bmatrix}0\\-1\\0\end{bmatrix} + \begin{bmatrix}0\\0\\-1\end{bmatrix} = \begin{bmatrix}-1\\-1\\-1\end{bmatrix}.$$

A different approach would be to find the 3×3 matrix A and point $\mathbf{p}$ which describe the affine map. Construct a coordinate system from the $\mathbf{p}_i$ tetrahedron. One way to do this is to choose $\mathbf{p}_1$ as the origin[1] and the three axes are defined as $\mathbf{p}_i - \mathbf{p}_1$ for $i = 2, 3, 4$. The coordinate system of the $\mathbf{p}'_i$ tetrahedron must be based on the same indices. Once we have defined A and $\mathbf{p}$ then we will be able to map $\mathbf{x}$ by this map:

$$\mathbf{x}' = A[\mathbf{x} - \mathbf{p}_1] + \mathbf{p}'_1$$

Thus the point $\mathbf{p} = \mathbf{p}'_1$. In order to determine A, let's write down some known relationships. Referring to Sketch 130, we know

$$A[\mathbf{p}_2 - \mathbf{p}_1] = \mathbf{p}'_2 - \mathbf{p}'_1,$$
$$A[\mathbf{p}_3 - \mathbf{p}_1] = \mathbf{p}'_3 - \mathbf{p}'_1,$$
$$A[\mathbf{p}_4 - \mathbf{p}_1] = \mathbf{p}'_4 - \mathbf{p}'_1,$$

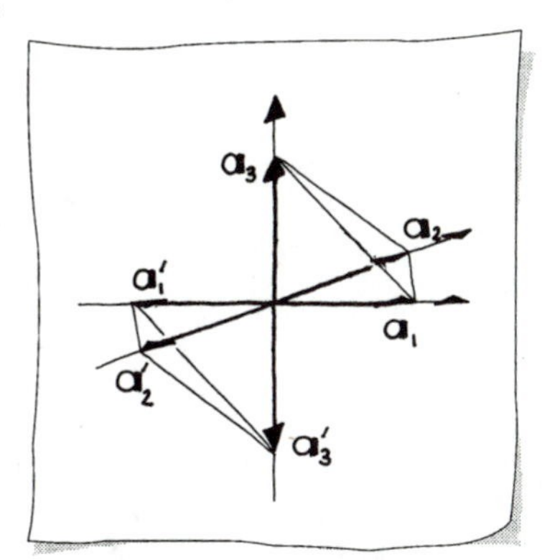

Sketch 130
The relationship between tetrahedra.

which may be written matrix form as

$$A\begin{bmatrix}\mathbf{p}_2 - \mathbf{p}_1 & \mathbf{p}_3 - \mathbf{p}_1 & \mathbf{p}_4 - \mathbf{p}_1\end{bmatrix} = \begin{bmatrix}\mathbf{p}'_2 - \mathbf{p}'_1 & \mathbf{p}'_3 - \mathbf{p}'_1 & \mathbf{p}'_4 - \mathbf{p}'_1\end{bmatrix}. \quad (13.5)$$

Thus

$$A = \begin{bmatrix}\mathbf{p}'_2 - \mathbf{p}'_1 & \mathbf{p}'_3 - \mathbf{p}'_1 & \mathbf{p}'_4 - \mathbf{p}'_1\end{bmatrix}\begin{bmatrix}\mathbf{p}_2 - \mathbf{p}_1 & \mathbf{p}_3 - \mathbf{p}_1 & \mathbf{p}_4 - \mathbf{p}_1\end{bmatrix}^{-1}, \quad (13.6)$$

and A is defined.

[1] Any of the four $\mathbf{p}_i$ would do, so for the sake of concreteness, we pick the first one.

Example 13.2 Revisiting the previous example, we now want to construct the matrix A. By selecting $\mathbf{p}_1$ as the origin for the $\mathbf{p}_i$ tetrahedron coordinate system there is no translation; $\mathbf{p}_1$ is the origin in the $[\mathbf{e}_1, \mathbf{e}_2, \mathbf{e}_3]$-system and $\mathbf{p}'_1 = \mathbf{p}_1$. We now compute A. (A is the product matrix in the bottom right position):

$$\begin{array}{ccc|ccc} & & & 1 & 0 & 0 \\ & & & 0 & 1 & 0 \\ & & & 0 & 0 & 1 \\ \hline -1 & 0 & 0 & -1 & 0 & 0 \\ 0 & -1 & 0 & 0 & -1 & 0 \\ 0 & 0 & -1 & 0 & 0 & -1 \end{array}$$

In order to compute $\mathbf{x}'$, we have

$$\mathbf{x}' = \begin{bmatrix} -1 & 0 & 0 \\ 0 & -1 & 0 \\ 0 & 0 & -1 \end{bmatrix} \begin{bmatrix} 1 \\ 1 \\ 1 \end{bmatrix} = \begin{bmatrix} -1 \\ -1 \\ -1 \end{bmatrix}.$$

This is the same result as in the previous example.

13.4 Projections

Take any object made of wires outside; let the sun shine on it, and you can observe a shadow. This shadow is the *parallel projection* of your object onto a plane. Everything we draw is a projection of necessity — paper is 2D, after all, whereas most interesting objects are 3D. Figure 13.4 gives an example. Also see Figure 12.5.

Projections reduce dimensionality; as basic linear maps, we encountered them in Sections 4.8 and 12.5. As affine maps, they map 3D points onto a plane. In most cases, we are interested in the case of these planes being the coordinate planes. All we have to do then is set one of the point's coordinates to zero. These basic projections are called *orthographic*: the simulated light ray hits the projection plane at a right angle.

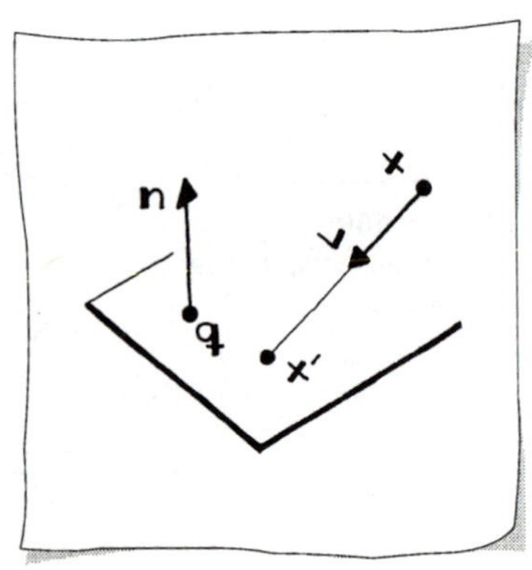

Sketch 131
Projecting a point on a plane.

It is a little more interesting to use directions which are at arbitrary angles to the projection plane. These are called *oblique projections*. Let $\mathbf{x}$ be the 3D point to be projected, let $\mathbf{n} \cdot [\mathbf{q} - \mathbf{x}] = 0$ be the projection plane, and let $\mathbf{v}$ indicate the projection direction (see Sketch 131).

We have already encountered this problem in Section 11.3 where it is called line/plane intersection. There, we established that $\mathbf{x}'$, the image of

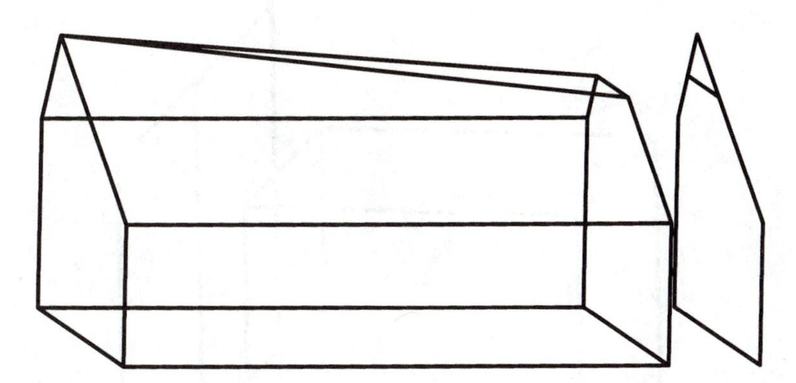

Figure 13.4.
Projections: a parallel projection of a 3D barn.

$\mathbf{x}$ under the projection is given by (11.6), which we repeat here:

$$\mathbf{x}' = \mathbf{x} + \frac{[\mathbf{q} - \mathbf{x}] \cdot \mathbf{n}}{\mathbf{v} \cdot \mathbf{n}} \mathbf{v}. \tag{13.7}$$

Figure 13.5 illustrates a simple oblique projection of a cube defined over $[-1, 1]$ in each coordinate with

$$\mathbf{n} = \begin{bmatrix} 1 \\ 0 \\ 0 \end{bmatrix} \quad \mathbf{q} = \begin{bmatrix} 2 \\ 0 \\ 0 \end{bmatrix} \quad \mathbf{v} = \begin{bmatrix} 1/\sqrt{2} \\ 1/\sqrt{2} \\ 0 \end{bmatrix}.$$

Figure 13.6 creates a projection plane that is not one of the coordinate planes; specifically,

$$\mathbf{n} = \begin{bmatrix} 1/\sqrt{3} \\ 1/\sqrt{3} \\ 1/\sqrt{3} \end{bmatrix} \quad \mathbf{q} = \begin{bmatrix} 4 \\ 0 \\ 0 \end{bmatrix} \quad \mathbf{v} = \begin{bmatrix} 1/\sqrt{3} \\ 1/\sqrt{3} \\ 1/\sqrt{3} \end{bmatrix}.$$

Finally, Figure 13.7 creates a general oblique projection with

$$\mathbf{n} = \begin{bmatrix} 1/\sqrt{3} \\ 1/\sqrt{3} \\ 1/\sqrt{3} \end{bmatrix} \quad \mathbf{q} = \begin{bmatrix} 4 \\ 0 \\ 0 \end{bmatrix} \quad \mathbf{v} = \begin{bmatrix} 1/\sqrt{2} \\ 1/\sqrt{2} \\ 0 \end{bmatrix}.$$

Revisiting the helix, Figure 13.8 is a projection with the same $\mathbf{n}$ and $\mathbf{v}$ as in the previous figure.

How do we write this as an affine map? Without much effort, we find

$$\mathbf{x}' = \mathbf{x} - \frac{\mathbf{n} \cdot \mathbf{x}}{\mathbf{v} \cdot \mathbf{n}} \mathbf{v} + \frac{\mathbf{q} \cdot \mathbf{n}}{\mathbf{v} \cdot \mathbf{n}} \mathbf{v}.$$

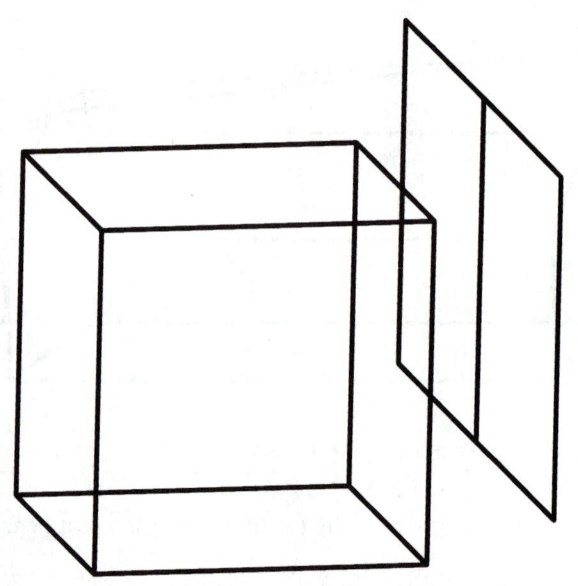

Figure 13.5.
Projections: a cube projected in a coordinate plane with an oblique angle.

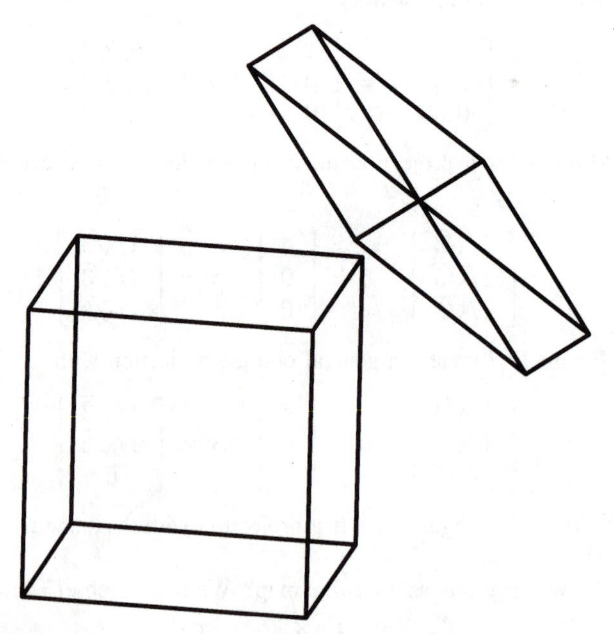

Figure 13.6.
Projections: a cube projected in an arbitrary plane with a right angle.

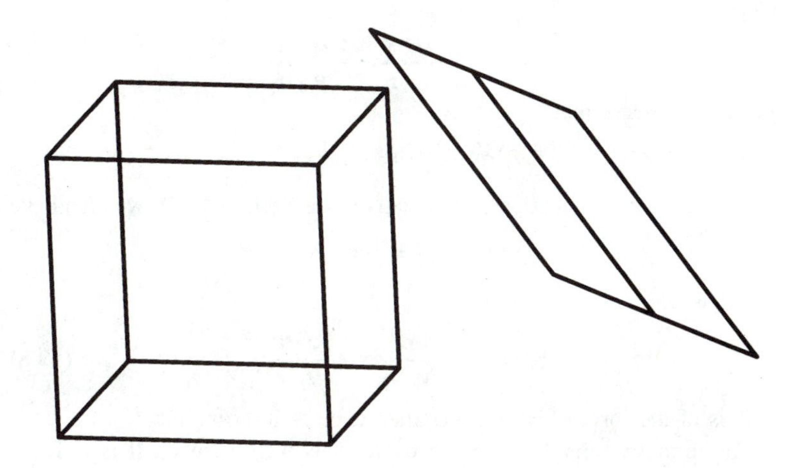

Figure 13.7.
Projections: a cube projected in an arbitrary plane with an oblique angle.

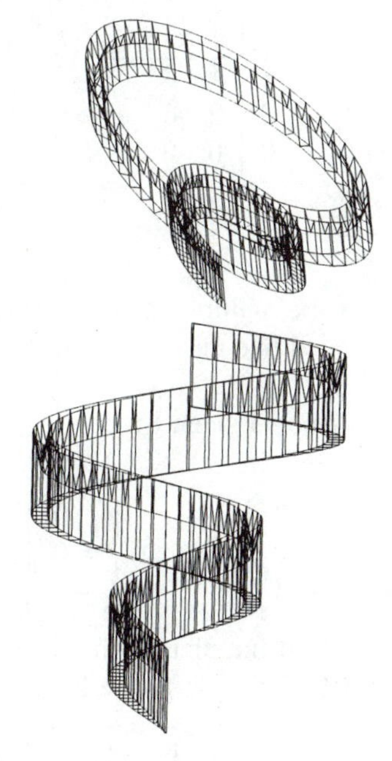

Figure 13.8.
Projections: a helix projected in an arbitrary plane with an oblique angle.

We know that we may write dot products in matrix form (see Section 4.11):

$$\mathbf{x}' = \mathbf{x} - \frac{\mathbf{n}^{\mathrm{T}}\mathbf{x}}{\mathbf{v}\cdot\mathbf{n}}\mathbf{v} + \frac{\mathbf{q}\cdot\mathbf{n}}{\mathbf{v}\cdot\mathbf{n}}\mathbf{v}.$$

Next, we observe that

$$[\mathbf{n}^{\mathrm{T}}\cdot\mathbf{x}]\mathbf{v} = \mathbf{v}[\mathbf{n}^{\mathrm{T}}\mathbf{x}].$$

Since matrix multiplication is associative (see Section 4.12), we also have

$$\mathbf{v}[\mathbf{n}^{\mathrm{T}}\mathbf{x}] = [\mathbf{v}\mathbf{n}^{\mathrm{T}}]\mathbf{x},$$

and thus

$$\mathbf{x}' = [I - \frac{\mathbf{v}\mathbf{n}^{\mathrm{T}}}{\mathbf{v}\cdot\mathbf{n}}]\mathbf{x} + \frac{\mathbf{q}\cdot\mathbf{n}}{\mathbf{v}\cdot\mathbf{n}}\mathbf{v}. \tag{13.8}$$

This is of the form $\mathbf{x}' = A\mathbf{x} + \mathbf{p}$ and hence is an affine map![2]

The term $\mathbf{v}\mathbf{n}^{\mathrm{T}}$ might appear odd, yet it is well-defined. It is a 3×3 matrix, as in the following example.

Example 13.3

	1	3	3
1	1	3	3
2	2	6	6
0	0	0	0

All rows of this matrix are multiples of each other; so are all columns. Matrices which are generated like this are called *dyadic*; their rank is one.

Example 13.4 Let a plane be given by $x_1 + x_2 + x_3 - 1 = 0$, a point $\mathbf{x}$ and a direction $\mathbf{v}$ by

$$\mathbf{x} = \begin{bmatrix} 3 \\ 2 \\ 4 \end{bmatrix}, \quad \mathbf{v} = \begin{bmatrix} 0 \\ 0 \\ -1 \end{bmatrix}.$$

If we project $\mathbf{x}$ along $\mathbf{v}$ onto the plane, what is $\mathbf{x}'$? First, we need the plane's normalized normal. Calling it $\mathbf{n}$, we have

$$\mathbf{n} = \frac{1}{\sqrt{3}}\begin{bmatrix} 1 \\ 1 \\ 1 \end{bmatrix}.$$

[2]Technically we should add the origin in order for $\mathbf{p}$ to be a point.

Now choose a point $\mathbf{q}$ in the plane. Let's choose

$$\mathbf{q} = \begin{bmatrix} 1 \\ 0 \\ 0 \end{bmatrix}$$

for simplicity. Now we are ready to calculate the quantities in (13.8):

$$\mathbf{v} \cdot \mathbf{n} = -1/\sqrt{3},$$

$$\frac{\mathbf{v}\mathbf{n}^{\mathrm{T}}}{-1/\sqrt{3}} = \begin{array}{r|rrr} & 1 & 1 & 1 \\ \hline 0 & 0 & 0 & 0 \\ 0 & 0 & 0 & 0 \\ -1 & 1 & 1 & 1 \end{array},$$

$$\frac{\mathbf{q} \cdot \mathbf{n}}{\mathbf{v} \cdot \mathbf{n}} \mathbf{v} = \begin{bmatrix} 0 \\ 0 \\ 1 \end{bmatrix}.$$

Putting all the pieces together:

$$\mathbf{x}' = \left(I - \begin{bmatrix} 0 & 0 & 0 \\ 0 & 0 & 0 \\ 1 & 1 & 1 \end{bmatrix} \right) \begin{bmatrix} 3 \\ 2 \\ 4 \end{bmatrix} + \begin{bmatrix} 0 \\ 0 \\ 1 \end{bmatrix} = \begin{bmatrix} 3 \\ 2 \\ -4 \end{bmatrix}.$$

Just to double check, enter $\mathbf{x}'$ into the plane equation

$$3 + 2 - 4 - 1 = 0,$$

and we see that

$$\begin{bmatrix} 3 \\ 2 \\ 4 \end{bmatrix} = \begin{bmatrix} 3 \\ 2 \\ -4 \end{bmatrix} + 8 \begin{bmatrix} 0 \\ 0 \\ 1 \end{bmatrix},$$

which together verify that this is the correct point.

Sketch 132 should convince you that this is indeed the correct answer.

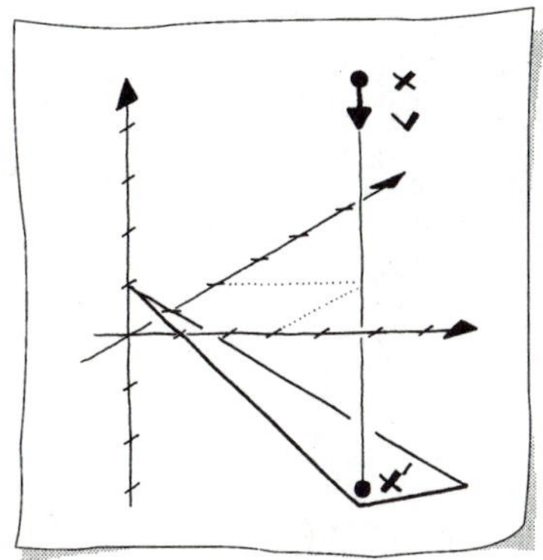

Sketch 132
A projection example.

Which of the two possibilities, (13.7) or the affine map (13.8) should you use? Clearly (13.7) is more straightforward and less involved. Yet in some computer graphics or CAD system environments, it may be desirable to have all maps in a unified format, i.e., $A\mathbf{x} + \mathbf{p}$.

13.5 Homogeneous Coordinates and Perspective Maps

There is a way to condense the form $\mathbf{x}' = A\mathbf{x} + \mathbf{p}$ of an affine map into just one matrix multiplication

$$\underline{\mathbf{x}}' = M\underline{\mathbf{x}}. \tag{13.9}$$

This is achieved by setting

$$M = \begin{bmatrix} a_{1,1} & a_{1,2} & a_{1,3} & p_1 \\ a_{2,1} & a_{2,2} & a_{2,3} & p_2 \\ a_{3,1} & a_{3,2} & a_{3,3} & p_3 \\ 0 & 0 & 0 & 1 \end{bmatrix}$$

and

$$\underline{\mathbf{x}} = \begin{bmatrix} x_1 \\ x_2 \\ x_3 \\ 1 \end{bmatrix}, \quad \underline{\mathbf{x}}' = \begin{bmatrix} x'_1 \\ x'_2 \\ x'_3 \\ 1 \end{bmatrix}.$$

The 4D point $\underline{\mathbf{x}}$ is called the *homogeneous form* of the affine point $\mathbf{x}$. You should verify for yourself that (13.9) is indeed the same affine map as before!

The homogeneous form is more general than just adding a fourth coordinate $x_4 = 1$ to a point. If, perhaps as the result of some computation, the fourth coordinate does not equal one, one gets from the homogeneous point $\underline{\mathbf{x}}$ to its affine counterpart $\mathbf{x}$ by dividing through by x_4. Thus one affine point has infinitely many homogeneous representations!

Example 13.5 (The symbol $\approx$ should be read "corresponds to".)

$$\begin{bmatrix} 1 \\ -1 \\ 3 \end{bmatrix} \approx \begin{bmatrix} 10 \\ -10 \\ 30 \\ 10 \end{bmatrix} \approx \begin{bmatrix} -2 \\ 2 \\ -6 \\ -2 \end{bmatrix}.$$

This example shows two homogeneous representations of one affine point.

Using the homogeneous matrix form of (13.9), the matrix M for the point into a plane projection from (13.8) becomes

$$
\left[\begin{array}{ccc|c}
\begin{bmatrix} \mathbf{v}\cdot\mathbf{n} & 0 & 0 \\ 0 & \mathbf{v}\cdot\mathbf{n} & 0 \\ 0 & 0 & \mathbf{v}\cdot\mathbf{n} \end{bmatrix} - \mathbf{v}\mathbf{n}^{\mathrm{T}} & & & (\mathbf{q}\cdot\mathbf{n})\mathbf{v} \\
\hline
0 \quad 0 \quad 0 & & & \mathbf{v}\cdot\mathbf{n}
\end{array}\right].
$$

Here, the element $m_{4,4} = \mathbf{v}\cdot\mathbf{n}$. Thus $\underline{x}_4 = \mathbf{v}\cdot\mathbf{n}$, and we will have to divide $\underline{\mathbf{x}}$'s coordinates by $\underline{x}_4$ in order to obtain the corresponding affine point.

A simple change in our equations will lead us from parallel projections onto a plane to *perspective projections*. Instead of using a constant direction $\mathbf{v}$ for all projections, now the direction depends on the point $\mathbf{x}$. More precisely, let it be the line from $\mathbf{x}$ to the origin of our coordinate system. Then, as shown in Sketch 133, $\mathbf{v} = -\mathbf{x}$, and (13.7) becomes

$$\mathbf{x}' = \mathbf{x} + \frac{[\mathbf{q}-\mathbf{x}]\cdot\mathbf{n}}{\mathbf{x}\cdot\mathbf{n}}\mathbf{x},$$

which quickly simplifies to

$$\mathbf{x}' = \frac{\mathbf{q}\cdot\mathbf{n}}{\mathbf{x}\cdot\mathbf{n}}\mathbf{x}. \tag{13.10}$$

In homogeneous form, this is described by the following matrix

$$
M: \left[\begin{array}{ccc|c}
I[\mathbf{q}\cdot\mathbf{n}] & & & \mathbf{o} \\
\hline
0 & 0 & 0 & \mathbf{x}\cdot\mathbf{n}
\end{array}\right].
$$

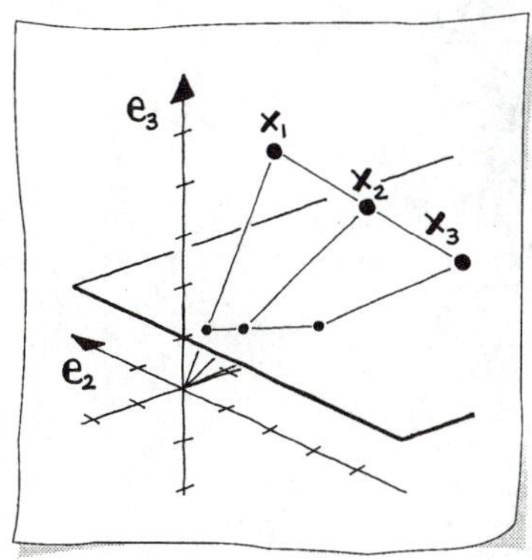

Sketch 133
Perspective projection.

Perspective projections are not affine maps anymore! To see this, a simple example will suffice.

Example 13.6 Take the plane $x_3 = 1$; let

$$\mathbf{q} = \begin{bmatrix} 0 \\ 0 \\ 1 \end{bmatrix}$$

be a point on the plane. Now $\mathbf{q}\cdot\mathbf{n} = 1$ and $\mathbf{x}\cdot\mathbf{n} = x_3$, resulting in the map

$$\mathbf{x}' = \frac{1}{x_3}\mathbf{x}.$$

Figure 13.9.
Parallel projection: A 3D helix and two orthographic projections on the left and bottom walls of the bounding cube — not visible due to the orthographic projection used for the whole scene.

Take the three points

$$\mathbf{x}_1 = \begin{bmatrix} 2 \\ 0 \\ 4 \end{bmatrix}, \quad \mathbf{x}_2 = \begin{bmatrix} 3 \\ -1 \\ 3 \end{bmatrix}, \quad \mathbf{x}_3 = \begin{bmatrix} 4 \\ -2 \\ 2 \end{bmatrix}.$$

This example is illustrated in Sketch 133. Note that $\mathbf{x}_2 = \frac{1}{2}\mathbf{x}_1 + \frac{1}{2}\mathbf{x}_3$, i.e., $\mathbf{x}_2$ is the midpoint of $\mathbf{x}_1$ and $\mathbf{x}_3$.

Their images are

$$\mathbf{x}'_1 = \begin{bmatrix} 1/2 \\ 0 \\ 1 \end{bmatrix}, \quad \mathbf{x}'_2 = \begin{bmatrix} 1 \\ -1/3 \\ 1 \end{bmatrix}, \quad \mathbf{x}'_3 = \begin{bmatrix} 2 \\ -1 \\ 1 \end{bmatrix}.$$

The perspective map destroyed the midpoint relation! For now $\mathbf{x}'_2 = \frac{2}{3}\mathbf{x}'_1 + \frac{1}{3}\mathbf{x}'_3$.

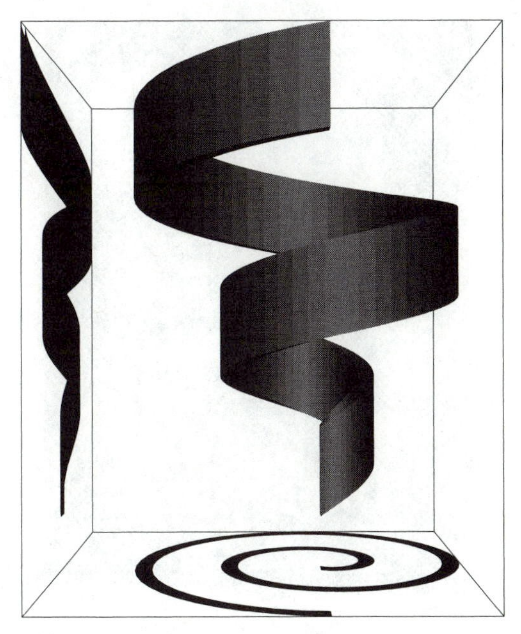

Figure 13.10.
Perspective projection: A 3D helix and two orthographic projections on the left and bottom walls of the bounding cube — visible due to the perspective projection used for the whole scene.

Thus the ratio of three points is changed by perspective maps. As a consequence, two parallel lines will not be mapped to parallel lines. Because of this effect, perspective maps are a good model for how we perceive 3D space around us. Parallel lines do seemingly intersect in a distance, and are thus not *perceived* as being parallel! Figure 13.9 is a parallel projection and Figure 13.10 illustrates the same geometry with a perspective projection. Notice in the perspective image, the sides of the bounding cube that move into the page are no longer parallel.

The study of perspective goes back to the fourteenth century — before that, artists simply could not draw realistic 3D images. One of the foremost researchers in the area of perspective maps was A. Dürer; see Figure 13.11 for one of his exploratory drawings.[3]

[3]From *The Complete Woodcuts of Albrecht Dürer,* edited by W. Durth, Dover Publications Inc., New York, 1963. A website with material on Dürer: http://www.bilkent.edu.tr/wm/paint/auth/durer/.

Figure 13.11.
Perspective maps: a woodcut by A. Dürer.

13.6 Exercises

We'll use four points

$$\mathbf{x}_1 = \begin{bmatrix} 1 \\ 0 \\ 0 \end{bmatrix}, \quad \mathbf{x}_2 = \begin{bmatrix} 0 \\ 1 \\ 0 \end{bmatrix}, \quad \mathbf{x}_3 = \begin{bmatrix} 0 \\ 0 \\ -1 \end{bmatrix}, \quad \mathbf{x}_4 = \begin{bmatrix} 0 \\ 0 \\ 1 \end{bmatrix}$$

and four points

$$\mathbf{y}_1 = \begin{bmatrix} -1 \\ 0 \\ 0 \end{bmatrix}, \quad \mathbf{y}_2 = \begin{bmatrix} 0 \\ -1 \\ 0 \end{bmatrix}, \quad \mathbf{y}_3 = \begin{bmatrix} 0 \\ 0 \\ -1 \end{bmatrix}, \quad \mathbf{y}_4 = \begin{bmatrix} 0 \\ 0 \\ 1 \end{bmatrix},$$

and also the plane through $\mathbf{q}$ with normal $\mathbf{n}$:

$$\mathbf{q} = \begin{bmatrix} 1 \\ 0 \\ 0 \end{bmatrix}, \quad \mathbf{n} = \frac{1}{5}\begin{bmatrix} 3 \\ 0 \\ 4 \end{bmatrix}.$$

1. Using a direction

$$\mathbf{v} = \frac{1}{4}\begin{bmatrix} 2 \\ 0 \\ 2 \end{bmatrix},$$

what are the images of the $\mathbf{x}_i$ when projected onto the plane with this direction?

2. Using the same $\mathbf{v}$ as in the previous problem, what are the images of the $\mathbf{y}_i$?
3. What are the images of the $\mathbf{x}_i$ when projected onto the plane by a perspective projection through the origin?
4. What are the images of the $\mathbf{y}_i$ when projected onto the plane by a perspective projection through the origin?
5. Compute the centroid $\mathbf{c}$ of the $\mathbf{x}_i$ and then the centroid $\mathbf{c}'$ of their perspective images (previous Exercise). Is $\mathbf{c}'$ the image of $\mathbf{c}$ under the perspective map?
6. An affine map $\mathbf{x}_i \rightarrow \mathbf{y}_i; i = 1, 2, 3, 4$ is uniquely defined. What is it?
7. What is the image of

$$\mathbf{p} = \begin{bmatrix} 1 \\ 1 \\ 1 \end{bmatrix}$$

under the map from the previous problem? Use two ways to compute it.
8. What are the geometric properties of the affine map from the last two problems?
9. We claimed that (13.8) reduces to (13.10). This necessitates that

$$[I - \frac{\mathbf{v}\mathbf{n}^{\mathrm{T}}}{\mathbf{n} \cdot \mathbf{v}}]\mathbf{x} = \mathbf{0}.$$

Show that this is indeed true.

General Linear Systems

14

Figure 14.1.
Linear systems: a shoe last is constructed that fits data points extracted by a coordinate measuring machine from a physical model.

In Chapter 5, we studied linear systems of two equations in two unknowns. A whole chapter for such a humble task seems like a bit of overkill — its main purpose was really to lay the groundwork for this chapter.

Linear systems arise in virtually every area of science and engineering — some are as big as 1,000,000 equations in as many unknowns. Such huge systems require more sophisticated treatment than the methods introduced here. They *will* allow you to solve systems with several thousand equations without a problem. Figure 14.1 illustrates a surface fitting problem that necessitated solving a system with approximately 300 equations.

This chapter explains the basic ideas underlying linear systems. Readers eager for hands-on experience should download linear system solvers from the net. The most prominent collection of routines is LINPACK, which was written in FORTRAN. The new generation of LINPACK is CLAPACK, and is written in C and can be found at http://www.netlib.org/clapack/.[1]

14.1 The Problem

A linear system is a set of equations like this:

$$\begin{aligned} 3u_1 - 2u_2 - 10u_3 + u_4 &= 0 \\ u_1 - u_3 &= 4 \\ u_1 + u_2 - 2u_3 + 3u_4 &= 1 \\ u_2 + 2u_4 &= -4. \end{aligned}$$

The unknowns are the numbers $u_1, \ldots, u_4$. There are as many equations as there are unknowns, four in this example.

We rewrite this system in matrix form:

$$\begin{bmatrix} 3 & -2 & -10 & 1 \\ 1 & 0 & -1 & 0 \\ 1 & 1 & -2 & 3 \\ 0 & 1 & 0 & 2 \end{bmatrix} \begin{bmatrix} u_1 \\ u_2 \\ u_3 \\ u_4 \end{bmatrix} = \begin{bmatrix} 0 \\ 4 \\ 1 \\ -4 \end{bmatrix}$$

Our example was a 4×4 linear system. A general, $n \times n$ linear system looks like this:

$$\begin{aligned} a_{1,1}u_1 + a_{1,2}u_2 + \ldots + a_{1,n}u_n &= b_1 \\ a_{2,1}u_1 + a_{2,2}u_2 + \ldots + a_{2,n}u_n &= b_2 \\ &\vdots \\ a_{n,1}u_1 + a_{n,2}u_2 + \ldots + a_{n,n}u_n &= b_n. \end{aligned}$$

In matrix form, it becomes

$$\begin{bmatrix} a_{1,1} & a_{1,2} & \ldots & a_{1,n} \\ a_{2,1} & a_{2,2} & \ldots & a_{2,n} \\ & & \vdots & \\ a_{n,1} & a_{n,2} & \ldots & a_{n,n} \end{bmatrix} \begin{bmatrix} u_1 \\ u_2 \\ \vdots \\ u_n \end{bmatrix} = \begin{bmatrix} b_1 \\ b_2 \\ \vdots \\ b_n \end{bmatrix}, \quad (14.1)$$

or even shorter

$$A\mathbf{u} = \mathbf{b}.$$

[1]For more scientific computing software, go to http://www.netlib.org/.

The *coefficient matrix* A has n rows and n columns. For example, the first row is

$$a_{1,1}, a_{1,2}, \ldots, a_{1,n},$$

and the second column is

$$\begin{matrix} a_{1,2} \\ a_{2,2} \\ \vdots \\ a_{n,2}. \end{matrix}$$

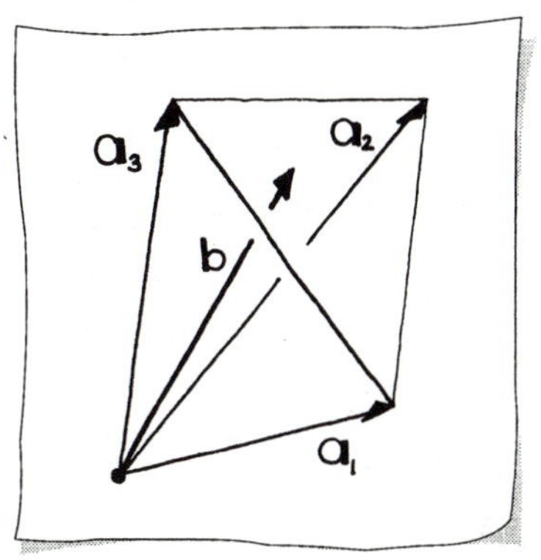

Sketch 134
A solvable 3×3 system.

Equation (14.1) is a compact way of writing n equations for the n unknowns $u_1, \ldots, u_n$. In the 2×2 case, such systems had nice geometric interpretations; in the general case, that interpretation needs n-dimensional linear spaces, and is not very intuitive. Still the methods that we developed for the 2×2 case can be gainfully employed here!

General linear systems defy geometric intuition, yet some underlying principles are of a geometric nature. It is best explained for the example $n = 3$. We are given a vector $\mathbf{b}$ and we try to write it as a linear combination of vectors $\mathbf{a}_1, \mathbf{a}_2, \mathbf{a}_3$. If the $\mathbf{a}_i$ are truly 3D, i.e., if they form a tetrahedron, then a unique solution may be found (see Sketch 134).

But if the three $\mathbf{a}_i$ all lie in a plane (i.e., if the volume formed by them is zero), then you cannot write $\mathbf{b}$ as a linear combination of them, unless it is itself in that 2D plane. In this case, you cannot expect uniqueness for your answer. Sketch 135 covers these cases.

In general, a linear system is uniquely solvable if the $\mathbf{a}_i$ have a nonzero n-dimensional volume. If they do not, they span a k-dimensional *subspace* (with $k < n$) — nonunique solutions only exist if $\mathbf{b}$ is itself in that subspace. A linear system is called *consistent* if at least one solution exits.

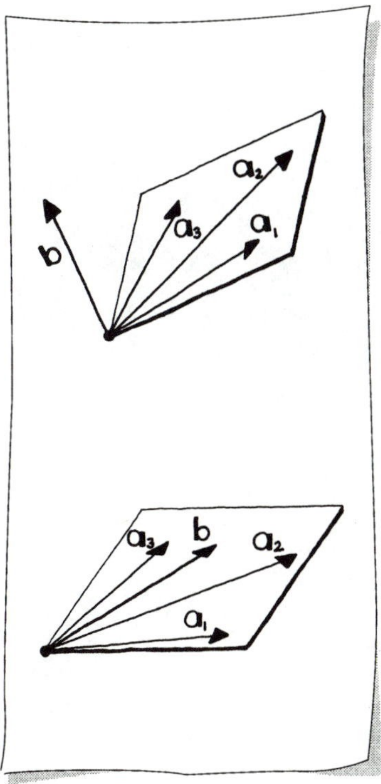

Sketch 135
Top: no solution;
bottom: nonunique solution.

14.2 The Solution via Gauss Elimination

The key to success in the 2×2 case was the application of a shear so that the matrix A was transformed to *upper triangular*. Then it was possible to apply *back substitution* to solve for the unknowns. The shear was constructed to map the first column vector of the matrix onto the $\mathbf{e}_1$-axis. Revisiting an example from Chapter 5, we have

$$\begin{bmatrix} 2 & 4 \\ 1 & 6 \end{bmatrix} \begin{bmatrix} u_1 \\ u_2 \end{bmatrix} = \begin{bmatrix} 4 \\ 4 \end{bmatrix}.$$

The shear used was

$$S_1 = \begin{bmatrix} 1 & 0 \\ -1/2 & 1 \end{bmatrix},$$

which when applied to the system as

$$S_1 A\mathbf{u} = S_1 \mathbf{b}$$

produced the system

$$\begin{bmatrix} 2 & 4 \\ 0 & 4 \end{bmatrix} \begin{bmatrix} u_1 \\ u_2 \end{bmatrix} = \begin{bmatrix} 4 \\ 2 \end{bmatrix}.$$

Algebraically, what this shear did was to change the *rows* of the system in the following manner:

$$\text{row}_1 \leftarrow \text{row}_1 \quad \text{and} \quad \text{row}_2 \leftarrow \text{row}_2 - \frac{1}{2}\text{row}_1.$$

Each of these are an *elementary row operation.*

Back substitution came next, with

$$u_2 = \frac{1}{4} \times 2 = \frac{1}{2}$$

and then

$$u_1 = \frac{1}{2}(4 - 4u_2) = 1.$$

The divisions in the back substitution equations are actually scalings, thus they could be re-written in terms of a scale matrix:

$$S_2 = \begin{bmatrix} 1/2 & 0 \\ 0 & 1/4 \end{bmatrix},$$

and then the system would be transformed via

$$S_2 S_1 A\mathbf{u} = S_2 S_1 \mathbf{b}.$$

The corresponding upper triangular matrix and system is

$$\begin{bmatrix} 1 & 2 \\ 0 & 1 \end{bmatrix} \begin{bmatrix} u_1 \\ u_2 \end{bmatrix} = \begin{bmatrix} 2 \\ 1/2 \end{bmatrix}.$$

Check for yourself that we get the same result.

Thus we see the geometric steps for solving a linear system have methodical algebraic interpretations. This algebraic approach is what will be followed for the rest of the chapter. For general linear systems the matrices, such as S_1 and S_2 above, are not actually constructed due to the speed and storage expense. Notice that the shear to zero one element in the matrix, only changed the elements in one row, thus it is unnecessary to manipulate the other row. This is an important observation for large systems.

In the general case, as in the 2×2 case, *pivoting* will be used. Recall for the 2×2 case this meant that the equations were reordered such that the matrix element $a_{1,1}$ is the largest one in the first column.

Example 14.1 Let's step through the necessary shears for a 3×3 linear system. The goal is to get it in upper triangular form so we may use back substitution to solve for the unknowns. The system is

$$\begin{bmatrix} 2 & -2 & 0 \\ 4 & 0 & -2 \\ 4 & 2 & -4 \end{bmatrix} \begin{bmatrix} u_1 \\ u_2 \\ u_3 \end{bmatrix} = \begin{bmatrix} 4 \\ -2 \\ 0 \end{bmatrix}.$$

The matrix element $a_{1,1}$ is not the largest in the first column, so we reorder:

$$\begin{bmatrix} 4 & 0 & -2 \\ 2 & -2 & 0 \\ 4 & 2 & -4 \end{bmatrix} \begin{bmatrix} u_1 \\ u_2 \\ u_3 \end{bmatrix} = \begin{bmatrix} -2 \\ 4 \\ 0 \end{bmatrix}.$$

To zero entries in the first column apply:

$$\mathrm{row}_2 \leftarrow \mathrm{row}_2 - \frac{1}{2}\mathrm{row}_1$$

$$\mathrm{row}_3 \leftarrow \mathrm{row}_3 - \mathrm{row}_1,$$

and the system becomes

$$\begin{bmatrix} 4 & 0 & -2 \\ 0 & -2 & 1 \\ 0 & 2 & -2 \end{bmatrix} \begin{bmatrix} u_1 \\ u_2 \\ u_3 \end{bmatrix} = \begin{bmatrix} -2 \\ 5 \\ 2 \end{bmatrix}.$$

Now the first column consists of only zeroes except for $a_{1,1}$, meaning that it is lined up with the $\mathbf{e}_1$–axis.

Now work on the second column vector. First check if pivoting is necessary; this means checking that $a_{2,2}$ is the largest in absolute value of all values in the second column that are below the diagonal. No pivoting is necessary. To zero the last element in this vector apply

$$\mathrm{row}_3 \leftarrow \mathrm{row}_3 + \mathrm{row}_2,$$

which produces

$$\begin{bmatrix} 4 & 0 & -2 \\ 0 & -2 & 1 \\ 0 & 0 & -1 \end{bmatrix} \begin{bmatrix} u_1 \\ u_2 \\ u_3 \end{bmatrix} = \begin{bmatrix} -2 \\ 5 \\ 7 \end{bmatrix}.$$

By chance, the second column is aligned with $\mathbf{e}_2$ because $a_{1,2} = 0$. If this extra zero had not appeared, then we would have mapped this 3D vector into the $[\mathbf{e}_1, \mathbf{e}_2]$-plane.

Now we are ready for back substitution:

$$\begin{aligned} u_3 &= \frac{1}{-1}(7) \\ u_2 &= \frac{1}{-2}(5 - u_3) \\ u_1 &= \frac{1}{4}(-2 + 2u_3). \end{aligned}$$

This implicitly incorporates a scaling matrix. We obtain the solution

$$\begin{bmatrix} u_1 \\ u_2 \\ u_3 \end{bmatrix} = \begin{bmatrix} -4 \\ -6 \\ -7 \end{bmatrix}.$$

It is usually a good idea to insert the solution into the original equations:

$$\begin{bmatrix} 2 & -2 & 0 \\ 4 & 0 & -2 \\ 4 & 2 & -4 \end{bmatrix} \begin{bmatrix} -4 \\ -6 \\ -7 \end{bmatrix} = \begin{bmatrix} 4 \\ -2 \\ 0 \end{bmatrix}.$$

It works!

Here is the algorithm for solving a general $n \times n$ system of linear equations:

Given: a coefficient matrix A and a right-hand side $\mathbf{b}$ describing a linear system

$$A\mathbf{u} = \mathbf{b},$$

which is short for the more detailed (14.1).

Wanted: the unknowns $u_1, \ldots, u_n$.

Algorithm:

For $j = 1, \ldots, n-1$: (note: j counts columns)

Pivoting step:

Find element in largest absolute value in column j that is below $a_{j,j}$: call it $a_{r,j}$.

Exchange equations r and j.

If $a_{j,j} = 0$, the system is not solvable.

Elimination step for column j:
For $i = j+1, \dots, n$: (elements below diagonal of column j)
Construct the *multiplier* $g_{i,j} = a_{i,j}/a_{j,j}$
For $k = j+1, \dots, n$ (each element in row i after column j)
$a_{i,k} = a_{i,k} - g_{i,j}a_{k,j}$
$b_i = b_i - g_{i,j}b_j$

After this loop, all elements below the diagonal have been set to zero – the matrix is now in *upper triangular* form. We call this transformed matrix $A^{\triangle}$.

Back substitution:
$u_n = b_n/a_{n,n}$
For $j = n-1, \dots, 1$
$u_j = \frac{1}{a_{j,j}}[b_j - a_{j,j+1}u_{j+1} - \dots - a_{j,n}u_n]$.

In a programming environment, it can be convenient to form an *augmented matrix* which is the matrix A augmented with the vector $\mathbf{b}$. Here is the idea for a 3×3 linear system:

$$\begin{bmatrix} a_{1,1} & a_{1,2} & a_{1,3} & b_1 \\ a_{2,1} & a_{2,2} & a_{2,3} & b_2 \\ a_{3,1} & a_{3,2} & a_{3,3} & b_3 \end{bmatrix}.$$

Then the k steps would run to $n+1$, and there would be no need for the extra line for the b_i element.

The operations in the elimination step above may also be written in matrix form. If A is the current matrix, then at step j, to produce zeroes under $a_{j,j}$ the matrix product G_jA is formed, where

$$G_j = \begin{bmatrix} 1 & & & & & & \\ & \ddots & & & & & \\ & & 1 & & & & \\ & & & 1 & & & \\ & & & -g_{j+1,j} & 1 & & \\ & & & \vdots & & \ddots & \\ & & & -g_{n,j} & & & 1 \end{bmatrix}. \qquad (14.2)$$

The elements $g_{i,j}$ of G_j are the multipliers. The matrix G_j is called a *Gauss matrix*. All entries except for the diagonal and the entries $g_{i,j}$ are zero. As we see with the algorithm, it is not necessary to explicitly form the Gauss matrix. In fact, it is more efficient with regard to speed and storage not to. Using the Gauss matrix blindly would result in many unnecessary calculations. In a programming environment, notice that it is possible to store the $g_{i,j}$ in the zero elements of A!

Example 14.2 We look at one more example, taken from [2]. Let the system be given by

$$\begin{bmatrix} 2 & 2 & 0 \\ 1 & 1 & 2 \\ 2 & 1 & 1 \end{bmatrix} \begin{bmatrix} u_1 \\ u_2 \\ u_3 \end{bmatrix} = \begin{bmatrix} 6 \\ 9 \\ 7 \end{bmatrix}$$

We start the algorithm with $j = 1$, and observe that no element in column 1 exceeds $a_{1,1}$ in absolute value – no pivoting is necessary at this step. Proceed with the elimination step for row 2 by constructing the multiplier

$$g_{2,1} = a_{2,1}/a_{1,1} = 1/2.$$

Change row 2 as follows:

$$\text{row}_2 \leftarrow \text{row}_2 - 1/2\text{row}_1.$$

Remember, this includes changing the element b_2. Similarly for row 3:

$$g_{3,1} = a_{3,1}/a_{1,1} = 2/2 = 1$$

then

$$\text{row}_3 \leftarrow \text{row}_3 - \text{row}_1.$$

Step $j = 1$ is complete and the linear system is now

$$\begin{bmatrix} 2 & 2 & 0 \\ 0 & 0 & 2 \\ 0 & -1 & 1 \end{bmatrix} \begin{bmatrix} u_1 \\ u_2 \\ u_3 \end{bmatrix} = \begin{bmatrix} 6 \\ 6 \\ 1 \end{bmatrix}.$$

Next is column 2, so $j = 2$. Observe that $a_{2,2} = 0$, whereas $a_{3,2} = -1$. We exchange equations 2 and 3 and the system becomes

$$\begin{bmatrix} 2 & 2 & 0 \\ 0 & -1 & 1 \\ 0 & 0 & 2 \end{bmatrix} \begin{bmatrix} u_1 \\ u_2 \\ u_3 \end{bmatrix} = \begin{bmatrix} 6 \\ 1 \\ 6 \end{bmatrix}.$$

If blindly following the algorithm above, we would proceed with the elimination for row 3 by forming the multiplier

$$g_{3,2} = a_{3,2}/a_{2,2} = 0/-1 = 0.$$

Then operate on the third row

$$\text{row}_3 \leftarrow \text{row}_3 - 0 \times \text{row}_2,$$

which doesn't change the row at all — ignoring numerical instabilities. Without putting a special check for a zero multiplier, this unnecessary work takes place. Tolerances are very important here.

Apply back substitution by first solving for the last unknown:

$$u_3 = 3.$$

Start the back substitution loop with $j = 2$:

$$u_2 = \frac{1}{-1}[1 - u_3] = 2,$$

and finally

$$u_1 = \frac{1}{2}[6 - 2u_2] = 1.$$

14.3 Determinants

When we take A to upper triangular form $A^{\triangle}$ we apply a sequence of shears to its initial column vectors. Shears do not change volumes; thus the row vectors of $A^{\triangle}$ span the same volume as did those of A. This volume is given by the product of the diagonal entries and is called the *determinant* of A:

$$\det A = a^{\triangle}_{1,1} \times \ldots \times a^{\triangle}_{n,n}.$$

In general, this is the best (and most stable) method for finding the determinant.

In the special case of a 3×3 matrix, it is convenient to write out the determinant explicitly. The formula is nearly impossible to remember, but the following trick is not. Copy the first two columns after the last column. Then form the product of the three "diagonals" and add them. Then, form the product of the three "anti-diagonals" and subtract them. The three "plus" products may be written as:

$$\begin{array}{ccccc} a_{1,1} & a_{1,2} & a_{1,3} & \square & \square \\ \square & a_{2,2} & a_{2,3} & a_{2,1} & \square \\ \square & \square & a_{3,3} & a_{3,1} & a_{3,2} \end{array}$$

and the three "minus" products as:

$$\begin{array}{ccccc} \square & \square & a_{1,3} & a_{1,1} & a_{1,2} \\ \square & a_{2,2} & a_{2,3} & a_{2,1} & \square \\ a_{3,1} & a_{3,2} & a_{3,3} & \square & \square \end{array}$$

The complete formula for the 3×3 determinant is

$$\begin{aligned}\det A = \quad & a_{1,1}a_{2,2}a_{3,3} + a_{1,2}a_{2,3}a_{3,1} + a_{1,3}a_{2,1}a_{3,2} \\ & - a_{3,1}a_{2,2}a_{1,3} - a_{3,2}a_{2,3}a_{1,1} - a_{3,3}a_{2,1}a_{1,2}\end{aligned}$$

Example 14.3 What is the volume spanned by the three vectors

$$\mathbf{a}_1 = \begin{bmatrix} 4 \\ 0 \\ 0 \end{bmatrix}, \quad \mathbf{a}_2 = \begin{bmatrix} -1 \\ 4 \\ 4 \end{bmatrix}, \quad \mathbf{a}_3 = \begin{bmatrix} 0.1 \\ -0.1 \\ 0.1 \end{bmatrix}?$$

All we have to do is to compute

$$\det[\mathbf{a}_1, \mathbf{a}_2, \mathbf{a}_3] = 4 \times 4 \times 0.1 - (-0.1) \times 4 \times 4 = 3.2.$$

In this computation, we did not write down zero terms.

For a more geometric derivation of 3×3 determinants, see Section 10.5.

14.4 Iterative Methods

In applications such as Finite Element Methods (FEM) in the context of the solution of fluid flow problems, scientists are faced with linear systems with many thousand equations. The above method would work, but would be far too slow. Typically, huge linear systems have one advantage: the coefficient matrix only has ten or fewer nonzero entries per row. Thus a 100,000 by 100,000 system would only have 1,000,000 nonzero entries, compared to 10,000,000,000 matrix elements!

In these cases, one does not store the whole matrix, but only its nonzero entries, together with their i, j location.

The solution to such systems is often obtained by *iterative methods*.

Example 14.4 Let the system (taken from Johnson and Riess [15]) be given by

$$\begin{bmatrix} 4 & 1 & 0 \\ 2 & 5 & 1 \\ -1 & 2 & 4 \end{bmatrix} \begin{bmatrix} u_1 \\ u_2 \\ u_3 \end{bmatrix} = \begin{bmatrix} 1 \\ 0 \\ 3 \end{bmatrix}.$$

An iterative method starts from a guess for the solution and then refines it until it *is* the solution. Let's take $u_1^{(1)} = u_2^{(1)} = u_3^{(1)} = 1$ for our first guess. We call this vector $\mathbf{u}^{(1)}$ and note that it clearly is not the solution to our system: $A\mathbf{u}^{(1)} \neq \mathbf{b}$.

A better guess ought to be obtained by using the current guess and solving the first equation for a new $u_1^{(2)}$, the second for a new $u_2^{(2)}$, and so on. This gives us

$$\begin{aligned} 4u_1^{(2)} + 1 &= 1 \\ 2 + 5u_2^{(2)} + 1 &= 0 \\ -1 + 2 + 4u_3^{(2)} &= 3 \end{aligned}$$

and thus

$$\begin{bmatrix} u_1^{(2)} \\ u_2^{(2)} \\ u_3^{(2)} \end{bmatrix} = \begin{bmatrix} 0 \\ -0.6 \\ 0.5 \end{bmatrix}.$$

The next iteration becomes

$$\begin{aligned} 4u_1^{(3)} - 0.6 &= 1 \\ 5u_2^{(3)} + 0.5 &= 0 \\ 1.2 + 4u_3^{(3)} &= 3 \end{aligned}$$

and thus

$$\begin{bmatrix} u_1^{(3)} \\ u_2^{(3)} \\ u_3^{(3)} \end{bmatrix} = \begin{bmatrix} 0.4 \\ -0.1 \\ 1.05 \end{bmatrix}.$$

After a few more iterations, we will be close enough to the true solution

$$\mathbf{u} = \begin{bmatrix} 0.333 \\ -0.333 \\ 1.0 \end{bmatrix}.$$

Try one more for yourself.

This iterative method is known as *Gauss-Jacobi iteration.* It always converges if the coefficient matrix is *diagonally dominant.* A matrix has this property if for every row, the sum of off-diagonal elements (in absolute value) is less than the diagonal element:

$$|a_{i,1}| + \ldots + |a_{i,i-1}| + |a_{i,i+1}| + \ldots + |a_{i,n}| < |a_{i,i}|.$$

In the above example, we were dealing with a diagonally dominant matrix.

14.5 Overdetermined Systems

Sometimes systems arise that have more equations than unknowns, such as

$$\begin{aligned} 4u_1 + u_2 &= 1 \\ -u_1 + 4u_2 + 2u_3 &= -1 \\ 4u_3 &= 0 \\ u_1 + u_2 + u_3 &= 2. \end{aligned}$$

This is a system of four equations in three unknowns. In matrix form:

$$\begin{bmatrix} 4 & 1 & 0 \\ -1 & 4 & 2 \\ 0 & 0 & 4 \\ 1 & 1 & 1 \end{bmatrix} \begin{bmatrix} u_1 \\ u_2 \\ u_3 \end{bmatrix} = \begin{bmatrix} 1 \\ -1 \\ 0 \\ 2 \end{bmatrix}.$$

Systems like this will in general not have solutions. But there is a recipe for finding an *approximate solution.* As usual, write the system as

$$A\mathbf{u} = \mathbf{b},$$

with a matrix that has more rows than columns. Simply multiply both sides by A^{T}:

$$A^{\mathrm{T}}A\mathbf{u} = A^{\mathrm{T}}\mathbf{b}. \tag{14.3}$$

This is a linear system with a square matrix $A^{\mathrm{T}}A$! Even more, that matrix is symmetric.

The solution to the new system (14.3) (when it has one) is the one that minimizes the *error*

$$\|A\mathbf{u} - \mathbf{b}\|.$$

It is called the *least squares solution* of the original system.

Example 14.5 Returning to the system above, the least squares solution is the solution of the linear system

$$\begin{bmatrix} 18 & 1 & -1 \\ 1 & 18 & 9 \\ -1 & 9 & 21 \end{bmatrix} \begin{bmatrix} u_1 \\ u_2 \\ u_3 \end{bmatrix} = \begin{bmatrix} 7 \\ -2 \\ 0 \end{bmatrix}.$$

Example 14.6 As a second example, consider the problem of fitting a straight line to a set of 2D data points. Let the points be given by

$$\mathbf{p}_2 = \begin{bmatrix} -2 \\ -2 \end{bmatrix}, \quad \mathbf{p}_1 = \begin{bmatrix} 0 \\ 0 \end{bmatrix}, \quad \mathbf{p}_3 = \begin{bmatrix} 3 \\ 0 \end{bmatrix}, \quad \mathbf{p}_4 = \begin{bmatrix} 0 \\ 1 \end{bmatrix}.$$

We want to find a line of the form

$$x_2 = ax_1 + b,$$

that is close to the given points.[2] For the solution, see Figure 14.2. Ideally, we would like all of our data points to be on the line. This would require each of the following equations to be satisfied for one pair of a and b:

$$\begin{aligned} -2a + b &= -2, \\ b &= 0, \\ 3a + b &= 0, \\ b &= 1. \end{aligned}$$

This is an overdetermined linear system for the unknowns a and b – obvious from looking at the second and fourth equations. Now our overdetermined linear system is of the form

$$\begin{bmatrix} -2 & 1 \\ 0 & 1 \\ 3 & 1 \\ 0 & 1 \end{bmatrix} \begin{bmatrix} a \\ b \end{bmatrix} = \begin{bmatrix} -2 \\ 0 \\ 0 \\ 1 \end{bmatrix}.$$

For the least squares system, we obtain

$$\begin{bmatrix} 13 & 1 \\ 1 & 4 \end{bmatrix} \begin{bmatrix} a \\ b \end{bmatrix} = \begin{bmatrix} 4 \\ -1 \end{bmatrix}.$$

It has the solution

$$\begin{bmatrix} a \\ b \end{bmatrix} = \begin{bmatrix} 0.34 \\ -0.34 \end{bmatrix}.$$

Thus our desired straight line (see Figure 14.2) is given by

$$x_2 = 0.34x_x - 0.34.$$

[2]For reasons beyond the scope of this text, we have formulated the line in the explicit form rather than the more favored parametric or implicit forms.

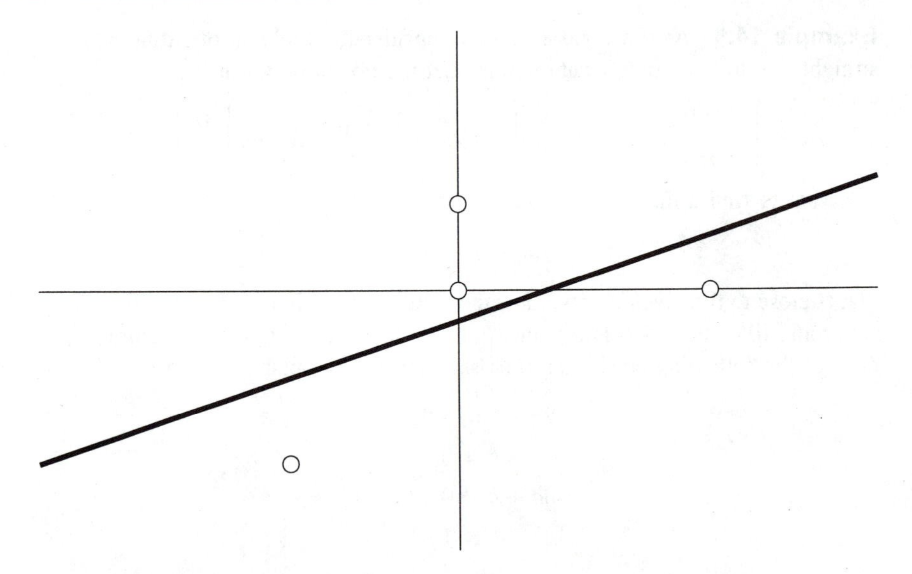

Figure 14.2.
Least squares: fitting a straight line to a set of points.

14.6 Inverse Matrices

The *inverse* of a square matrix A is the matrix that "undoes" A's action, i.e., the combined action of A and A^{-1} is the identity:

$$AA^{-1} = I. \tag{14.4}$$

Example 14.7 The following scheme shows a matrix A multiplied by its inverse A^{-1}. The matrix A is on the left, A^{-1} is on top, and the result of the multiplication, the identity, is on the lower right.

$$\begin{array}{ccc|ccc}
 & & & 1 & 0 & -1 \\
 & & & 3 & 1 & -3 \\
 & & & 1 & 2 & -2 \\
\hline
-4 & 2 & -1 & 1 & 0 & 0 \\
-3 & 1 & 0 & 0 & 1 & 0 \\
-5 & 2 & -1 & 0 & 0 & 1
\end{array}$$

How do we find the inverse of a matrix? In much the same way as we did in the 2×2 case in Section 5.5, write

$$A \begin{bmatrix} \bar{\mathbf{a}}_1 & \dots & \bar{\mathbf{a}}_n \end{bmatrix} = \begin{bmatrix} \mathbf{e}_1 & \dots & \mathbf{e}_n \end{bmatrix}. \tag{14.5}$$

Here, the matrices are $n \times n$, and the vectors $\bar{\mathbf{a}}_i$ as well as the $\mathbf{e}_i$ are vectors with n components. The vector $\mathbf{e}_i$ has all nonzero entries except for its ith component; it equals 1.

We may now interpret (14.5) as n linear systems:

$$A\bar{\mathbf{a}}_1 = \mathbf{e}_1, \dots, A\bar{\mathbf{a}}_n = \mathbf{e}_n. \tag{14.6}$$

Each of these may be solved as per Section 14.2. Most economically, an LU decomposition should be used here (see Section 14.7).

The inverse of a matrix A only exists if the action of A does not reduce dimensionality, as in a projection. This means that all columns of A must be linearly independent. There is a simple way to see if a matrix A is invertible; just perform Gauss elimination for the first of the linear systems in (14.6). If you are able to transform A to upper triangular with all nonzero diagonal elements, then A is invertible. Otherwise, it is said to be *singular.*

A singular matrix reduces dimensionality. An invertible (hence square!) matrix does not do this; it is said to have *rank* n, or *full rank.* If it reduces dimensionality by k, then it has rank $n - k$.

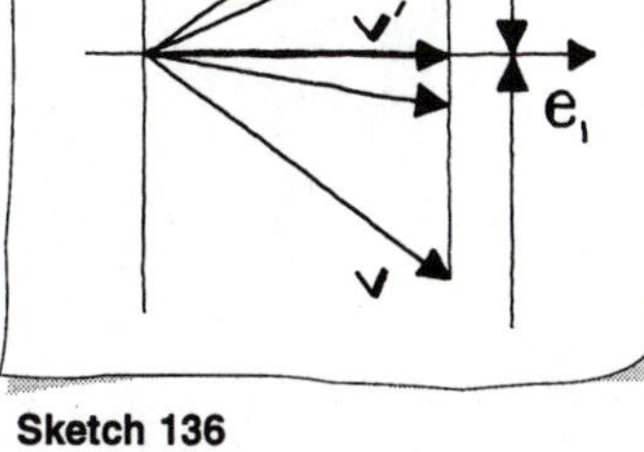

Sketch 136
Several vectors are projected onto the same image.

Example 14.8 The 4×4 matrix

$$\begin{bmatrix} 1 & 3 & -3 & 0 \\ 0 & 3 & 3 & 1 \\ 0 & 0 & 0 & 0 \\ 0 & 0 & 0 & 0 \end{bmatrix}$$

has rank 2, while

$$\begin{bmatrix} 1 & 3 & -3 & 0 \\ 0 & 3 & 3 & 1 \\ 0 & 0 & -1 & 0 \\ 0 & 0 & 0 & 0 \end{bmatrix}$$

has rank 3.

The $n \times n$ identity matrix has rank n; the zero matrix has rank 0.

An example of a matrix which does not have full rank is a projection. All of its image vectors $\mathbf{v}'$ may be the result of many different operations $A\mathbf{v}$, as Sketch 136 shows.

Example 14.9 Let us compute the inverse of the $n \times n$ matrix G_j as defined in (14.2):

$$G_j^{-1} = \begin{bmatrix} 1 & & & & & & \\ & \ddots & & & & & \\ & & 1 & & & & \\ & & & 1 & & & \\ & & & g_{j+1,j} & 1 & & \\ & & & \vdots & & \ddots & \\ & & & g_{n,j} & & & 1 \end{bmatrix}.$$

That is simple! To make some geometric sense of this, you should realize that G_j is a shear, and so is G_j^{-1}, except it "undoes" G_j.

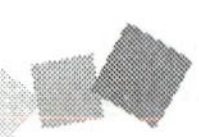

Here is another interesting property of the inverse of a matrix. Suppose $k \neq 0$ and kA is an invertible matrix, then

$$(kA)^{-1} = \frac{1}{k}A^{-1}.$$

And yet another: If two matrices A and B are invertible, then the product AB is invertible too.

Inverse matrices are primarily a theoretical concept. They suggest to solve a linear system $A\mathbf{v} = \mathbf{b}$ by computing A^{-1} and then to set $\mathbf{v} = A^{-1}\mathbf{b}$. Don't do that! It is a very expensive way to solve a linear system; simple Gauss elimination is much cheaper.

14.7 The LU Decomposition

Gauss elimination has two major parts: transforming the system to upper triangular form, and then back substitution. The creation of the upper triangular matrix may be written in terms of matrix multiplies using Gauss matrices G_j. If we denote the final upper triangular matrix[3] by U, then we have

$$G_{n-1} \cdot \ldots \cdot G_1 \cdot A = U. \tag{14.7}$$

It follows that

$$A = G_1^{-1} \cdot \ldots \cdot G_{n-1}^{-1} U.$$

[3]This matrix U was called $A^{\triangle}$ in Section 14.2.

The neat thing about the product $G_1^{-1} \cdot \ldots \cdot G_{n-1}^{-1}$ is that it is a lower triangular matrix with elements $g_{i,j}$:

$$G_1^{-1} \cdot \ldots \cdot G_{n-1}^{-1} = \begin{bmatrix} 1 & & & \\ g_{2,1} & 1 & & \\ \vdots & \ddots & \ddots & \\ g_{n,1} & \cdots & g_{n,n-1} & 1 \end{bmatrix}.$$

We denote this product by L (for lower triangular). Thus

$$A = LU, \tag{14.8}$$

which is known as the LU decomposition of A. It is also called the factorization of A. Every invertible matrix A has such a decomposition, although it may be necessary to employ pivoting during its computation.

If we denote the elements of L by $l_{i,j}$ (keeping in mind that $l_{i,i} = 1$) and those of U by $u_{i,j}$, the elements of A may be rewritten as

$$a_{i,j} = l_{i,1}u_{1,j} + \ldots + l_{i,j}u_{j,j}; \quad j < i,$$

i.e., for those $a_{i,j}$ below A's diagonal. For those on or above the diagonal, we get

$$a_{i,j} = l_{i,1}u_{i,j} + \ldots + l_{i,i-1}u_{i-1,j} + u_{i,j}; \quad j \geq i.$$

This leads to

$$l_{i,j} = \frac{1}{u_{j,j}}(a_{i,j} - l_{i,1}u_{1,j} - \ldots - l_{i,j-1}u_{j-1,j}); \quad j < i \tag{14.9}$$

and

$$u_{i,j} = a_{i,j} - l_{i,1}u_{1,j} - \ldots - l_{i,i-1}u_{i-1,j}; \quad j \geq i. \tag{14.10}$$

If A has a decompostion $A = LU$, then the system can be written as

$$LU\mathbf{u} = \mathbf{b}. \tag{14.11}$$

The matrix vector product $U\mathbf{u}$ results in a vector; call this $\mathbf{y}$. Re-examining (14.11), it becomes a two step problem. First solve

$$L\mathbf{y} = \mathbf{b}, \tag{14.12}$$

then solve

$$U\mathbf{u} = \mathbf{y}. \tag{14.13}$$

Hence we have if $U\mathbf{u} = \mathbf{y}$, then $LU\mathbf{u} = L\mathbf{y} = \mathbf{b}$. The two systems in (14.12) and (14.13) are triangular and easy to solve. (Forward substitution is used for L.)

Here is a more direct method for forming L and U, rather than through Gauss elimination. This then is the method of LU-decomposition.

Algorithm:

Calculate the nonzero elements of L and U:
For $k = 1, \ldots, n$
$u_{k,k} = a_{k,k} - l_{k,1}u_{1,k} - \ldots - l_{k,k-1}u_{k-1,k}$
For $i = k+1, \ldots, n$
$l_{i,k} = \frac{1}{u_{k,k}}[a_{i,k} - l_{i,1}u_{1,k} - \ldots - l_{i,k-1}u_{k-1,k}]$
For $j = k+1, \ldots, n$
$u_{k,j} = a_{k,j} - l_{k,1}u_{1,j} - \ldots - l_{k,k-1}u_{k-1,j}$

Using forward substitution solve $\mathbf{Ly} = \mathbf{b}$.
Using back substitution solve $\mathbf{Uu} = \mathbf{y}$.

The $u_{k,k}$ term is equivalent to the pivot element in Gauss elimination. Pivoting makes the bookkeeping trickier; see [15] for a detailed description.

Example 14.10 Decompose A into LU, where

$$A = \begin{bmatrix} 2 & 2 & 4 \\ -1 & 2 & -3 \\ 1 & 2 & 2 \end{bmatrix}.$$

Following the steps in the algorithm above:

$$\begin{aligned}
u_{1,1} &= 2 \\
l_{2,1} &= a_{2,1}/u_{1,1} = -1/2 \\
l_{3,1} &= a_{3,1}/u_{1,1} = 1/2 \\
u_{1,2} &= a_{1,2} = 2 \\
u_{2,2} &= a_{2,2} - l_{2,1}u_{1,2} = 2 + 1 = 3 \\
l_{3,2} &= \frac{1}{u_{2,2}}[a_{3,2} - l_{3,1}u_{1,2}] = \frac{1}{3}[2 - 1] = 1/3 \; u_{1,3} = 4 \\
u_{2,3} &= a_{2,3} - l_{2,1}u_{1,3} = -3 + 2 = -1 \\
u_{3,3} &= a_{3,3} - l_{3,1}u_{1,3} - l_{3,2}u_{2,3} = 2 - 2 + 1/3 = 1/3
\end{aligned}$$

Check that this produced valid entries for L and U:

$$
\begin{array}{ccc|ccc}
 & & & 2 & 2 & 4 \\
 & & & 0 & 3 & -1 \\
 & & & 0 & 0 & 1/3 \\
\hline
1 & 0 & 0 & 2 & 2 & 4 \\
-1/2 & 1 & 0 & -1 & 2 & -3 \\
1/2 & 1/3 & 1 & 1 & 2 & 2
\end{array}
$$

Finally, the major benefit of the LU-decomposition: speed. In cases where one has to solve multiple linear systems with the same coefficient matrix,[4] the LU-decomposition is a big timesaver. We perform it once, and then perform the forward and backward substitutions (14.12) and (14.13) for each right-hand side. This is significantly less work than performing a complete Gauss elimination every time over!

14.8 Exercises

1. Solve the linear system $A\mathbf{v} = \mathbf{b}$ where

$$A = \begin{bmatrix} 1 & 0 & -1 & 2 \\ 0 & 0 & 1 & -2 \\ 2 & 0 & 0 & 1 \\ 1 & 1 & 1 & 0 \end{bmatrix}, \quad \text{and} \quad \mathbf{b} = \begin{bmatrix} -1 \\ 2 \\ 1 \\ -3 \end{bmatrix}.$$

2. Solve the linear system $A\mathbf{v} = \mathbf{b}$ where

$$A = \begin{bmatrix} 0 & 0 & 1 \\ 1 & 0 & 0 \\ 1 & 1 & 1 \end{bmatrix}, \quad \text{and} \quad \mathbf{b} = \begin{bmatrix} -1 \\ 0 \\ -1 \end{bmatrix}.$$

3. Find the inverse of the matrix from the previous problem.

4. Let five points be given by

$$\mathbf{p}_2 = \begin{bmatrix} 1 \\ -1 \end{bmatrix}, \mathbf{p}_1 = \begin{bmatrix} 0 \\ 0 \end{bmatrix}, \mathbf{p}_3 = \begin{bmatrix} -3 \\ 0 \end{bmatrix}, \mathbf{p}_4 = \begin{bmatrix} 3 \\ -1 \end{bmatrix}, \mathbf{p}_5 = \begin{bmatrix} 0 \\ 1 \end{bmatrix}.$$

Find the least squares approximation using a straight line.

5. Carry out three iterations for the linear system

$$A = \begin{bmatrix} 4 & 1 & 0 & 1 \\ 1 & 4 & 1 & 0 \\ 0 & 1 & 4 & 1 \\ 1 & 0 & 1 & 4 \end{bmatrix}, \quad \text{and} \quad \mathbf{b} = \begin{bmatrix} 0 \\ 1 \\ 1 \\ 0 \end{bmatrix},$$

[4]Finding the inverse of a matrix is an example.

starting with the initial guess

$$\mathbf{u}^{(1)} = \begin{bmatrix} 1 \\ 1 \\ 1 \\ 1 \end{bmatrix}.$$

After each iteration, check that the *residual vector* $||A\mathbf{u}^{(i)} - \mathbf{b}||$ decreases in size!

6. Restate the Gauss elimination algorithm in pseudocode with pivoting. Do not actually exchange rows, but rather use an ordering vector.

7. Calculate the determinant of

$$A = \begin{bmatrix} 3 & 0 & 1 \\ 1 & 2 & 0 \\ 1 & 1 & 1 \end{bmatrix}.$$

8. Calculate the LU decomposition of the matrix in the previous problem.

15 Putting Lines Together: Polylines and Polygons

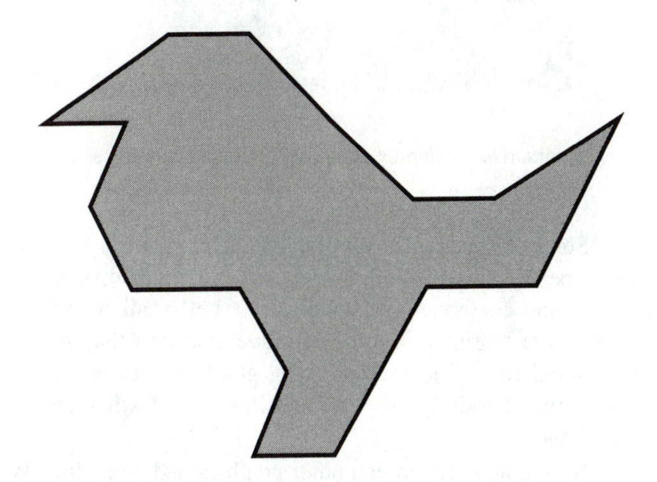

Figure 15.1.
Polygon: straight line segments forming a bird shape.

Figure 15.1 shows a *polygon*. It is the outline of a shape, drawn with straight line segments. Since such shapes are all a printer or plotter can draw, just about every computer-generated drawing consists of polygons. If we add an "eye" to the bird-shaped polygon from Figure 15.1, and if we apply a sequence of rotations and translations to it, then we arrive at Figure 15.2 — it turns out copies of our special bird polygon can cover the whole plane! This technique is also present in the Escher illustration, Figure 6.8.

15.1 Polylines

Straight line segments, called *edges*, connecting an ordered set of *vertices* constitute a *polyline*. The first and last vertices are not necessarily

Figure 15.2.
Mixing maps: a pattern is created by composing rotations and translations.

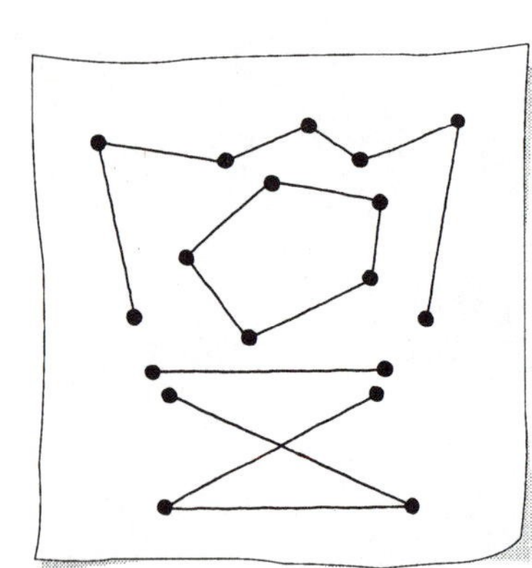

Sketch 137
2D polyline examples.

connected. Some 2D examples are illustrated in Sketch 137, however a polyline can be 3D too. Since the vertices are ordered, the edges are oriented and can be thought of as vectors. Let's call them *edge vectors*. Polylines are a primary output primitive, and thus they are included in graphics standards. One example of a graphics standard is the *GKS* (Graphical Kernel System); this is a specification of what belongs in a graphics package.

Polylines have many uses in computer graphics and modeling. Whether in 2D or 3D, they are typically used to outline a shape. The power of polylines to reveal a shape is illustrated in Figure 15.3 in the display of a 3D surface. The surface is *evaluated* in an organized fashion so that the points can be logically connected as polylines, giving the observer a feeling of the "flow" of the surface. In modeling, a polyline is often used to approximate a complex curve or data, which in turn makes analysis easier and less costly. An example of this is illustrated in Figure 3.4.

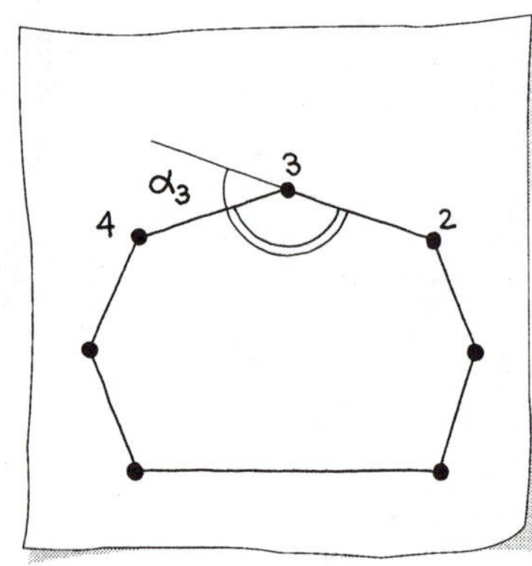

Sketch 138
Interior and exterior angles.

15.2 Polygons

When the first and last vertices of a polyline are connected, it is called a *polygon*. Normally, a polygon is thought to enclose an area. For this reason, we will consider planar polygons only. Just as with polylines, polygons constitute an ordered set of vertices and we will continue to use the term edge vectors. Thus a polygon with n edges is given by an ordered

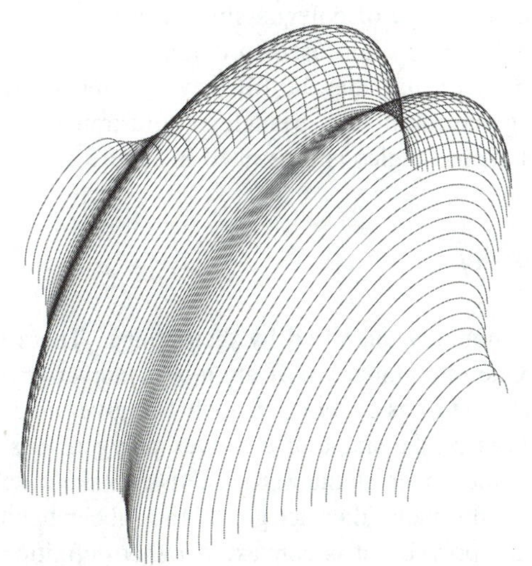

Figure 15.3.
Polylines: the display of a 3D surface.

Sketch 139
A minmax box is a polygon.

set of points

$$\mathbf{p}_1, \mathbf{p}_2, \ldots, \mathbf{p}_n$$

and has edge vectors $\mathbf{v}_i = \mathbf{p}_{i+1} - \mathbf{p}_i; i = 1, \ldots, n$. Note that the edge vectors sum to zero![1]

If you look at the edge vectors carefully, you'll discover that $\mathbf{v}_n = \mathbf{p}_{n+1} - \mathbf{p}_n$, but there is no vertex $\mathbf{p}_{n+1}$! This apparent problem is resolved by defining $\mathbf{p}_{n+1} = \mathbf{p}_1$, a convention which is called *cyclic numbering*. We'll use this convention throughout, and will not mention it every time. We also add one more, topological, characterization of polygons: the number of vertices equals the number of edges.

Since a polygon is closed, it divides the plane into two parts: a finite part, the polygon's *interior*, and an infinite part, the polygon's *exterior*.

As you traverse a polygon, you follow the path determined by the vertices and edge vectors. Between vertices, you'll move along straight lines (the edges), but at the vertices, you'll have to perform a rotation before resuming another straight line path. The angle α_i by which you rotate at vertex $\mathbf{p}_i$ is called the *turning angle* or *exterior angle* at $\mathbf{p}_i$. The *interior angle* is then given by $\pi - \alpha_i$ (see Sketch 138).

Polygons are used a lot! For instance, in Chapter 1 we discussed the *extents* of geometry in a 2D coordinate system. Another name for these extents is a *minmax box*, see Sketch 139. It is a special polygon, namely

Sketch 140
Convex (left) and nonconvex (right) polygons.

[1]To be precise, they sum to the *zero vector*.

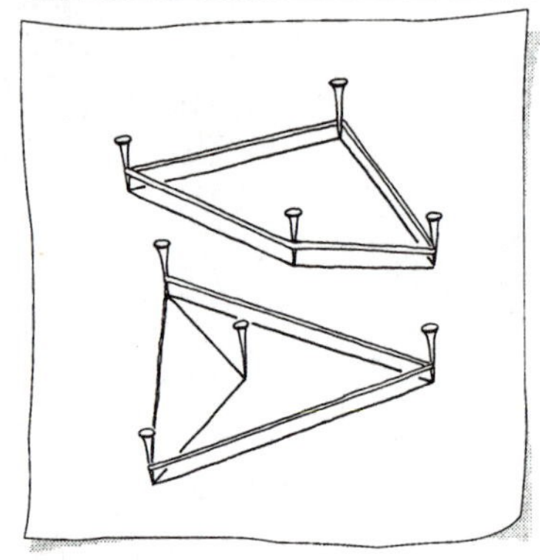

Sketch 141
Rubberband test for convexity of a polygon.

a rectangle. Another type of polygon studied in Chapter 8 is the triangle. This type of polygon is often used to define a *polygonal mesh* of a 3D model. The triangles may then be filled with color to produce a shaded image. A polygonal mesh from triangles is illustrated in Figures 8.3 and 8.4, and one from rectangles is illustrated in Figure 10.2.

15.3 Convexity

Polygons are commonly classified by their shape. There are many ways of doing this. One important classification is as *convex* or *nonconvex*. The latter is also referred to as *concave*. Sketch 140 gives an example of each. How do you describe the shape of a convex polygon? As in Sketch 141, stick nails into the paper at the vertices. Now take a rubber band and stretch it around the nails, then let go. If the rubber band shape follows the outline of the polygon, it is convex. Another definition: take any two points in the polygon (including on the edges) and connect them with a straight line. If the line never leaves the polygon, then it is convex. This must work for *all* possible pairs of points!

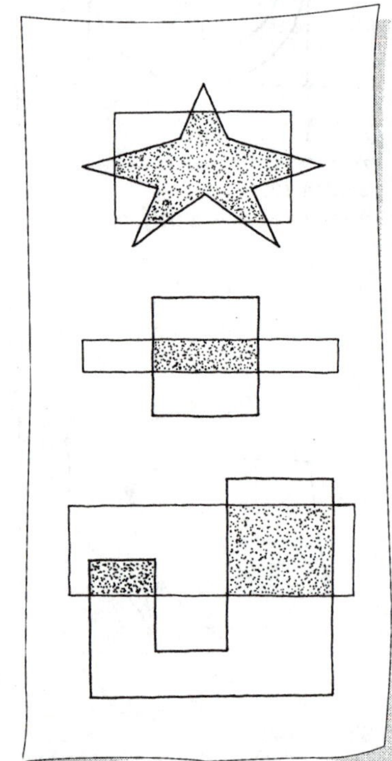

Sketch 142
Polygon clipping.

The issue of convexity is important, because algorithms which involve polygons can be simplified if the polygons are known to be convex. This is true for algorithms for the problem of *polygon clipping*. This problem starts with two polygons, and the goal is to find the intersection of the polygon areas. Some examples are illustrated in Sketch 142. The intersection area isdefined in terms of one or more polygons. If both polygons are convex, then the result will be just one convex polygon. However, if even one polygon is not convex then the result might be two or more, possibly disjoint or nonconvex, polygons. Thus nonconvex polygons need more record keeping in order to properly define the intersection area(s). Not all algorithms are designed for nonconvex polygons. See [10, 11] for a detailed description of clipping algorithms.

An n-sided convex polygon has a sum of interior angles I equal to

$$I = (n - 2)\pi. \tag{15.1}$$

To see this, take one polygon vertex and form triangles with the other vertices, as illustrated in Sketch 143. This forms $n - 2$ triangles. The sum of interior angles of a triangle is known to be π. Thus we get the above result.

The sum of the exterior angles of a convex polygon is easily found with this result. Each interior and exterior angle sums to π. Suppose the ith interior angle is α_i radians, then the exterior angle is $\pi - \alpha_i$ radians. Sum over all angles, and the exterior angle sum E is

$$E = n\pi - (n-2)\pi = 2\pi. \quad (15.2)$$

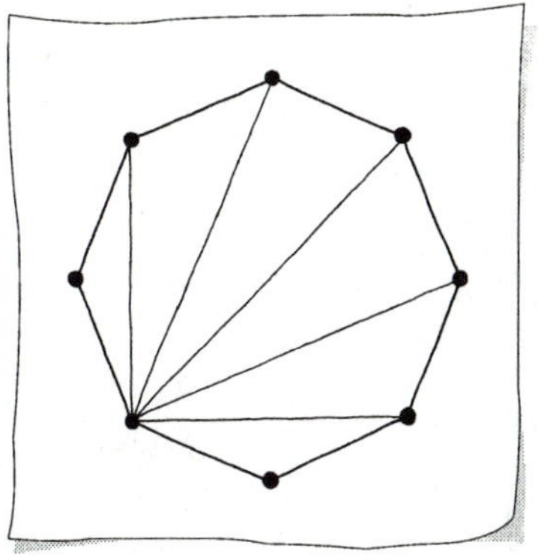

Sketch 143
Sum of interior angles using triangles.

To test if an n-sided polygon is convex, we'll use the *barycenter* of the vertices $\mathbf{p}_1, \ldots, \mathbf{p}_n$ The barycenter $\mathbf{b}$ is a special barycentric combination (see Sections 8.1 and 8.3):

$$\mathbf{b} = \frac{1}{n}(\mathbf{p}_1 + \ldots + \mathbf{p}_n).$$

It is the center of gravity of the vertices. We need to construct the implicit line equation for each edge vector in a consistent manner. If the polygon is convex, then the point $\mathbf{b}$ will be on the "same" side of every line. The implicit equation will result in all positive or all negative values. (This will not work for some unusual — nonsimple — polygons, see the Exercises and Section 15.5.)

Another test for convexity is to check if there is a *re-entrant angle*. This is an interior angle which is greater than π.

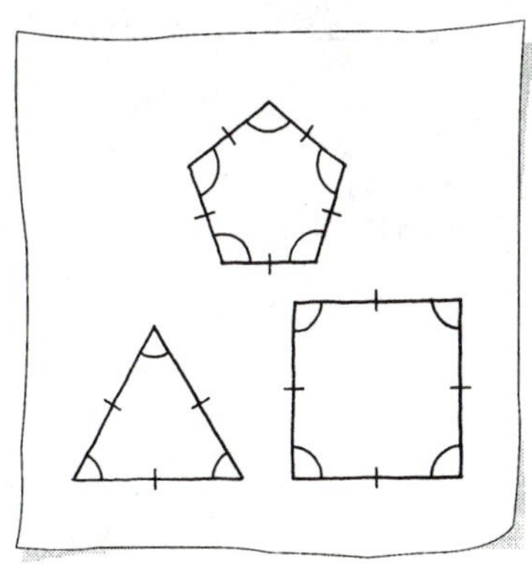

Sketch 144
Regular polygons.

15.4 Types of Polygons

There are a variety of special polygons. First, we introduce two terms to help describe these polygons:

- *equilateral* means that all sides are of equal length, and
- *equiangular* means that all *interior angles* at the vertices are equal.

In the following illustrations, edges with the same number of tick marks are of equal length and angles with the same number of arc markings are equal.

A very special polygon is the *regular polygon*: it is equilateral and equiangular. Examples are illustrated in Sketch 144. This polygon is also referred to as an *n-gon*, indicating it has n edges. We list the names of the "classical" n-gons::

- a 3-gon is a equilateral triangle,
- a 4-gon is a square,
- a 5-gon is a pentagon,
- a 6-gon is a hexagon, and
- an 8-gon is a octagon

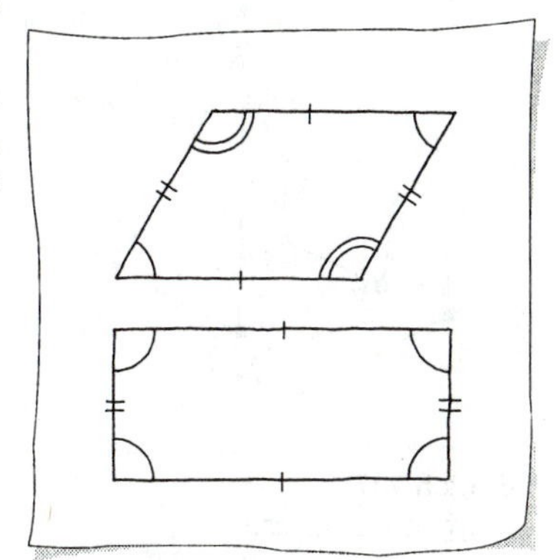

Sketch 145
Rhombus and rectangle.

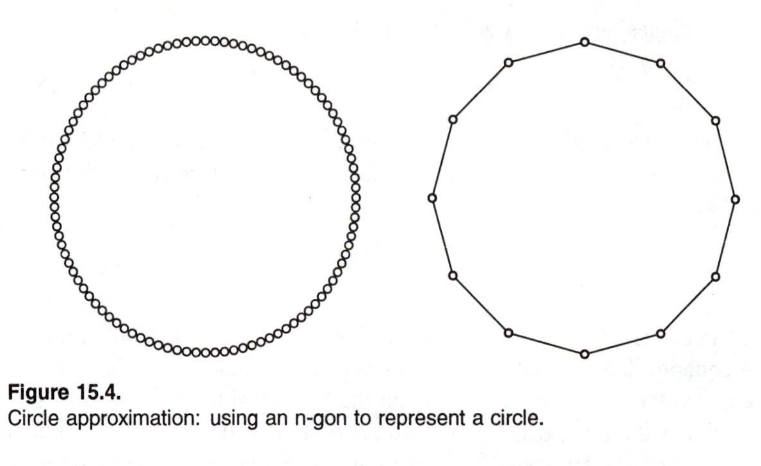

Figure 15.4.
Circle approximation: using an n-gon to represent a circle.

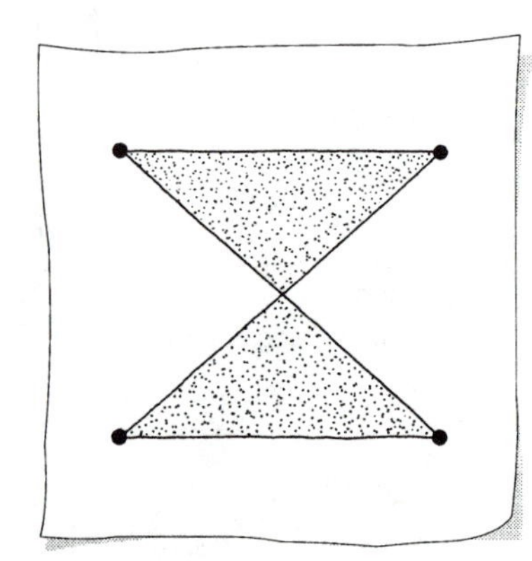

Sketch 146
Nonsimple polygon.

An n-gon is commonly used to approximate a circle in computergraphics, as illustrated in Figure 15.4.

A *rhombus* is equilateral but not equiangular, whereas a *rectangle* is equiangular but not equilateral. These are illustrated in Sketch 145.

15.5 Unusual Polygons

Most often, applications deal with *simple* polygons, as opposed to *nonsimple* polygons. A nonsimple polygon, as illustrated in Sketch 146, is characterized by edges intersecting other than at the vertices. Topology is the reason nonsimple polygons can cause havoc in some algorithms. For convex and nonconvex simple polygons, as you traverse along the boundary of the polygon, the interior remains on one side. This is not the case for nonsimple polygons. At the mid-edge intersections, the interior switches sides. In more concrete terms, recall how the implicit line equation could be used to determine if a point is on the line, and more generally, which side it is on. Suppose you have developed an algorithm that associates the + side of the line with being inside the polygon. This rule will work fine if the polygon is simple; however not otherwise.

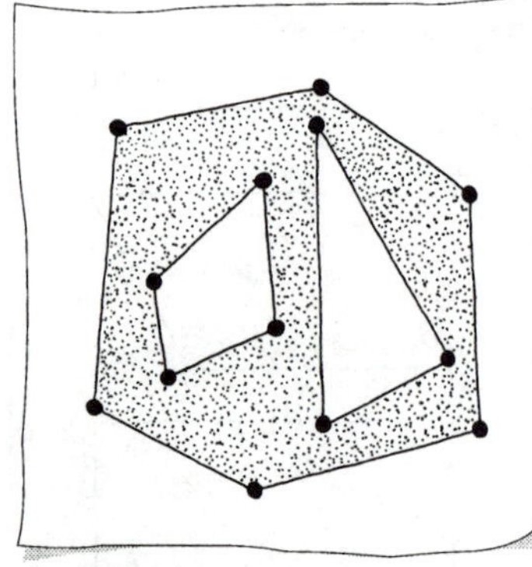

Sketch 147
Polygon with holes.

Sometimes nonsimple polygons can arise due to an error. The polygon clipping algorithms, as discussed in Section 15.3, involve sorting vertices to form the final polygon. If this sorting goes haywire, then you could end up with a nonsimple polygon rather than a simple one as desired.

In applications, it is not uncommon to encounter polygons with holes. Such a polygon is illustrated in Sketch 147. As you see, this is actually more than one polygon. An example of this, illustrated in Figure 15.5, is a special CAD/CAM surface called a *trimmed surface*. The polygons define

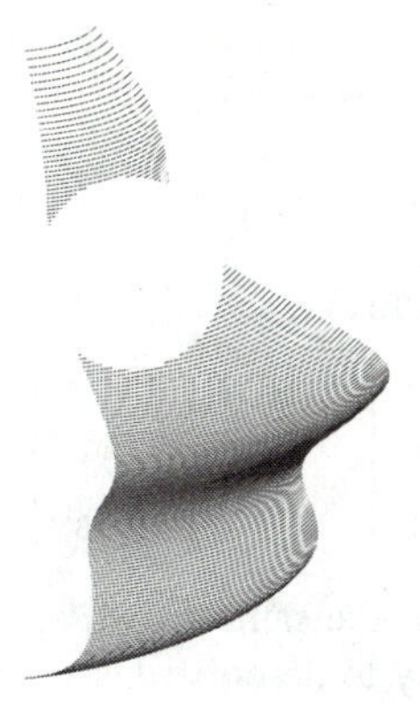

Figure 15.5.
Trimmed surface: an application of polygons with holes.

parts of the material to be cut or punched out. This allows other parts to fit to this one.

For trimmed surfaces and other CAD/CAM applications, a certain convention is accepted in order to make more sense out of this multi-polygon geometry. The polygons must be oriented a special way.The *visible region*, or the region that is not cut out, is to the "left". As a result, the outer boundary is oriented counterclockwise and the inner boundaries are oriented clockwise. More on the visible region in Section 15.9.

15.6 Turning Angles and Winding Numbers

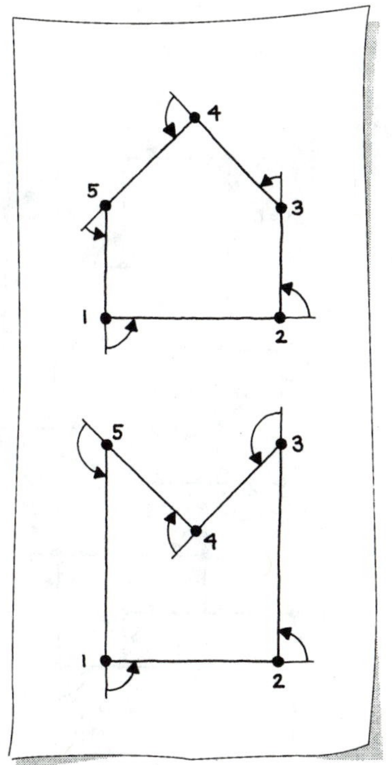

Sketch 148
Turning angles.

The *turning angle* of a polygon or polyline is essentially another name for the *exterior angle*, which is illustrated in Sketch 138. Sketch 148 illustrates the turning angles for a convex and a nonconvex polygon. Here the difference between a turning angle and an exterior angle is illustrated. The turning angle has an orientation as well as an angle measure. All turning angles for a convex polygon have the same orientation, which is not the case for a nonconvex polygon. This fact will allow us to easily differentiate the two types of polygons.

Here is an application of the turning angle. Suppose for now that a given polygon is 2D and lives in the $[\mathbf{e}_1, \mathbf{e}_2]$-plane. Its n vertices are labeled

$$\mathbf{p}_1, \mathbf{p}_2, \ldots \mathbf{p}_n.$$

We want to know if the polygon is convex. We only need to look at the orientation of the turning angles, not the actual angles. Recall that the cross product of two 2D vectors will produce a vector which points "in" or "out", that is in the $+\mathbf{e}_3$ or $-\mathbf{e}_3$ direction. Therefore, by taking the cross product of successive edge vectors

$$\mathbf{u}_i = (\mathbf{p}_{i+1} - \mathbf{p}_i) \wedge (\mathbf{p}_{i+2} - \mathbf{p}_{i+1}), \tag{15.3}$$

we'll encounter $\mathbf{u}_i$ of the form

$$\begin{bmatrix} 0 \\ 0 \\ u_3 \end{bmatrix} \quad \text{or} \quad \begin{bmatrix} 0 \\ 0 \\ -u_3 \end{bmatrix}.$$

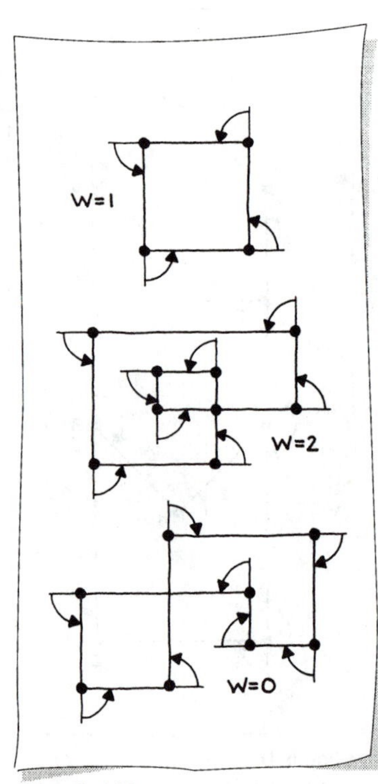

Sketch 149
Winding numbers.

If the sign of the u_3 value is the same for all angles, then the polygon is convex. A mathematical way to describe this is by using the scalar triple product (see Section 10.5). The turning angle orientation is determined by the scalar

$$u_3 = \mathbf{e}_3 \cdot ((\mathbf{p}_{i+1} - \mathbf{p}_i) \wedge (\mathbf{p}_{i+2} - \mathbf{p}_{i+1})).$$

Notice that the sign is dependent upon the traversal direction of the polygon, but only a change of sign is important.

If the polygon lies in an arbitrary plane, having a normal $\mathbf{n}$, then the above convex/concave test is changed only a bit. The cross product in (15.3) produces a vector $\mathbf{u}_i$ that has direction $\pm\mathbf{n}$. Now we need the dot product, $\mathbf{n} \cdot \mathbf{u}_i$ to extract a signed scalar value.

If we actually computed the turning angle at each vertex, we could form an accumulated value called the *total turning angle*. Recall from (15.2) that the total turning angle for a convex polygon is 2π. For a polygon that is not known to be convex, assign a sign using the scalar triple product as above, to each angle measurement. The sum E will then be used to compute the *winding number* of the polygon. The winding number W is

$$W = \frac{E}{2\pi}.$$

Thus for a convex polygon, the winding number is one. Sketch 149 illustrates a few examples. A convex polygon is essentially one loop. A polygon can have more than one loop, with different orientation: clockwise versus counterclockwise. The winding number gets decremented for each clockwise loop and incremented for each counterclockwise loop, or vice versa depending on how you assign signs to your angles.

15.7 Area

A simple method for calculating the area of a polygon is to use the *signed area* of a triangle as per Section 4.9. For a convex polygon, as in Sketch

150, the sign of the area is not an issue. Simply choose one vertex and form triangles to the other vertices. The sum of the areas of the triangles is the area of the convex polygon. For a nonconvex polygon, things work pretty much the same way, although one must take some more care. Sketch 151 illustrates the procedure for a nonconvex polygon. The triangles are formed just as with a convex polygon, however now the ordering of the vertices of the triangles comes into play: order with increasing index. Use the right hand rule to study which areas are positive and which are negative in the sketch.

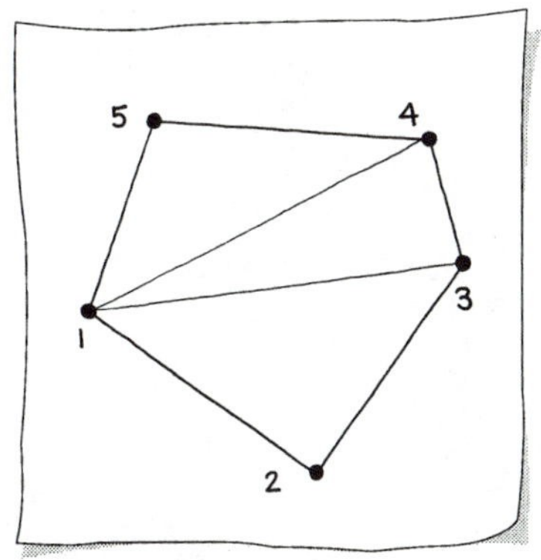

Sketch 150
Area of a convex polygon.

Thus we see for a convex or nonconvex polygon with vertices $\mathbf{p}_i$ for $i = 1, \ldots, n$ in a plane with normal $\mathbf{n}$, the area A can be computed as the sum of areas $A = a_1 + a_2 + \ldots a_{n-1}$, with

$$a_j = 1/2[\mathbf{n} \cdot ((\mathbf{p}_{j+1} - \mathbf{p}_1) \wedge (\mathbf{p}_{j+2} - \mathbf{p}_1))] \qquad (15.4)$$

for $j = 1, \ldots, n-2$. If (15.4) is expanded for all j, then some of the terms cancel and we are left with the area of a polygon being $A = b_1 + b_2 + \ldots b_{n-1}$, where

$$b_k = 1/2[\mathbf{n} \cdot (\mathbf{p}_k \wedge \mathbf{p}_{k+1})]. \qquad (15.5)$$

for $k = 1, \ldots n$. Recall $\mathbf{p}_{n+1} = \mathbf{p}_1$. Equation (15.5) seems to have lost all geometric meaning because it involves the cross product of point pairs; but actually we are using the cross products of the points minus the origin. It is preferred over (15.4) because it is more efficient.

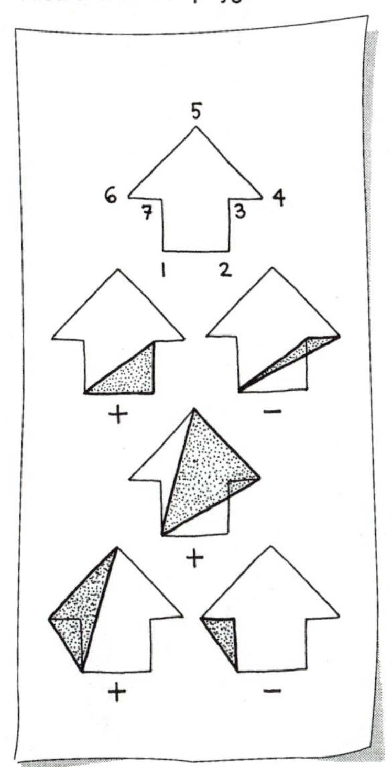

Sketch 151
Area of a nonconvex polygon.

Example 15.1 Take the four coplanar 3D points

$$\mathbf{p}_1 = \begin{bmatrix} 0 \\ 2 \\ 0 \end{bmatrix}, \quad \mathbf{p}_2 = \begin{bmatrix} 3 \\ 0 \\ 0 \end{bmatrix}, \quad \mathbf{p}_3 = \begin{bmatrix} 0 \\ 2 \\ 3 \end{bmatrix}, \quad \mathbf{p}_4 = \begin{bmatrix} 3 \\ 0 \\ 3 \end{bmatrix}.$$

Apply formula (15.4) and (15.5), and obtain $A = 15$. You may also realize that our simple example polygon is just a rectangle, and so you have a second way to check the area!

For points in the $[\mathbf{e}_1, \mathbf{e}_2]$-plane, there is a neat way to write (15.5) due to M. Stone, see [18]. Let

$$\mathbf{p}_i = \begin{bmatrix} p_i \\ q_i \end{bmatrix}.$$

Then

$$A = \frac{1}{2} \begin{vmatrix} p_1 & p_2 & \cdots & p_n & p_1 \\ q_1 & q_2 & \cdots & q_n & q_1 \end{vmatrix},$$

where the generalized determinant is computed by adding the products of all "downward" diagonals, and subtracting the products of all "upward" diagonals.

Example 15.2 Let

$$\mathbf{p}_1 = \begin{bmatrix} 0 \\ 0 \end{bmatrix}, \quad \mathbf{p}_2 = \begin{bmatrix} 1 \\ 0 \end{bmatrix}, \quad \mathbf{p}_3 = \begin{bmatrix} 1 \\ 1 \end{bmatrix}, \quad \mathbf{p}_4 = \begin{bmatrix} 0 \\ 1 \end{bmatrix}.$$

We have

$$A = \frac{1}{2}\begin{vmatrix} 0 & 1 & 1 & 0 & 0 \\ 0 & 0 & 1 & 1 & 0 \end{vmatrix} = \frac{1}{2}[0+1+1+0-0-0-0-0] = 1.$$

Since our polygon was a square, this is as expected.

But now take

$$\mathbf{p}_1 = \begin{bmatrix} 0 \\ 0 \end{bmatrix}, \quad \mathbf{p}_2 = \begin{bmatrix} 1 \\ 1 \end{bmatrix}, \quad \mathbf{p}_3 = \begin{bmatrix} 0 \\ 1 \end{bmatrix}, \quad \mathbf{p}_4 = \begin{bmatrix} 1 \\ 0 \end{bmatrix}.$$

This is a nonsimple polygon! Its area computes to

$$A = \frac{1}{2}\begin{vmatrix} 0 & 1 & 0 & 1 & 0 \\ 0 & 1 & 1 & 1 & 0 \end{vmatrix} = \frac{1}{2}[0+1+0+0-0-0-1-0] = 0.$$

Draw a sketch and convince yourself this is correct!

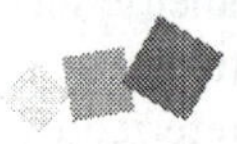

15.8 Planarity Test

Suppose someone sends you a CAD file that contains a polygon. For your application, the polygon must be 2D, however it is oriented arbitrarily in 3D. How do you verify that the data points are *coplanar*?

There are many ways to solve this problem, although some solutions have clear advantages over the others. Some considerations when comparing algorithms include

- numerical stability,
- speed,
- ability to define a meaningful tolerance,
- size of data set, and
- maintainability of the algorithm.

The order of importance is arguable.

Let's look at three possible methods to solve this planarity test and then compare them.

- Volume test: Choose the first polygon vertex as a base point. Form vectors to the next three vertices. Use the scalar triple product to calculate the volume spanned by these three vectors. If it is less than a given tolerance, then the four points are coplanar. Continue for all other sets.

- Plane test: Construct the plane through the first three vertices. Check if all of the other vertices lie in this plane, within a given tolerance.

- Average normal test: Find the centroid $\mathbf{c}$ of all points. Compute all normals $\mathbf{n}_i = [\mathbf{c} - \mathbf{p}_i] \wedge [\mathbf{c} - \mathbf{p}_{i+1}]$. Check if all angles formed by two subsequent normals are below a given angle tolerance.

If we compare these three methods, we see that they employ different kinds of tolerances: for volumes, distances, and angles. Which of these is preferable must depend on the application at hand. Clearly the plane test is the fastest of the three; yet it has a problem if the first three vertices are close to being coplanar.

15.9 Inside or Outside?

Another important concept for 2D polygons is the *inside/outside test* or *visibility test*. The problem is this: Given a polygon in the $[\mathbf{e}_1, \mathbf{e}_2]$-plane and a point $\mathbf{p}$, determine if the point lies inside the polygon.

One obvious application for this is polygon fill for raster device software, e.g., as with PostScript. Each pixel must be checked to see if it is in the polygon and should be colored. The inside/outside test is also encountered in CAD with *trimmed surfaces*, which were introduced in Section 15.5. With both applications it is not uncommon to have a polygon with one or more holes, as illustrated in Sketch 147. In the PostScript fill application, nonsimple polygons are not unusual either.[2]

We will present two similar algorithms, producing different results in some special cases. However we choose to solve this problem, we will want to incorporate a *trivial reject* test. This simply means that if a point is "obviously" not in the polygon, then we output that result immediately, i.e., with a minimal amount of calculation. In this problem, trivial reject refers to constructing a *minmax box* around the polygon. If a point lies

[2] As an example, take Figure 3.5. The lines inside the bounding rectangle are the edges of a nonsimple polygon.

Figure 15.6.
Even-Odd Rule: applied to polygon fill.

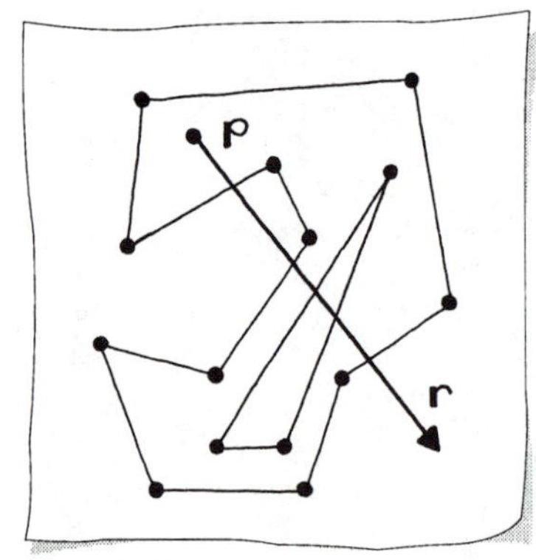

Sketch 152
Even-Odd rule.

outside of this minmax box then it may be trivially rejected. As you see, this involves simple comparison of $\mathbf{e}_1$ and $\mathbf{e}_2$ coordinates.

15.9.1 Even-Odd Rule. From the point $\mathbf{p}$, construct a line in parametric form with vector $\mathbf{r}$ in any direction. The parametric line is

$$\mathbf{l}(t) = \mathbf{p} + t\mathbf{r}.$$

This is illustrated in Sketch 152. Count the number of intersections this line has with the polygon edges for $t \geq 0$ only. This is why the vector is sometimes referred to as a *ray*. The number of intersection will be odd if $\mathbf{p}$ is inside and even if $\mathbf{p}$ is outside. Figure 15.6 illustrates the results of this rule with the polygon fill application.

It can happen that $\mathbf{l}(t)$ coincides with an edge of the polygon or passes through a vertex. Either a more elaborate counting scheme must be developed, or you can choose a different $\mathbf{r}$. As a rule, it is better to not choose $\mathbf{r}$ parallel to the $\mathbf{e}_1$- or $\mathbf{e}_2$-axis, because the polygons often have edges parallel to these axes.

15.9.2 Non-Zero Winding Number. In Section 15.6 the winding number was introduced. Here is another use for it. This method proceeds

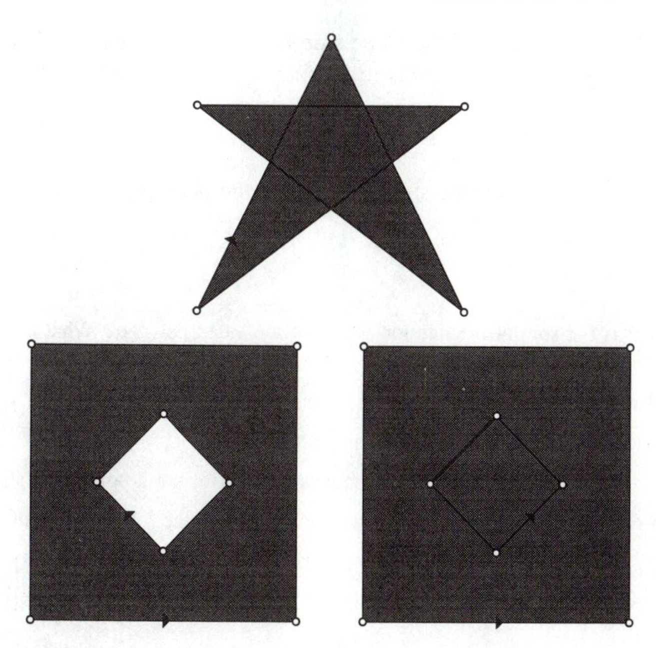

Figure 15.7.
Non-Zero Winding Rule: applied to polygon fill.

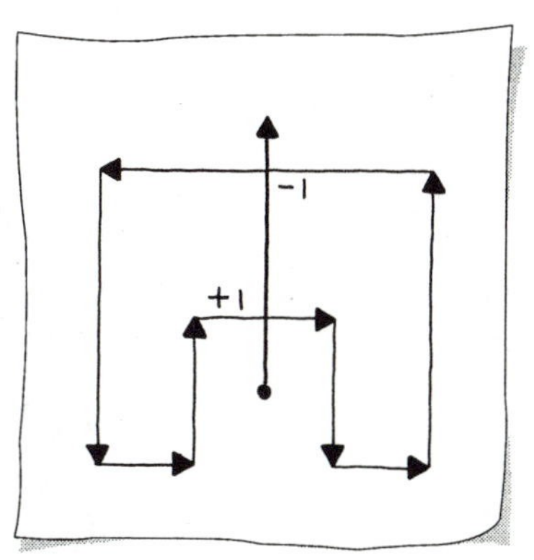

Sketch 153
Non-Zero winding number rule.

similarly to the even-odd rule. Construct a parametric line at the point **p** and intersect the polygon edges. Again, only consider those intersections for $t \geq 0$. The counting method depends on the orientation of the polygon edges. Start with a winding number of zero. Following Sketch 153, if a polygon edge is oriented "right to left" then add one to the winding number. If a polygon edge is oriented "left to right" then subtract one from the winding number. If the final result is zero then the point is outside the polygon. Figure 15.7 illustrates the results of this rule with the same polygons used in the even-odd rule. As with the previous rule, if you encounter edges head-on, then choose a different ray.

The differences in the algorithms are interesting. The PostScript language uses the non-zero winding number rule as the default. The authors (of the PostScript) language feel that this produces better results for the polygon fill application, but the even-odd rule is available with a special command. PostScript must deal with the most general (and crazy) polygons. In the trimmed surface application, the polygons must be simple and polygons cannot intersect; therefore either algorithm is suitable.

If you happen to know that you are dealing only with *convex polygons*, another inside/outside test is available. Check which side of the edges the point $\mathbf{p}$ is on. If it is on the same side for all edges, then $\mathbf{p}$ is inside the polygon. All you have to do is to compute all determinants of the form

$$\left| \; [\mathbf{p} - \mathbf{p}_i] \quad [\mathbf{p} - \mathbf{p}_{i+1}] \; \right|.$$

If they are all of the same sign, $\mathbf{p}$ is inside the polygon.

15.10 Exercises

1. What is the sum of the interior angles of a 6-sided polygon? What is the sum of the exterior angles?
2. What type of polygon is equiangular and equilateral?
3. Which polygon is equilateral but not equiangular?
4. Develop an algorithm which determines whether or not a polygon is simple.
5. Calculate the winding number of the polygon with the following vertices.

$$\mathbf{p}_1 = \begin{bmatrix} 0 \\ 0 \end{bmatrix} \quad \mathbf{p}_2 = \begin{bmatrix} -2 \\ 0 \end{bmatrix} \quad \mathbf{p}_3 = \begin{bmatrix} -2 \\ 2 \end{bmatrix} \quad \mathbf{p}_4 = \begin{bmatrix} 0 \\ 2 \end{bmatrix}$$

$$\mathbf{p}_5 = \begin{bmatrix} 2 \\ 2 \end{bmatrix} \quad \mathbf{p}_6 = \begin{bmatrix} 2 \\ -2 \end{bmatrix} \quad \mathbf{p}_7 = \begin{bmatrix} 3 \\ -2 \end{bmatrix} \quad \mathbf{p}_8 = \begin{bmatrix} 3 \\ -1 \end{bmatrix} \quad \mathbf{p}_9 = \begin{bmatrix} 0 \\ -1 \end{bmatrix}$$

6. Compute the area of the polygon with the following vertices.

$$\mathbf{p}_1 = \begin{bmatrix} -1 \\ 0 \end{bmatrix} \quad \mathbf{p}_2 = \begin{bmatrix} 0 \\ 1 \end{bmatrix} \quad \mathbf{p}_3 = \begin{bmatrix} 1 \\ 0 \end{bmatrix}$$

$$\mathbf{p}_4 = \begin{bmatrix} 1 \\ 2 \end{bmatrix} \quad \mathbf{p}_5 = \begin{bmatrix} -1 \\ 2 \end{bmatrix}$$

   Use both methods from Section 15.7.

7. Give an example of a nonsimple polygon that will pass the test for convexity, which uses the barycenter from Section 15.3.

16 Curves

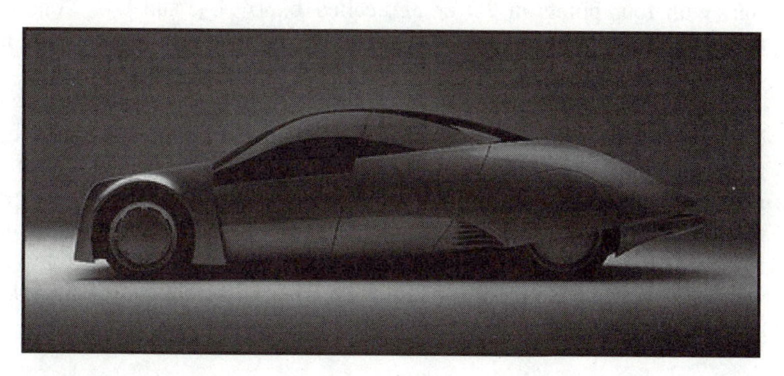

Figure 16.1.
Car design: curves are used to design cars such as the the Ford Synergy 2010 concept car. (Source: www.ford.com/concept)

Earlier in this book, we mentioned that all letters that you see here were designed by a font designer, and then put into a font library. The font designer's main tool is a cubic curve, also called a cubic Bézier curve. Such curves are handy for font design, but they were initially invented for car design. This happened in France in the early 1960's at Rénault and Citroën in Paris. These techniques are still in use today, as illustrated in Figure 16.1. We will briefly outline this kind of curve, and also apply previous geometric concepts to the study of curves in general. This type of work is called *Geometric Modeling* or *Computer Aided Geometric Design*, see [5] or [12].

16.1 Parametric Curves

You will recall that one way to write a straight line was the *parametric* form:

$$\mathbf{x}(t) = (1-t)\mathbf{a} + t\mathbf{b}.$$

If we interpret t as time, then this says at time $t = 0$, a moving point is at $\mathbf{a}$. It moves towards $\mathbf{b}$, and reaches it at time $t = 1$.

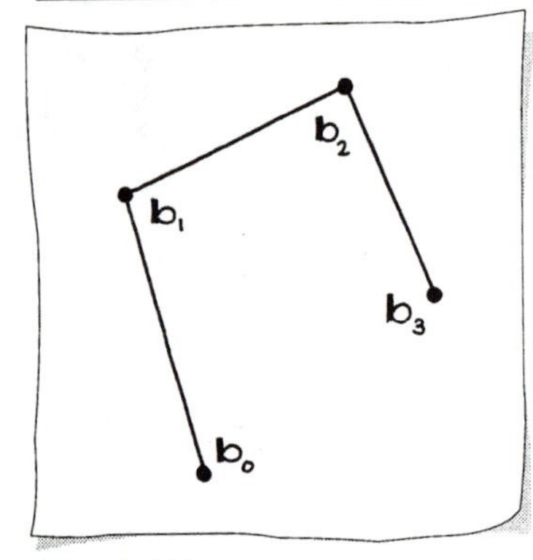

Sketch 154
A Bézier polygon.

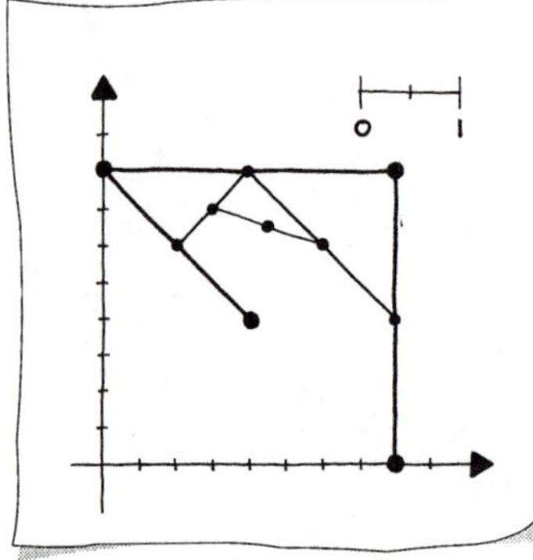

Sketch 155
The de Casteljau algorithm.

Let us be a bit more ambitious now and study motion along *curves*, i.e., paths that do not have to be straight. The simplest example is that of driving a car along a road. At time $t = 0$, you start, you follow the road, and at time $t = 1$, you have arrived somewhere. It does not really matter what kind of units we use to measure time; the $t = 0$ and $t = 1$ may just be viewed as a normalization of an arbitrary time interval.

We will now attack the problem of modeling curves, and we will choose a particularly simple way of doing this, namely cubic *Bézier curves*. We start with four points in 2D or 3D, called $\mathbf{b}_0, \mathbf{b}_1, \mathbf{b}_2$, and $\mathbf{b}_3$. Connect them with straight lines as shown in Sketch 154. The resulting polygon is called a *Bézier polygon.*

Here is how you generate a curve from it. Pick a parameter value t between 0 and 1. Find the corresponding point on each polygon leg by linear interpolation. This gives you three points $\mathbf{b}_0^1, \mathbf{b}_1^1, \mathbf{b}_2^1$. They form a polygon themselves — repeat the linear interpolation process and you get two points $\mathbf{b}_0^2, \mathbf{b}_1^2$. Repeat one more time, and you have a point $\mathbf{b}_0^3$. This is the point on the Bézier curve defined by $\mathbf{b}_0, \mathbf{b}_1, \mathbf{b}_2, \mathbf{b}_3$ at the parameter value t. The recursive process of applying linear interpolation is called the *de Casteljau algorithm*, and it is shown in Sketch 155.

Example 16.1 A numerical counterpart to Sketch 155 follows. Let the polygon be given by

$$\mathbf{b}_0 = \begin{bmatrix} 4 \\ 4 \end{bmatrix}, \quad \mathbf{b}_1 = \begin{bmatrix} 0 \\ 8 \end{bmatrix}, \quad \mathbf{b}_2 = \begin{bmatrix} 8 \\ 8 \end{bmatrix}, \quad \mathbf{b}_3 = \begin{bmatrix} 8 \\ 0 \end{bmatrix}.$$

For simplicity, let $t = 1/2$. Linear interpolation is then nothing but finding midpoints, and we have

$$\mathbf{b}_0^1 = \frac{1}{2}\mathbf{b}_0 + \frac{1}{2}\mathbf{b}_1 = \begin{bmatrix} 2 \\ 6 \end{bmatrix},$$

$$\mathbf{b}_1^1 = \frac{1}{2}\mathbf{b}_1 + \frac{1}{2}\mathbf{b}_2 = \begin{bmatrix} 4 \\ 8 \end{bmatrix},$$

$$\mathbf{b}_2^1 = \frac{1}{2}\mathbf{b}_2 + \frac{1}{2}\mathbf{b}_3 = \begin{bmatrix} 8 \\ 4 \end{bmatrix}.$$

Next,

$$\mathbf{b}_0^2 = \frac{1}{2}\mathbf{b}_0^1 + \frac{1}{2}\mathbf{b}_1^1 = \begin{bmatrix} 3 \\ 7 \end{bmatrix},$$

$$\mathbf{b}_1^2 = \frac{1}{2}\mathbf{b}_1^1 + \frac{1}{2}\mathbf{b}_2^1 = \begin{bmatrix} 6 \\ 6 \end{bmatrix},$$

and finally

$$\mathbf{b}_0^3 = \frac{1}{2}\mathbf{b}_0^2 + \frac{1}{2}\mathbf{b}_1^2 = \begin{bmatrix} \frac{9}{2} \\ \frac{13}{2} \end{bmatrix}.$$

This is the point on the curve corresponding to $t = 1/2$.

In general, we have

$$\begin{aligned}\mathbf{b}_0^3 &= (1-t)\mathbf{b}_0^2 + t\mathbf{b}_1^2 \\ &= (1-t)[(1-t)\mathbf{b}_0^1 + t\mathbf{b}_1^1] + t[(1-t)\mathbf{b}_1^1 + t\mathbf{b}_2^1] \\ &= (1-t)[(1-t)[(1-t)\mathbf{b}_0 + t\mathbf{b}_1] + t[(1-t)\mathbf{b}_1 + t\mathbf{b}_2] + t[(1-t)[(1-t)\mathbf{b}_1 + t\mathbf{b}_2] \\ &\quad + t[(1-t)\mathbf{b}_2 + t\mathbf{b}_3]].\end{aligned}$$

After some re-assembly, this becomes

$$\mathbf{b}_0^3(t) = (1-t)^3\mathbf{b}_0 + 3(1-t)^2t\mathbf{b}_1 + 3(1-t)t^2\mathbf{b}_2 + t^3\mathbf{b}_3. \qquad (16.1)$$

This is the general form of a cubic Bézier curve. As t traces out values between 0 and 1, the point $\mathbf{b}_0^3(t)$ traces out a curve. See Figure 16.2 for some examples. From now on, we will also use the shorter $\mathbf{b}(t)$ instead of $\mathbf{b}_0^3(t)$.

The original curve was given by the control polygon

$$\mathbf{b}_0, \mathbf{b}_1, \mathbf{b}_2, \mathbf{b}_3.$$

Inspection of Sketch 155 suggests that the two curve segments generated by $\mathbf{b}(t)$ also have control polygons. This is indeed so; the curve segment

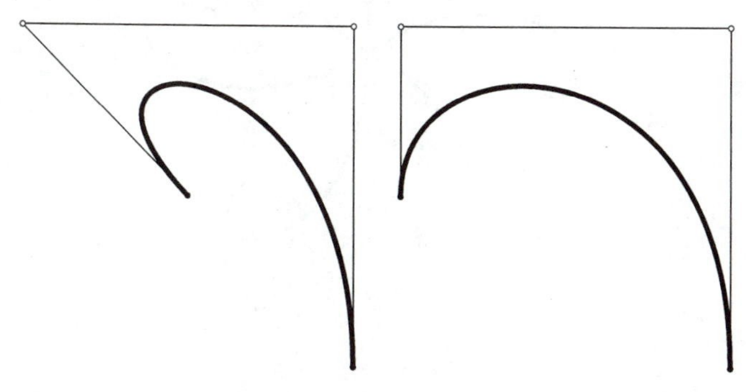

Figure 16.2.
Bézier curves: two examples.

from $\mathbf{b}_0$ to $\mathbf{b}(t)$ has

$$\mathbf{b}_0, \mathbf{b}_0^1, \mathbf{b}_0^2, \mathbf{b}_0^3$$

as its control polygon. The segment from $\mathbf{b}(t)$ to $\mathbf{b}_3$ has

$$\mathbf{b}_0^3, \mathbf{b}_1^2, \mathbf{b}_2^1, \mathbf{b}_3$$

as its control polygon. This process: generating two Bézier curves from one, is called *subdivision.*

16.2 Properties of Bézier Curves

From inspection of the examples, but also from (16.1), we see that the curve passes through the first and last control points:

$$\mathbf{b}(0) = \mathbf{b}_0, \quad \mathbf{b}(1) = \mathbf{b}_3. \tag{16.2}$$

If we map the control polygon using an affine map, then the curve undergoes the same transformation, as shown in Figure 16.3. This can be seen by inspecting the examples, but also by observing that the cubic

Figure 16.3.
Bézier curves: as the control polygon rotates, so does the curve.

Figure 16.4.
Bézier curves: the curve lies in the convex hull of the control polygon.

coefficients of the control points in (16.1) sum to one. This can be seen as follows:

$$(1-t)^3 + 3(1-t)^2 t + 3(1-t)t^2 + t^3 = [(1-t)+t]^3 = 1.$$

Thus every point on the curve is a *barycentric combination* of the control points. Such relationships are not changed under affine maps, as per Section 6.1.

The curve also lies in the convex hull of the control polygon — a fact called the *convex hull property* . This can be seen by observing that the coefficients of the control points in (16.1) are nonnegative (recall that we only consider values of t between 0 and 1). It follows that every point on the curve is a *convex combination* of the control points, and hence is inside their *convex hull.* For a definition, see Section 8.4; for an illustration, see Figure 16.4.

Clearly the control polygon is inside its minmax box.[1] Because of the convex hull property, we also know that the curve is inside this box — a property that has numerous applications. See Figure 16.5 for an illustration.

[1] Recall that the minmax box of a polygon is the smallest rectangle with edges parallel to the coordinate axes that contains the polygon.

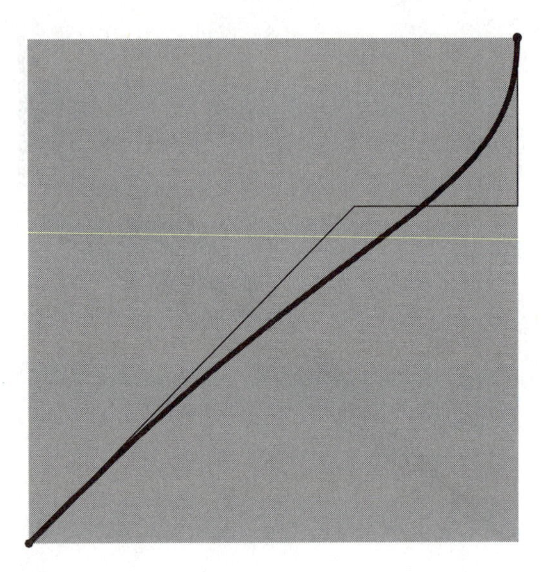

Figure 16.5.
Bézier curves: the curve lies inside the minmax box of the control polygon.

16.3 The Matrix Form

As a preparation for what is to follow, let us rewrite (16.1) using the formalism of dot products. It then looks like this:

$$\mathbf{b}(t) = \begin{bmatrix} \mathbf{b}_0 & \mathbf{b}_1 & \mathbf{b}_2 & \mathbf{b}_3 \end{bmatrix} \begin{bmatrix} (1-t)^3 \\ 3(1-t)^2 t \\ 3(1-t)t^2 \\ t^3 \end{bmatrix}. \qquad (16.3)$$

Most people think of polynomials as combinations of the *monomials*; they are $1, t, t^2, t^3$ for the cubic case. Our expression (16.1) may be rewritten into this form:

$$\mathbf{b}(t) = \mathbf{b}_0 + 3t(\mathbf{b}_1 - \mathbf{b}_0) + 3t^2(\mathbf{b}_2 - 2\mathbf{b}_1 + \mathbf{b}_0) + t^3(\mathbf{b}_3 - 3\mathbf{b}_2 + 3\mathbf{b}_1 - \mathbf{b}_0).$$

This allows a more concise formulation using matrices:

$$\mathbf{b}(t) = \begin{bmatrix} \mathbf{b}_0 & \mathbf{b}_1 & \mathbf{b}_2 & \mathbf{b}_3 \end{bmatrix} \begin{bmatrix} 1 & -3 & 3 & -1 \\ 0 & 3 & -6 & 3 \\ 0 & 0 & 3 & -3 \\ 0 & 0 & 0 & 1 \end{bmatrix} \begin{bmatrix} 1 \\ t \\ t^2 \\ t^3 \end{bmatrix}. \qquad (16.4)$$

This is the matrix form of a Bézier curve.

Equation (16.4) shows how to write a Bézier curve in monomial form. A curve in monomial form looks like this:

$$\mathbf{b}(t) = \mathbf{a}_0 + \mathbf{a}_1 t + \mathbf{a}_2 t^2 + \mathbf{a}_3 t^3.$$

Rewritten using the dot product form, this becomes

$$\mathbf{b}(t) = \begin{bmatrix} \mathbf{a}_0 & \mathbf{a}_1 & \mathbf{a}_2 & \mathbf{a}_3 \end{bmatrix} \begin{bmatrix} 1 \\ t \\ t^2 \\ t^3 \end{bmatrix}.$$

Thus the monomial coefficients $\mathbf{a}_i$ are defined as

$$\begin{bmatrix} \mathbf{a}_0 & \mathbf{a}_1 & \mathbf{a}_2 & \mathbf{a}_3 \end{bmatrix} = \begin{bmatrix} \mathbf{b}_0 & \mathbf{b}_1 & \mathbf{b}_2 & \mathbf{b}_3 \end{bmatrix} \begin{bmatrix} 1 & -3 & 3 & -1 \\ 0 & 3 & -6 & 3 \\ 0 & 0 & 3 & -3 \\ 0 & 0 & 0 & 1 \end{bmatrix}. \tag{16.5}$$

How about the inverse process: If we are given a curve in monomial form, how can we write it as a Bézier curve? Simply rearrange (16.5) to solve for the $\mathbf{b}_i$:

$$\begin{bmatrix} \mathbf{b}_0 & \mathbf{b}_1 & \mathbf{b}_2 & \mathbf{b}_3 \end{bmatrix} = \begin{bmatrix} \mathbf{a}_0 & \mathbf{a}_1 & \mathbf{a}_2 & \mathbf{a}_3 \end{bmatrix} \begin{bmatrix} 1 & -3 & 3 & -1 \\ 0 & 3 & -6 & 3 \\ 0 & 0 & 3 & -3 \\ 0 & 0 & 0 & 1 \end{bmatrix}^{-1}.$$

A matrix inversion is all that is needed here!

Notice that the square matrix in this equation is nonsingular. Because of its nonsingularity, we can conclude that any cubic curve can be written in either the Bézier or the monomial form.

16.4 Derivatives

Equation (16.1) consists of two (in 2D) or three (in 3D) cubic equations in t. We can take the derivative in each of the components:

$$\frac{\mathrm{d}\mathbf{b}(t)}{\mathrm{d}t} = -3(1-t)^2\mathbf{b}_0 - 6(1-t)t\mathbf{b}_1 + 3(1-t)^2\mathbf{b}_1 - 3t^2\mathbf{b}_2 + 6(1-t)t\mathbf{b}_2 + 3t^2\mathbf{b}_3.$$

Rearranging, and using the abbreviation $\frac{\mathrm{d}\mathbf{b}(t)}{\mathrm{d}t} = \dot{\mathbf{b}}(t)$, we have

$$\dot{\mathbf{b}}(t) = 3(1-t)^2[\mathbf{b}_1 - \mathbf{b}_0] + 6(1-t)t[\mathbf{b}_2 - \mathbf{b}_1] + 3t^2[\mathbf{b}_3 - \mathbf{b}_2]. \tag{16.6}$$

As expected, the derivative of a degree three curve is one of degree two.[2]

For $t = 0$, we obtain

$$\dot{\mathbf{b}}(0) = 3[\mathbf{b}_1 - \mathbf{b}_0],$$

and, similarly, for $t = 1$,

$$\dot{\mathbf{b}}(1) = 3[\mathbf{b}_3 - \mathbf{b}_2].$$

In words, the control polygon is tangent to the curve at the curve's endpoints. This is not a surprising statement if you check the example figures!

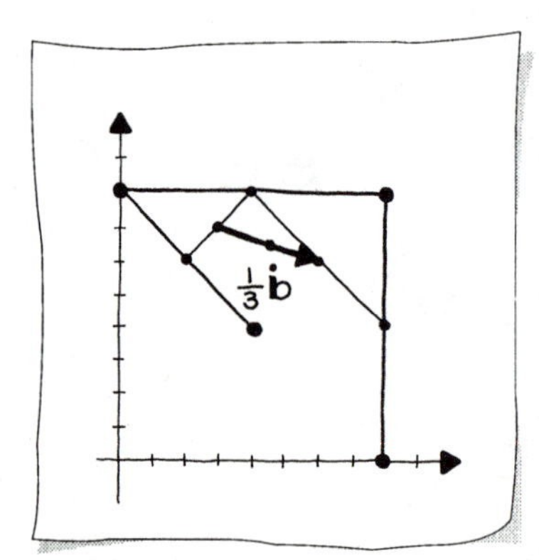

Sketch 156
A derivative vector.

Example 16.2 Let us compute the derivative of the example curve from Section 16.1 for $t = 1/2$. We obtain

$$\dot{\mathbf{b}}(\frac{1}{2}) = 3\cdot\frac{1}{4}[\begin{bmatrix} 0 \\ 8 \end{bmatrix} - \begin{bmatrix} 4 \\ 4 \end{bmatrix}] + 6\cdot\frac{1}{4}[\begin{bmatrix} 8 \\ 8 \end{bmatrix} - \begin{bmatrix} 0 \\ 8 \end{bmatrix}] + 3\cdot\frac{1}{4}[\begin{bmatrix} 8 \\ 0 \end{bmatrix} - \begin{bmatrix} 8 \\ 8 \end{bmatrix}],$$

which yields

$$\dot{\mathbf{b}}(\frac{1}{2}) = \begin{bmatrix} 9 \\ -3 \end{bmatrix}.$$

See Sketch 156 for an illustration.

Note that the derivative of a curve is a *vector*. It is tangent to the curve — apparent from our example, but nothing we want to prove here. A convenient way to think about the derivative is by interpreting it as a *velocity vector*. If you interpret the parameter t as time, and you think of traversing the curve such that at time t you have reached $\mathbf{b}(t)$, then the derivative measures your velocity. The larger the magnitude of the tangent vector, the faster you move.

If we rotate the control polygon, the curve will follow, and so will all of its derivative vectors. In calculus, a "horizontal tangent" has a special meaning; it indicates an extreme value of a function. Not here: the very notion of an extreme value is meaningless for parametric curves since the term "horizontal tangent" depends on the curve's orientation and is not a property of the curve itself.

We may take the derivative of (16.6) with respect to t. We then have the *second derivative*. It is given by

$$\ddot{\mathbf{b}}(t) = -6(1-t)[\mathbf{b}_1 - \mathbf{b}_0] - 6t[\mathbf{b}_2 - \mathbf{b}_1] + 6(1-t)[\mathbf{b}_2 - \mathbf{b}_1] + 6t[\mathbf{b}_3 - \mathbf{b}_2]$$

[2]Note that the derivative curve does not have control *points* anymore, but rather *control vectors*!

and may be rearranged to

$$\ddot{\mathbf{b}}(t) = 6(1-t)[\mathbf{b}_2 - 2\mathbf{b}_1 + \mathbf{b}_0] + 6t[\mathbf{b}_3 - 2\mathbf{b}_2 + \mathbf{b}_1]. \quad (16.7)$$

The second derivative at $\mathbf{b}_0$ (see Sketch 157) is particularly simple — it is given by

$$\ddot{\mathbf{b}}(0) = 6[\mathbf{b}_2 - 2\mathbf{b}_1 + \mathbf{b}_0].$$

Loosely speaking, we may interpret the second derivative $\ddot{\mathbf{b}}(t)$ as acceleration when traversing the curve.

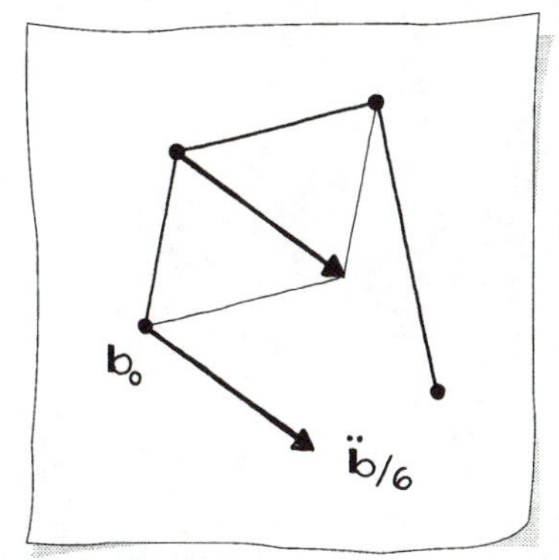

Sketch 157
A second derivative vector.

16.5 Composite Curves

A Bézier curve is a handsome tool, but one such curve would rarely suffice for describing much of any shape! For "real" shapes, we have to be able to line up many cubic Bézier curves. In order to define a smooth overall curve, these pieces must join smoothly.

This is easily achieved. Let $\mathbf{b}_0, \mathbf{b}_1, \mathbf{b}_2, \mathbf{b}_3$ and $\mathbf{c}_0, \mathbf{c}_1, \mathbf{c}_2, \mathbf{c}_3$ be the control polygons of two Bézier curves with a common point $\mathbf{b}_3 = \mathbf{c}_0$ (see Sketch 158), If the two curves are to have the same tangent vector direction at $\mathbf{b}_3 = \mathbf{c}_0$, then all that is required is

$$\mathbf{c}_1 - \mathbf{c}_0 = c[\mathbf{b}_3 - \mathbf{b}_2] \quad (16.8)$$

for some positive real number c, meaning that the three points $\mathbf{b}_2, \mathbf{b}_3 = \mathbf{c}_0, \mathbf{c}_1$ are collinear.

If we use this rule to piece curve segments together, we can design many 2D and 3D shapes; Figure 16.6 gives an example.

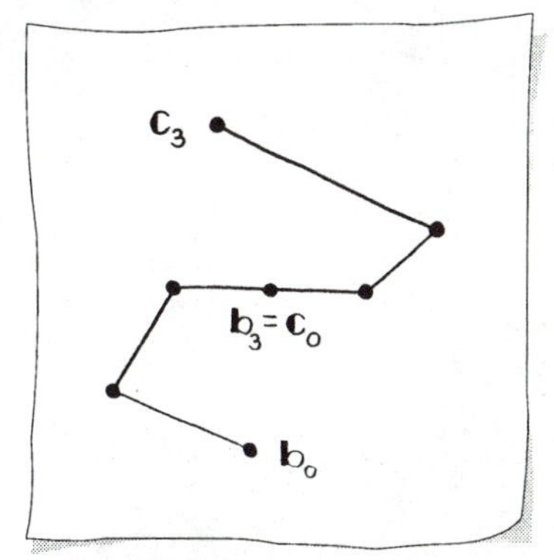

Sketch 158
Smoothly joining Bézier curves.

16.6 The Geometry of Planar Curves

The geometry of planar curves is centered around one concept: their *curvature*. It is easily understood if you imagine driving a car along a road. For simplicity, let's assume you are driving with constant speed. If the road does not curve, i.e., it is straight, you will not have to tilt your steering wheel. When the road does curve, you will have to tilt the steering wheel, and more so if the road curves rapidly. The curviness of the road (our model of a curve) is thus proportional to the tilting of the steering wheel.

Returning to the more abstract concept of a curve, let us sample its tangents at various points (see Sketch 159). Where the curve bends sharply, i.e., where its curvature is high, successive tangents differ from each other significantly. In areas where the curve is relatively flat, or where its curvature is low, successive tangents are almost identical. Curvature may thus

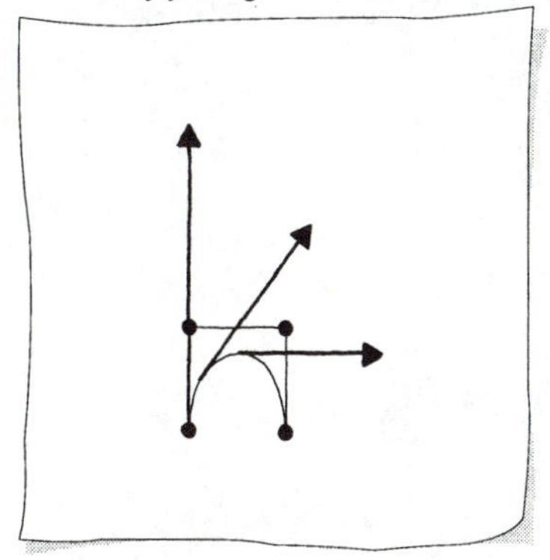

Sketch 159
Tangents on a curve.

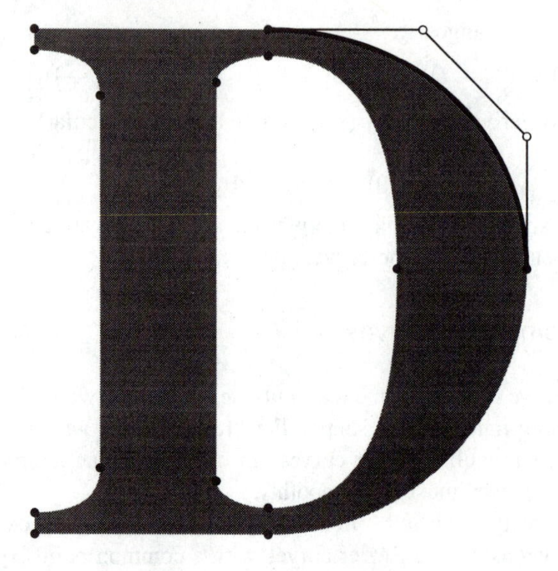

Figure 16.6.
Bézier curves: the letter "D" as a collection of cubic Bézier curves. Only one Bézier curve is shown.

be defined as *rate of change of tangents*. (In terms of our car example, the rate of change of tangents is proportional to the tilt of the steering wheel.)

Since the tangent is determined by the curve's first derivative, its rate of change should be determined by the second derivative. This is indeed so, but the actual formula for curvature is a bit more complex than can be derived in the context of this book. We denote the curvature of the curve at $\mathbf{b}(t)$ by κ; it is given by

$$\kappa(t) = \frac{\|\dot{\mathbf{b}} \wedge \ddot{\mathbf{b}}\|}{\|\dot{\mathbf{b}}\|^3}. \tag{16.9}$$

This formula holds for both 2D and 3D curves. In the 2D case, it may be rewritten as

$$\kappa(t) = \frac{\left| \, \dot{\mathbf{b}} \quad \ddot{\mathbf{b}} \, \right|}{\|\dot{\mathbf{b}}\|^3} \tag{16.10}$$

with the use of a 2×2 determinant. Since determinants may be positive or negative, curvature in 2D is *signed*. A point where $\kappa = 0$ is called an *inflection point*: the 2D curvature changes sign here. In Figure 16.7, the inflection point is marked.

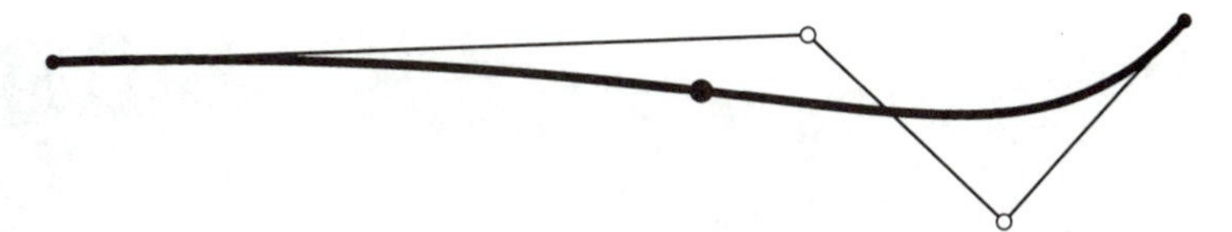

Figure 16.7.
Bézier curves: an inflection point is marked on the curve.

In calculus, you learned that a curve has an inflection point if the second derivative vanishes. For parametric curves, the situation is different. An inflection point occurs when the first and second derivative vectors are parallel, or linearly dependent. This can lead to the curious effect of a cubic with *two* inflection points. It is illustrated in Figure 16.8.

16.7 Moving along a Curve

Take a look at Figure 16.9. You will see the letter "B" sliding along a curve. If the curve is given in Bézier form, how can that effect be achieved?

The answer can be seen in Sketch 160. If you want to position an object, such as the letter "B," at a point on a curve, all you need to know is the point and the curve's tangent there.

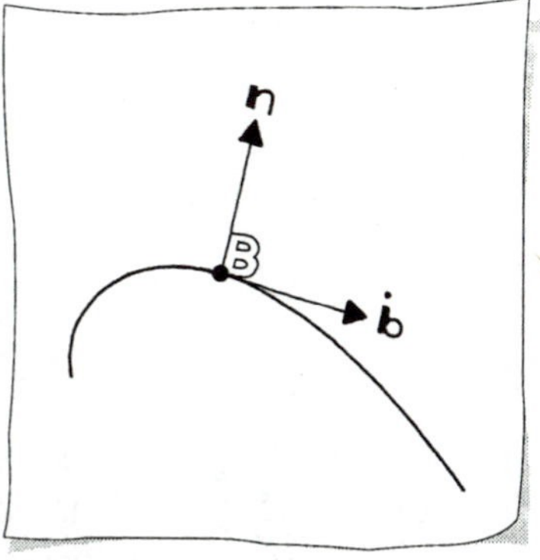

Sketch 160
Sliding along a curve.

Figure 16.8.
Bézier curves: a cubic with two inflection points.

Figure 16.9.
Curve motions: a letter is moved along a curve.

If $\dot{\mathbf{b}}$ is the tangent, then simply define $\mathbf{n}$ to be a vector perpendicular to it.[3] Using the local coordinate system with origin $\mathbf{b}(t)$ and $[\dot{\mathbf{b}}, \mathbf{n}]$-axes, you can position any object as per Section 4.1.

The same story is far trickier in 3D! If you had a point on the curve and its tangent, the exact location of your object would not be fixed; it could still rotate around the tangent. Yet there is a unique way to position objects along a 3D curve. At every point on the curve, we may define a *local coordinate system* as follows.

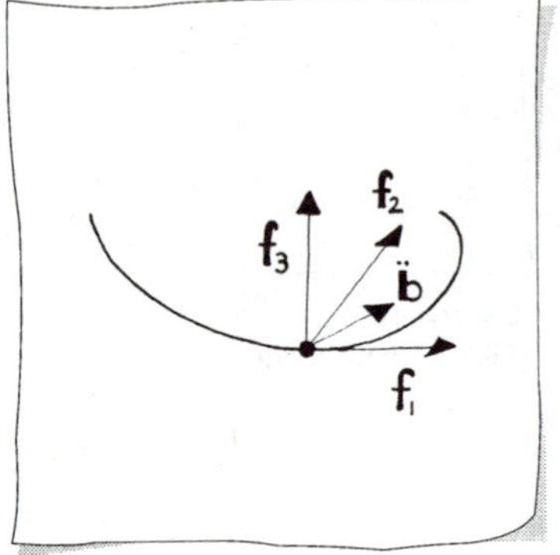

Sketch 161
A Frenet frame.

Let the point on the curve be $\mathbf{b}(t)$; we now want to set up a local coordinate system defined by three vectors $\mathbf{f}_1, \mathbf{f}_2, \mathbf{f}_3$. Following the 2D example, we set $\mathbf{f}_1$ to be in the tangent direction; $\mathbf{f}_1 = \dot{\mathbf{b}}(t)$. If the curve does not have an inflection point at t, then $\dot{\mathbf{b}}(t)$ and $\ddot{\mathbf{b}}(t)$ will not be collinear. This means that they span a plane, and that plane's normal is given by $\dot{\mathbf{b}}(t) \wedge \ddot{\mathbf{b}}(t)$. See Sketch 161 for some visual information. We make the plane's normal one of our local coordinate axes, namely $\mathbf{f}_3$. The plane, by the way, has a name: it is called the *osculating plane* at $\mathbf{x}(t)$. Since we have two coordinate axes, namely $\mathbf{f}_1$ and $\mathbf{f}_3$, it is not hard to come up with the remaining axis, we just set $\mathbf{f}_2 = \mathbf{f}_1 \wedge \mathbf{f}_3$. Thus for every point on the curve (as long as it is not an inflection point), there exists an orthogonal coordinate system. It is customary to use coordinate axes of unit length, and then we have

$$\mathbf{f}_1 = \frac{\dot{\mathbf{b}}(t)}{\|\dot{\mathbf{b}}(t)\|}, \tag{16.11}$$

$$\mathbf{f}_3 = \frac{\dot{\mathbf{b}}(t) \wedge \ddot{\mathbf{b}}(t)}{\|\dot{\mathbf{b}}(t) \wedge \ddot{\mathbf{b}}(t)\|}, \tag{16.12}$$

$$\mathbf{f}_2 = \mathbf{f}_1 \wedge \mathbf{f}_3. \tag{16.13}$$

This system with local origin $\mathbf{b}(t)$ and normalized axes $\mathbf{f}_1, \mathbf{f}_2, \mathbf{f}_3$ is called the *Frenet frame* of the curve at $\mathbf{b}(t)$. Equipped with the tool of Frenet frames, we may now position objects along a 3D curve! See Figure 16.10.

[3] If $\dot{\mathbf{b}} = \begin{bmatrix} \dot{b}_1 \\ \dot{b}_2 \end{bmatrix}$ then $\mathbf{n} = \begin{bmatrix} -\dot{b}_2 \\ \dot{b}_1 \end{bmatrix}$.

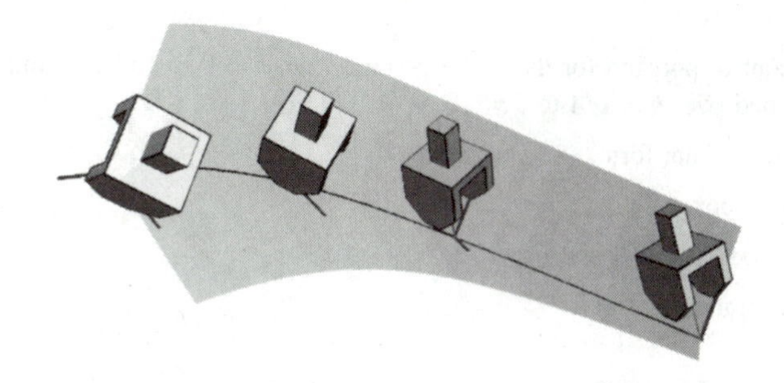

Figure 16.10.
Curve motions: a robot arm is moved along a curve. (Courtesy of M. Wagner, Arizona State University.)

Let us now work out exactly how to carry out our object-positioning plan. The object is given in some local coordinate system with axes $\mathbf{u}_1, \mathbf{u}_2, \mathbf{u}_3$. Any point of the object has coordinates

$$\mathbf{u} = \begin{bmatrix} u_1 \\ u_2 \\ u_3 \end{bmatrix}.$$

It is mapped to

$$\mathbf{x}(t, \mathbf{u}) = \mathbf{x}(t) + u_1\mathbf{f}_1 + u_2\mathbf{f}_2 + u_3\mathbf{f}_3.$$

A typical application is *robot motion*. Robots are used extensively in automotive assembly lines; one job is to grab a part and move it to its destination inside the car body. This movement happens along well-defined curves. While the car part is being moved, it has to be oriented into its correct position — exactly the process described in this section!

16.8 Exercises

Let a cubic Bézier curve be given by the control polygon

$$\mathbf{b}_0 = \begin{bmatrix} 0 \\ 0 \\ 0 \end{bmatrix}, \quad \mathbf{b}_1 = \begin{bmatrix} 4 \\ 0 \\ 0 \end{bmatrix}, \quad \mathbf{b}_2 = \begin{bmatrix} 4 \\ 4 \\ 0 \end{bmatrix}, \quad \mathbf{b}_3 = \begin{bmatrix} 4 \\ 4 \\ 4 \end{bmatrix}.$$

1. Sketch this 3D curve manually.
2. Using the de Casteljau algorithm, evaluate it for $t = 1/4$.
3. Evaluate the first and second derivative for $t = 1/4$. Add these vectors to the sketch from exercise 1.
4. What is the control polygon for the curve defined from $t = 0$ to $t = 1/4$ and the curve defined over $t = 1/4$ to $t = 1$?
5. Rewrite it in monomial form.
6. Find its minmax box.
7. Find its curvatures at $t = 0$ and $t = 1/2$.
8. Find its Frenet frame for $t = 1$.

A PostScript Tutorial

The figures in this book are created using the PostScript language. This is a mini-guide to PostScript with the purpose of giving you enough information so you can alter these images or create similar ones yourself.

PostScript is a page description language.[1] A PostScript program tells the output device (printer or previewer) how to format a printed page. Most laser printers today understand the PostScript language. A previewer, such as Ghostview or xpsview, allows you to view your document without printing — a great way to save paper.

Before proceeding, check if you have a previewer available. If not, there are free versions available. Ghostview for instance can be found at

http://www.cs.wisc.edu/ghost/index.html.

In case you would like more in-depth information about PostScript, see [14, 13] or check out

http://www.cs.indiana.edu/docproject/programming/postscript/

for more help.

A.1 A Warm-Up Example

Let's go through the PostScript file that generates Figure A.1.

```
%!
newpath
  200 200 moveto
  300 200 lineto
  300 300 lineto
  200 300 lineto
  200 200 lineto
stroke

showpage
```

[1] Its origins are in the late seventies, when it was developed at Evans & Sutherland, Xerox, and finally by Adobe Systems, who now owns it.

Figure A.1.
A simple PostScript example.

Figure A.1 shows the result of this program: we have drawn a box on a standard size page. (In this figure, the outline of the page is shown for clarity; everything is reduced to fit here.)

The first line of the file is "%!". Nothing else belongs on this line, and there are no extra spaces on the line. This command tells the printer that what is to come is in the PostScript language.

The actual drawing is done with the `newpath` command. Move to the starting position with `moveto`. Record the path with `lineto`. Finally, indicate that the path should be drawn with `stroke`. These commands simulate the movement of a "virtual pen" in a fairly obvious way.

PostScript uses *prefix notation* for its commands. So if you want to move your "pen" to position $(100, 200)$, the command is `100 200 moveto`.

Finally, you need to invoke the `showpage` command to cause PostScript to actually print your figure. This is important!

Now for some variation:

```
%!
% This is a comment line because it begins with a percent sign
% draw the box
newpath
   200 200 moveto
   300 200 lineto
   300 300 lineto
   200 300 lineto
   200 200 lineto
stroke

80 80 translate
0.5 setgray

newpath
   200 200 moveto
   300 200 lineto
   300 300 lineto
   200 300 lineto
   200 200 lineto
fill

showpage
```

This is illustrated in Figure A.2. The `80 80 translate` command moves the origin of your current coordinate system by 80 units in both the $\mathbf{e}_1$- and $\mathbf{e}_2$-directions. The `0.5 setgray` command causes the next item to be drawn in a gray shade (0 is black, 1 is white).

What follows is the same square as before; instead of `stroke`, however, there is a `fill`. This fills the square with the specified gray scale. Note how the second square is drawn on top of the first one!

Another handy basic geometry element is a circle. To create a circle, use the command

```
250 250 50 0 360 arc stroke
```

This generates a circle centered at $(250, 250)$ with radius 50. The `0 360 arc` indicates we want the whole circle. To create a white filled circle change the command to

```
1.0 setgray
250 250 50 0 360 arc fill
```

Figure A.3 illustrates all these additions.

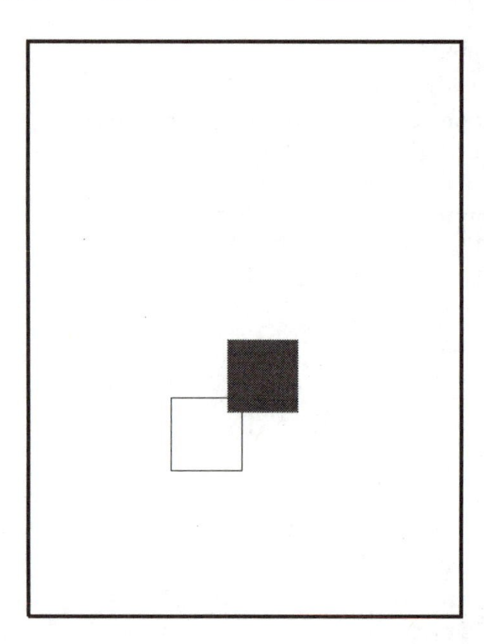

Figure A.2.
Another simple PostScript example.

A.2 Overview

The PostScript language has basic graphics operators: shapes such as line or arc, painting, text, bitmapped images, and coordinate transformation. Variables and calculations are also in its repertoire; it is a powerful tool. Nearly all the figures for this book are available as PostScript (text) files at the book's web site. To illustrate the aspects of the file format, we'll return to some of these figures.

We use three basic scenarios for generating PostScript files.

1. A program (such as C) generates geometry, then opens a file and writes PostScript commands to plot the geometry. This type of file is typically filled simply with move, draw, and style (gray scale or line width) commands. Example file: `Bez_ex.ps` which is displayed in Section 16.2.

2. The PostScript language is used to generate geometry. Using a text editor, the PostScript program is typed in. This program might even use control structures such as "for loops." Example file: `D_trans.ps` which is displayed in Section 6.2.

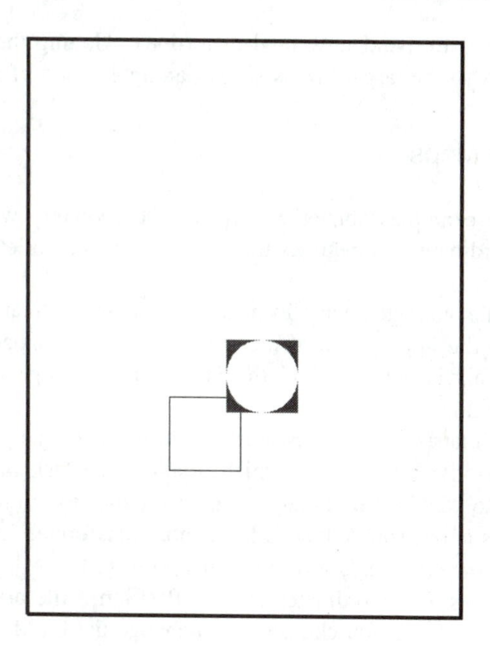

Figure A.3.
Yet another simple PostScript example.

3. A screen dump creating a bit mapped image is made part of a PostScript file. This is a "save' option in the Netscape Web browser, for example. Example file: `Flight.ps` which is displayed in Chapter 12.

There are also figures which draw from more than one of these scenarios.

Since PostScript tells the printer how to format a page, it is necessary for the move and draw commands to indicate locations on the piece of paper. For historical reasons, printers use a coordinate system based on *points*, abreviated as pt.

$$1 \text{ inch} \equiv 72\text{pt}.$$

PostScript follows suit. This means that an $8\frac{1}{2} \times 11$ inch page has the following extents,

$$\text{lower left}: \ (0,0) \qquad \text{upper right}: \ (612,792).$$

Whichever scenario from above is followed, it is always necessary to be sure that the PostScript commands are drawing in the appropriate area. Keep in mind you probably want a margin. Chapter 1 covers the basics of setting up the dimensions of the 'target box' (on the paper) and that of the 'source box' to avoid unwanted distortions. Getting the geometry in the source box to the target box is simply an application of affine maps!

A.3 Affine Maps

In our simple examples above, we created the geometry with respect to the paper coordinates. Sometimes this is inconvenient, so let's discuss the options.

If you create your geometry in a, say C, program, then it is an easy task to apply the appropriate affine map to put it in paper coordinates. However, if this is not the case, the PostScript language has the affine map tools builtin.[2]

There is a matrix (which describes an affine map, i.e., a linear map and a translation) in PostScript which assumes the task of taking "user coordinates" to "device coordinates". In accordance to the terminology of this book, this is our source box to target box transformation. This matrix is called the *current transformation matrix*, or CTM.

In other words, the coordinates in your PostScript file are always multiplied by the CTM. If you choose not to change the CTM, then your coordinates must live within the page coordinates, or "device coordinates". Here we give a few details on how to alter the CTM.

There are two "levels" of changing the CTM. The simplest, and most basic ones use the `scale`, `rotate`, and `translate` commands. As we know from Chapter 6, these are essentially the most basic affine operations.

Unless you have a complete understanding of the CTM, it is probably a good idea to use each of these commands only once in a PostScript file. It can get confusing! (See Section A.6.)

A scale command such as

```
72 72 scale
```

automatically changes one "unit" from being a *point* to being an inch. A translate command such as

```
2 2 translate
```

will translate the origin to coordinates $(2, 2)$. If this `translate` was preceded by the `scale` command, the effect is different. Try both options for yourself!

---

[2]There are many techniques for displaying 3D geometry in 2D; some "realistic" methods are not in the realms of affine geometry. A graphics text should be consulted to determine what is best.

A rotate command such as

```
45 rotate
```

will cause a rotation of the coordinate system by 45 degrees in a counter-clockwise direction.

These commands are used in the web site files, `D_scale.ps`, `D_trans.ps`, and `D_rot.ps`.

Instead of the commands above, a more general manipulation of the CTM is available, see Section A.6.

A.4 Variables

The figures in this book use variables quite often. This is a powerful tool that allows a piece of geometry to be defined once, and then affine maps can be applied to it to change its appearance on the page.

Let's use an example to illustrate. Take the file for Figure A.2. We can rewrite this as

```
%!

% define the box
/box {
    200 200 moveto
    300 200 lineto
    300 300 lineto
    200 300 lineto
    200 200 lineto
} def

newpath
box
stroke

80 80 translate
0.5 setgray

newpath
box
fill

showpage
```

The figure does not change. The box that was repeated is defined only once now. Notice the `/box {...} def` structure. This defines the variable `box`. It is then used without the forward slash.

A.5 Loops

The ability to create "for loops" is another very powerful tool. If you are not familiar with the prefix this might look odd. Let's look at the file `D_trans.ps`, which is displayed in Figure 6.3.

```
%!
%%BoundingBox: 90 100 375 300
/Times-Bold findfont
70 scalefont setfont
/printD{

 0 0 moveto
 (D) show
}def


100 100 translate

2.5 2.5 scale
.95 -.05 0{ setgray printD 3 1 translate} for

showpage
```

Before getting to the loop, let's look at a few other new things in the file. The `BoundingBox` command is not used by PostScript; this is to help in the placement of the figure in a LaTeX file. The `Times-Bold findfont` command allows us to access a particular set of fonts — we want to draw the letter D.

Now for the "for loop." The command above

```
.95 -0.5 0  {...} for
```

tells PostScript to start with the value 0.95 and decrement by 0.5 until it reaches 0. At each step it will execute the commands within the parenthesis. This allows the D to be printed 19 times in different gray scales and translated each time.

A.6 CTM

One of the most complicated figure files is Nocomm.ps, which is illustrated in Figure 4.12. Here is its listing.

```
%!
%%BoundingBox: 50 50  350 250
% show that matrix multiply does not commute

% define a set of unit vectors with varying gray scale.
% if a vector points at 0 or 90 degrees it gets painted
% in a special way.
/vectors

{       /inc 5 360 div def
        /deg 0 def
        /gray 0 def
        0 6 354
        {
            /gray gray inc add def
            gray setgray
            newpath
            0 0 moveto
            0 2 lineto
            100 2 lineto
            100 5 lineto
            120 0 lineto
            100 -5 lineto
            100 -2 lineto
            0 -2 lineto
            0 0 lineto
            deg 90 eq  deg 0 eq or
            { stroke } {fill} ifelse
            6 rotate
            /deg 6 deg add def
        } for
} def

%definition of the two matrices:
% rotate 90 deg
/mat1[0 1 -1 0  0 0] def
```

```
% shear in x-dir
/mat2[1  0  0.5  1 0 0] def

% inverse matrices defined - computed below
/mat1inv[0 0 0 0 0 0]def
/mat2inv[0 0 0 0 0 0]def

% create the inverse of the shear and rotation
mat1 mat1inv invertmatrix
mat2 mat2inv invertmatrix

% move to place on page and scale to fit
100 200 translate
0.35 0.35 scale

% plot the vectors without a transformation
     vectors

% plot vectors with rotation
     250 0 translate
     mat1 concat
     vectors

% undo the rotation
     mat1inv concat

% plot vectors with rotation then shear
     250 0 translate
     mat2 concat
     mat1 concat
     vectors

% undo shear then rotation -- order important!
     mat1inv concat
     mat2inv concat

% move to a new row and plot vector without a transf.
     -500 -300 translate
     vectors

% plot vectors with shear
     250 0 translate
```

```
    mat2 concat
    vectors

% undo shear
    mat2inv concat

% plot vectors with shear then rotation
    250 0 translate
    mat1 concat
    mat2 concat
    vectors

% undo rotation then shear (not needed but to illustrate)
    mat2inv concat
    mat1inv concat

showpage
```

This one is commented a bit more than the original! Hopefully the comments help you get a hang of the language. What makes this figure more complicated than others is that it manipulates the CTM directly rather than only using the `translate`, `scale`, and `rotate` commands.

Here is the tricky part. PostScript uses a transformation matrix for "left multiply." This book works on the basis of "right multiply".[3] Specifically, this book applies a transformation to a point $\mathbf{p}$ as

$$\mathbf{p}' = \begin{bmatrix} a & c \\ b & d \end{bmatrix} \begin{bmatrix} p_1 \\ p_2 \end{bmatrix},$$

whereas PostScript would apply the transformation as

$$\mathbf{p}' = \begin{bmatrix} p_1 & p_2 \end{bmatrix} \begin{bmatrix} a & b \\ c & d \end{bmatrix}.$$

So the PostScript CTM is the transpose of the matrix in this book.

A matrix definition in PostScript has the translation vector included. The matrix has the syntax

$$[a, b, c, d, t_x, t_y].$$

[3]The computer graphics community typically uses left multiply whereas mathematicians typically use right multiply.

From the listing above, let's look at the part where we want to rotate, then shear.

```
% plot vectors with rotation then shear
    250 0 translate
    mat2 concat
    mat1 concat
    vectors
```

From the notation in this book, we would write

$$\mathbf{p}' = SR\mathbf{p},$$

where S is a shear matrix and R is a rotation matrix. For left multiply, this becomes

$$\mathbf{p}' = \mathbf{p}^T RS.$$

So our transformation matrix T is $T = SR$, but the PostScript matrix is $T = RS$! This is reflected in the PostScript code segment by concatenating (multiplying) the S matrix `mat2` to the CTM first, then concatenating the R matrix.

This same idea is reflected in the procedure to "undo" the shear and rotation from the CTM. To restore the CTM we apply,

$$S^{-1}R^{-1}RS(CTM),$$

which means that the rotation gets taken off first, then the shear. This is the opposite order they were put on.

That should be enough detail to put you well on your way to becoming a PostScript expert!

Selected Problem Solutions B

Chapter 1

1.1.a) The triangle vertex $(0.1, 0.1)$ in the $[\mathbf{d}_1, \mathbf{d}_2]$-system is mapped to

$$x_1 = 0.9 \times 1 + 0.1 \times 3 = 1.2,$$
$$x_2 = 0.9 \times 2 + 0.1 \times 3 = 2.1.$$

The triangle vertex $(0.9, 0.2)$ in the $[\mathbf{d}_1, \mathbf{d}_2]$-system is mapped to

$$x_1 = 0.1 \times 1 + 0.9 \times 3 = 2.8,$$
$$x_2 = 0.8 \times 2 + 0.2 \times 3 = 2.2.$$

The triangle vertex $(0.4, 0.7)$ in the $[\mathbf{d}_1, \mathbf{d}_2]$-system is mapped to

$$x_1 = 0.6 \times 1 + 0.4 \times 3 = 1.8,$$
$$x_2 = 0.3 \times 2 + 0.7 \times 3 = 2.7.$$

1.1.b) The point $(2, 2)$ in the $[\mathbf{e}_1, \mathbf{e}_2]$-system is mapped to

$$u_1 = \frac{2-1}{3-1} = \frac{1}{2},$$
$$u_2 = \frac{2-2}{3-2} = 0.$$

1.2. The local coordinate $(0.5, 0, 0.7)$ is mapped to

$$x_1 = 1 + 0.5 \times 1 = 1.5,$$
$$x_2 = 1 + 0 \times 2 = 1,$$
$$x_3 = 1 + 0.7 \times 4 = 3.8.$$

Chapter 2

2.4. A triangle.

2.5. The length of the vector

$$\mathbf{v} = \begin{bmatrix} -4 \\ -3 \end{bmatrix} \quad \text{is}$$

$$\|\mathbf{v}\| = \sqrt{4^2 + 3^2} = 5.$$

2.7. The normalized vector

$$\frac{\mathbf{v}}{\|\mathbf{v}\|} = \begin{bmatrix} \frac{-4}{5} \\ \frac{-3}{5} \end{bmatrix}$$

2.9. The angle between the vectors

$$\begin{bmatrix} 5 \\ 5 \end{bmatrix} \quad \text{and} \quad \begin{bmatrix} 3 \\ -3 \end{bmatrix}$$

is 90° by inspection! Sketch it and this will be clear. Additionally, notice $5 \times 3 + 5 \times -3 = 0$.

Chapter 3

3.2. Form a vector

$$\mathbf{a} = \begin{bmatrix} 1 \\ 2 \end{bmatrix}$$

perpendicular to the line, then $c = 2$, which makes the equation of the line $x_1 + 2x_2 + 2 = 0$.

3.3 We'll only do one of the four points. The point

$$\begin{bmatrix} 0 \\ 0 \end{bmatrix}$$

is not on the line $x_1 + 2x_2 + 2 = 0$, since $0 + 2 \times 0 + 2 = 2 \neq 0$.

3.4. First compute $\|\mathbf{a}\| = \sqrt{5}$. The distance d of point

$$\begin{bmatrix} 0 \\ 0 \end{bmatrix}$$

from the line $x_1 + 2x_2 + 2 = 0$ is

$$d = \frac{0 + 2 \times 0 + 2}{\sqrt{5}} \approx 0.9.$$

Sketch the line and the point to convince yourself that this is reasonable.

3.7. The lines are parallel!

Chapter 4

4.2. The product $A\mathbf{v}$:

$$\begin{array}{cc|c} & & 2 \\ & & 3 \\ \hline 0 & -1 & -3 \\ 1 & 0 & 2 \end{array}.$$

The product $B\mathbf{v}$:

$$\begin{array}{cc|c} & & 2 \\ & & 3 \\ \hline 1 & -1 & -1 \\ -1 & 1/2 & -1/2 \end{array}.$$

4.3. The sum

$$A + B = \begin{bmatrix} 1 & -2 \\ 0 & 1/2 \end{bmatrix}.$$

The product $(A + B)\mathbf{v}$:

$$\begin{array}{cc|c} & & 2 \\ & & 3 \\ \hline 1 & -2 & -4 \\ 0 & 1/2 & 3/2 \end{array}$$

and

$$A\mathbf{v} + B\mathbf{v} = \begin{bmatrix} -4 \\ 3/2 \end{bmatrix}.$$

4.7. The determinant of A:

$$|A| = \begin{vmatrix} 0 & -1 \\ 1 & 0 \end{vmatrix} = 0 \times 0 - (-1) \times 1 = 1.$$

Chapter 5

5.1. The transformation takes the form

$$\begin{bmatrix} 2 & 6 \\ -3 & 0 \end{bmatrix} \begin{bmatrix} x_1 \\ x_2 \end{bmatrix} + \begin{bmatrix} 3 \\ 3 \end{bmatrix} = \begin{bmatrix} 6 \\ 3 \end{bmatrix}.$$

5.2. The linear system

$$\begin{bmatrix} 2 & 6 \\ -3 & 0 \end{bmatrix} \begin{bmatrix} x_1 \\ x_2 \end{bmatrix} = \begin{bmatrix} 3 \\ 0 \end{bmatrix},$$

has the solution

$$x_1 = \frac{\begin{vmatrix} 3 & 6 \\ 0 & 0 \end{vmatrix}}{\begin{vmatrix} 2 & 6 \\ -3 & 0 \end{vmatrix}} = 0 \qquad x_1 = \frac{\begin{vmatrix} 2 & 3 \\ -3 & 0 \end{vmatrix}}{\begin{vmatrix} 2 & 6 \\ -3 & 0 \end{vmatrix}} = \frac{1}{2}.$$

5.5. The inverse of the matrix A:

$$A^{-1} = \begin{bmatrix} 0 & -3/9 \\ 1/6 & 1/9 \end{bmatrix}.$$

Check that $AA^{-1} = I$!

5.11. The matrix is

$$A = \begin{bmatrix} 1 & 0 \\ 0 & -1 \end{bmatrix}.$$

Chapter 6

6.1. The point

$$\mathbf{q} = \begin{bmatrix} 2/3 \\ 4/3 \end{bmatrix}.$$

The transformed points:

$$\mathbf{r}' = \begin{bmatrix} 3 \\ 4 \end{bmatrix} \quad \mathbf{s}' = \begin{bmatrix} 11/2 \\ 6 \end{bmatrix} \quad \mathbf{q}' = \begin{bmatrix} 14/3 \\ 16/3 \end{bmatrix}.$$

The point $\mathbf{q}'$ is in fact equal to $1/3\mathbf{r}' + 2/3\mathbf{s}'$.

6.3. In order to rotate a point $\mathbf{x}$ around another point $\mathbf{r}$, construct the affine map

$$\mathbf{x}' = A(\mathbf{x} - \mathbf{r}) + \mathbf{r}.$$

In this exercise, we rotate 90°,

$$\mathbf{x} = \begin{bmatrix} -2 \\ -2 \end{bmatrix}, \quad \text{and} \quad \mathbf{r} = \begin{bmatrix} -2 \\ 2 \end{bmatrix}.$$

The matrix

$$A = \begin{bmatrix} 0 & -1 \\ 1 & 0 \end{bmatrix},$$

thus

$$\mathbf{x}' = \begin{bmatrix} 2 \\ 2 \end{bmatrix}.$$

Be sure to draw a sketch!

6.5. The point

$$\mathbf{x}' = \begin{bmatrix} 0 \\ 0 \end{bmatrix}.$$

Chapter 7

7.1. Form the matrix $A - \lambda I$. The characteristic equation is

$$\lambda^2 - 2\lambda - 3 = 0.$$

The eigenvalues are $\lambda_1 = -1$ and $\lambda_2 = 3$. The eigenvectors are

$$\mathbf{r}_1 = \begin{bmatrix} 1/\sqrt{2} \\ 1/\sqrt{2} \end{bmatrix} \quad \text{and} \quad \mathbf{r}_2 = \begin{bmatrix} 1/\sqrt{2} \\ -1/\sqrt{2} \end{bmatrix}.$$

7.3. Form $A^{\mathrm{T}}A$ to be

$$\begin{bmatrix} 1.49 & -0.79 \\ -0.79 & 1.09 \end{bmatrix}.$$

The characteristic equation $|A^{\mathrm{T}}A - \lambda I|$ is

$$\lambda^2 - 2.58\lambda + 1 = 0,$$

and its roots are $\lambda_1 = 2.10$ and $\lambda_2 = 0.475$. Thus the condition number of the matrix A is $\lambda_1/\lambda_2 \approx 4.42$.

Chapter 8

8.1.a) First of all, draw a sketch. Before calculating the barycentric coordinates (u, v, w) of the point

$$\begin{bmatrix} 0 \\ 1.5 \end{bmatrix},$$

notice that this point is on the edge formed by $\mathbf{p}_1$ and $\mathbf{p}_3$. Thus, the barycentric coordinate $v = 0$.

The problem now is simply to find u and w such that

$$\begin{bmatrix} 0 \\ 1.5 \end{bmatrix} = u \begin{bmatrix} 1 \\ 1 \end{bmatrix} + w \begin{bmatrix} -1 \\ 2 \end{bmatrix}$$

and $u + w = 1$. This is simple enough to see, without computing! The barycentric coordinates are $(1/2, 0, 1/2)$.

8.1.b) Add the point

$$\mathbf{p} = \begin{bmatrix} 0 \\ 0 \end{bmatrix}$$

to the sketch from the previous exercise. Notice that $\mathbf{p}$, $\mathbf{p}_1$, and $\mathbf{p}_2$ are collinear. Thus we know that $w = 0$.

The problem now is to find u and v such that

$$\begin{bmatrix} 0 \\ 0 \end{bmatrix} = u \begin{bmatrix} 1 \\ 1 \end{bmatrix} + v \begin{bmatrix} 2 \\ 2 \end{bmatrix}$$

and $u + v = 1$. This is easy to see: $u = 2$ and $v = -1$. If this wasn't obvious, you would calculate

$$u = \frac{\|\mathbf{p}_2 - \mathbf{p}\|}{\|\mathbf{p}_2 - \mathbf{p}_1\|},$$

then $v = 1 - u$.

Thus the barycentric coordinates of $\mathbf{p}$ are $(2, -1, 0)$.

If you were to write a subroutine to calculate the barycentric coordinates, you would not proceed as we did here. Instead, you would calculate the area of the triangle and two of the three sub-triangle areas. The third barycentric coordinate, say w, can be calculated as $1 - u - v$.

8.1.c) To calculate the barycentric coordinates (i_1, i_2, i_3) of the incenter, we need the lengths of the sides of the triangles:

$$s_1 = 3 \quad s_2 = \sqrt{5} \quad s_3 = \sqrt{2}.$$

The sum of the lengths, or circumference c, is approximately $c = 6.65$. Thus the barycentric coordinates are

$$(\frac{3}{6.65}, \frac{\sqrt{5}}{6.65}, \frac{\sqrt{2}}{6.65}) = (0.45, 0.34, 0.21).$$

Always double-check that the barycentric coordinates sum to one. Additionally, check that these barycentric coordinates result in a point in the correct location:

$$0.45\begin{bmatrix}1\\1\end{bmatrix}+0.34\begin{bmatrix}2\\2\end{bmatrix}+0.21\begin{bmatrix}-1\\2\end{bmatrix}=\begin{bmatrix}0.92\\1.55\end{bmatrix}.$$

Plot this point on your sketch, and this looks correct! Recall the incenter is the intersection of the three angle bisectors.

8.1.d) Referring to the circumcenter equations from Section 8.3, first calculate the dot products

$$d_1=\begin{bmatrix}1\\1\end{bmatrix}\cdot\begin{bmatrix}-2\\1\end{bmatrix}=-1$$
$$d_2=\begin{bmatrix}-1\\-1\end{bmatrix}\cdot\begin{bmatrix}-3\\0\end{bmatrix}=3$$
$$d_3=\begin{bmatrix}2\\-1\end{bmatrix}\cdot\begin{bmatrix}3\\0\end{bmatrix}=6,$$

then $D=18$. The barycentric coordinates (cc_1, cc_2, cc_3) of the circumcenter are

$$cc_1=-1\times 9/18=-1/2$$
$$cc_2=3\times 5/18=5/6$$
$$cc_3=6\times 2/18=2/3.$$

The circumcenter is

$$\frac{-1}{2}\begin{bmatrix}1\\1\end{bmatrix}+\frac{5}{6}\begin{bmatrix}2\\2\end{bmatrix}+\frac{2}{3}\begin{bmatrix}-1\\2\end{bmatrix}=\begin{bmatrix}0.5\\2.5\end{bmatrix}.$$

Plot this point on your sketch. Construct the perpendicular bisectors of each edge to verify.

Chapter 9

9.1. The matrix

$$A=\begin{bmatrix}1&1\\1&0\end{bmatrix},$$

thus the characteristic equation is

$$\lambda^2-\lambda-1=0,$$

which has roots $\lambda_1=1.62$ and $\lambda_2=-0.62$. Since there are two distinct roots of opposite sign, the conic is a hyperbola.

9.2. Translate by

$$\mathbf{v}=\begin{bmatrix}3\\-1\end{bmatrix}$$

and scale with the matrix

$$\begin{bmatrix}1/4&0\\0&1/8\end{bmatrix}.$$

Chapter 10

10.1. For the given vector

$$\mathbf{r} = \begin{bmatrix} 4 \\ 2 \\ 4 \end{bmatrix},$$

we have $\|\mathbf{r}\| = 6$. Then $\|2\mathbf{r}\| = 2\|\mathbf{r}\| = 12$.

10.3. The cross product of the vectors $\mathbf{v}$ and $\mathbf{w}$:

$$\mathbf{v} \wedge \mathbf{w} = \begin{bmatrix} 0 \\ -1 \\ 1 \end{bmatrix}.$$

10.5. The sine of the angle between $\mathbf{v}$ and $\mathbf{w}$:

$$\sin\theta = \frac{\|\mathbf{v} \wedge \mathbf{w}\|}{\|\mathbf{v}\|\|\mathbf{w}\|} = \frac{\sqrt{2}}{1 \times \sqrt{3}} = 0.82$$

This means that $\theta = 55°$. Draw a sketch to double-check this for yourself.

10.7. The point normal form of the plane through $\mathbf{p}$ with normal direction $\mathbf{r}$ is found by first defining the normal $\mathbf{n}$ by normalizing $\mathbf{r}$:

$$\mathbf{n} = \begin{bmatrix} 2/3 \\ 1/3 \\ 2/3 \end{bmatrix}.$$

The point normal form of the plane is

$$\mathbf{n}(\mathbf{x} - \mathbf{p}) = 0,$$

or

$$\frac{2}{3}x_1 + \frac{1}{3}x_2 + \frac{2}{3}x_3 - \frac{2}{3} = 0.$$

10.9. A parametric form of the plane P through the points $\mathbf{p}$, $\mathbf{q}$, and $\mathbf{r}$ is

$$\begin{aligned} P(s,t) &= \mathbf{p} + s(\mathbf{q} - \mathbf{p}) + t(\mathbf{r} - \mathbf{p}) \\ &= (1 - s - t)\mathbf{p} + s\mathbf{q} + t\mathbf{r}. \end{aligned}$$

10.11. The volume V formed by the vectors $\mathbf{v}, \mathbf{w}, \mathbf{u}$ can be computed as the scalar triple product

$$V = \mathbf{v} \cdot (\mathbf{w} \wedge \mathbf{u}).$$

This is invariant under cyclic permutations, thus we can also compute V as

$$V = \mathbf{u} \cdot (\mathbf{v} \wedge \mathbf{w}),$$

which allows us to reuse the cross product from exercise 1. Thus

$$V = \begin{bmatrix} 0 \\ 0 \\ 1 \end{bmatrix} \cdot \begin{bmatrix} 0 \\ -1 \\ 1 \end{bmatrix} = 1.$$

Chapter 11

11.1. Make a sketch! You will find that the line is parallel to the plane. The actual calculations would involve finding the parameter t on the line for the intersection:

$$t = \frac{(\mathbf{p} - \mathbf{q}) \cdot \mathbf{n}}{\mathbf{v} \cdot \mathbf{n}}.$$

In this exercise, $\mathbf{v} \cdot \mathbf{n} = 0$.

11.5. The vector $\mathbf{a}$ is projected to the vector

$$\mathbf{a}' = \mathbf{a} - \frac{\mathbf{a} \cdot \mathbf{n}}{\mathbf{v} \cdot \mathbf{n}} \mathbf{v}.$$

Chapter 12

12.2. The scale matrix is

$$\begin{bmatrix} 2 & 0 & 0 \\ 0 & 1/4 & 0 \\ 0 & 0 & -4 \end{bmatrix}.$$

This matrix changes the volume of a unit cube to be $2 \times 1/4 \times -4 = -2$.

12.4. The shear matrix is

$$\begin{bmatrix} 1 & 0 & -a/c \\ 0 & 1 & -b/c \\ 0 & 0 & 1 \end{bmatrix}.$$

Shears do not change volume, therefore the volume of the mapped unit cube is still 1.

12.6. To rotate about the vector

$$\begin{bmatrix} -1 \\ 0 \\ -1 \end{bmatrix},$$

first form the unit vector

$$\mathbf{a} = \begin{bmatrix} -1/\sqrt{2} \\ 0 \\ -1/\sqrt{2} \end{bmatrix}.$$

Then, following (12.8), the rotation matrix is

$$\begin{bmatrix} \frac{1}{2}(1 + \frac{\sqrt{2}}{2}) & 1/2 & \frac{1}{2}(1 - \frac{\sqrt{2}}{2}) \\ -1/2 & \sqrt{2}/2 & 1/2 \\ \frac{1}{2}(1 - \frac{\sqrt{2}}{2}) & -1/2 & \frac{1}{2}(1 + \frac{\sqrt{2}}{2}) \end{bmatrix}.$$

The matrices for rotating about an arbitrary vector are difficult to verify by inspection. One test is to check the vector about which we rotated. You'll find that

$$\begin{bmatrix} -1 \\ 0 \\ -1 \end{bmatrix} \rightarrow \begin{bmatrix} -1 \\ 0 \\ -1 \end{bmatrix},$$

which is precisely correct.

12.7. The resulting matrix is

$$\begin{bmatrix} 2 & 3 & -4 \\ 3 & 9 & -4 \\ -1 & -9 & 4 \end{bmatrix}.$$

Chapter 13

13.1. As always, draw a sketch when possible! Construct the parallel projection map defined in (13.8):

$$\mathbf{x}' = \begin{bmatrix} 8/14 & 0 & -8/14 \\ 0 & 1 & 0 \\ -6/14 & 0 & 6/14 \end{bmatrix} \mathbf{x} + \begin{bmatrix} 6/14 \\ 0 \\ 6/14 \end{bmatrix}.$$

The point

$$\mathbf{x}_1 = \begin{bmatrix} 1 \\ 0 \\ 0 \end{bmatrix}$$

is mapped to $\mathbf{x}_1' = \mathbf{x}_1$. By inspection of your sketch and the plane equation, this is clear since $\mathbf{x}_1$ is already in the plane.

The point

$$\mathbf{x}_2 = \begin{bmatrix} 0 \\ 1 \\ 0 \end{bmatrix}$$

is mapped to the point

$$\mathbf{x}_2' = \begin{bmatrix} 3/7 \\ 1 \\ 3/7 \end{bmatrix}.$$

Check that $\mathbf{x}_2'$ lies in the projection plane. This should seem reasonable too. The projection direction $\mathbf{v}$ is parallel to the $\mathbf{e}_2$-axis, thus this coordinate is unchanged. Additionally, $\mathbf{v}$ projects equally into the $\mathbf{e}_1$- and $\mathbf{e}_3$-axes.

Try the other points yourself, and rationalize each resulting point as we have done here.

13.3. Construct the perspective projection vector equation defined in (13.10). This will be different for each of the $\mathbf{x}_i$:

$$\mathbf{x}_1' = \frac{3/5}{3/5}\mathbf{x}_1 = \begin{bmatrix} 1 \\ 0 \\ 0 \end{bmatrix}$$

$$\mathbf{x}_2' = \frac{3/5}{0}\mathbf{x}_2$$

$$\mathbf{x}_3' = \frac{3/5}{-4/5}\mathbf{x}_3 = \begin{bmatrix} 0 \\ 0 \\ 3/4 \end{bmatrix}$$

$$\mathbf{x}_4' = \frac{3/5}{4/5}\mathbf{x}_4 = \begin{bmatrix} 0 \\ 0 \\ 3/4 \end{bmatrix}.$$

There is no solution for $\mathbf{x}_2'$ because $\mathbf{x}_2$ is projected parallel to the plane.

13.6. Construct the matrices in (13.6). For this problem, use the notation $A = YX^{-1}$, where

$$Y = \begin{bmatrix} \mathbf{y}_2 - \mathbf{y}_1 & \mathbf{y}_3 - \mathbf{y}_1 & \mathbf{y}_4 - \mathbf{y}_1 \end{bmatrix} = \begin{bmatrix} 1 & 1 & 1 \\ -1 & 0 & 0 \\ 0 & -1 & 1 \end{bmatrix}$$

and

$$X = \begin{bmatrix} \mathbf{x}_2 - \mathbf{x}_1 & \mathbf{x}_3 - \mathbf{x}_1 & \mathbf{x}_4 - \mathbf{x}_1 \end{bmatrix} = \begin{bmatrix} -1 & -1 & -1 \\ 1 & 0 & 0 \\ 0 & -1 & 1 \end{bmatrix}.$$

The first task is to find X^{-1}:

$$X^{-1} = \begin{bmatrix} 0 & 1 & 0 \\ -1/2 & -1/2 & -1/2 \\ -1/2 & -1/2 & 1/2 \end{bmatrix}.$$

Always check that $XX^{-1} = I$. Now A takes the form:

$$A = \begin{bmatrix} -1 & 0 & 0 \\ 0 & -1 & 0 \\ 0 & 0 & 1 \end{bmatrix},$$

which isn't surprising at all if you sketched the tetrahedra formed by the $\mathbf{x}_i$ and $\mathbf{y}_i$.

13.7. The point

$$\begin{bmatrix} 1 \\ 1 \\ 1 \end{bmatrix}$$

is mapped to

$$\begin{bmatrix} -1 \\ -1 \\ 1 \end{bmatrix}.$$

Chapter 14

14.1. The solution vector

$$\mathbf{v} = \begin{bmatrix} 1 \\ -4 \\ 0 \\ -1 \end{bmatrix}.$$

14.2. The solution vector

$$\mathbf{v} = \begin{bmatrix} 0 \\ 0 \\ -1 \end{bmatrix}.$$

14.4. Use the explicit form of the line, $x_2 = ax_1 + b$. The overdetermined system is

$$\begin{bmatrix} 1 & 1 \\ 0 & 1 \\ -3 & 1 \\ 3 & 1 \\ 0 & 1 \end{bmatrix} \begin{bmatrix} a \\ b \end{bmatrix} = \begin{bmatrix} -1 \\ 0 \\ 0 \\ -1 \\ 1 \end{bmatrix}.$$

We want to form a square matrix, and the solution should be the least squares solution. Following (14.3), the linear system becomes

$$\begin{bmatrix} 19 & 1 \\ 1 & 5 \end{bmatrix} \begin{bmatrix} a \\ b \end{bmatrix} = \begin{bmatrix} -4 \\ -1 \end{bmatrix}.$$

Thus the least squares line is $x_2 = -2.2x_1 + 0.24$. Sketch the data and the line to convince yourself.

14.7. The determinant is equal to 5.

Chapter 15

15.3. A rhombus is equilateral but not equiangular.

15.5. The winding number is 0.

15.6. The area is 3.

Chapter 16

16.2. The evaluation point at $t = 1/4$ is calculated as follows:

$$\begin{bmatrix} 1 \\ 0 \\ 0 \end{bmatrix} \begin{bmatrix} 4 \\ 0 \\ 0 \end{bmatrix} \begin{bmatrix} 4 \\ 4 \\ 0 \end{bmatrix} \begin{bmatrix} 4 \\ 4 \\ 4 \end{bmatrix}$$
$$\begin{bmatrix} 1 \\ 0 \\ 0 \end{bmatrix} \begin{bmatrix} 4 \\ 1 \\ 0 \end{bmatrix} \begin{bmatrix} 4 \\ 4 \\ 1 \end{bmatrix}$$
$$\begin{bmatrix} 7/4 \\ 1/4 \\ 0 \end{bmatrix} \begin{bmatrix} 4 \\ 7/4 \\ 1/4 \end{bmatrix}$$
$$\begin{bmatrix} 37/16 \\ 10/16 \\ 1/16 \end{bmatrix}.$$

16.3. The first derivative is

$$\dot{\mathbf{b}}(1/4) = 3\left(\begin{bmatrix} 4 \\ 7/4 \\ 1/4 \end{bmatrix} - \begin{bmatrix} 7/4 \\ 1/4 \\ 0 \end{bmatrix}\right) = 3\begin{bmatrix} 9/4 \\ 6/4 \\ 1/4 \end{bmatrix}.$$

The second derivative is

$$\ddot{\mathbf{b}}(1/4) = 3 \times 2\left(\begin{bmatrix} 4 \\ 4 \\ 1 \end{bmatrix} - 2\begin{bmatrix} 4 \\ 1 \\ 0 \end{bmatrix} + \begin{bmatrix} 1 \\ 0 \\ 0 \end{bmatrix}\right) = 6\begin{bmatrix} -3 \\ 2 \\ 1 \end{bmatrix}.$$

16.5. The monomial coefficients $\mathbf{a}_i$ are

$$\mathbf{a}_0 = \begin{bmatrix} 0 \\ 0 \\ 0 \end{bmatrix} \quad \mathbf{a}_1 = \begin{bmatrix} 12 \\ 0 \\ 0 \end{bmatrix} \quad \mathbf{a}_2 = \begin{bmatrix} -12 \\ 12 \\ 0 \end{bmatrix} \quad \mathbf{a}_3 = \begin{bmatrix} 4 \\ -8 \\ 4 \end{bmatrix}.$$

For additional insight, compare these vectors to the point and derivatives of the Bézier curve at $t = 0$.

16.8. The Frenet frame at $t = 1$ is computed by first finding the first and second derivatives of the curve there:

$$\dot{\mathbf{b}}(1) = 3\begin{bmatrix} 0 \\ 0 \\ 4 \end{bmatrix} \quad \ddot{\mathbf{b}}(1) = 6\begin{bmatrix} 0 \\ -4 \\ 4 \end{bmatrix}.$$

Because of the homogeneous property of the cross product, see Section 10.2, the 3 and 6 factors from the derivatives simply cancel in the calculation of the Frenet frame. The Frenet frame:

$$\mathbf{f}_1 = \begin{bmatrix} 0 \\ 0 \\ 1 \end{bmatrix} \quad \mathbf{f}_2 = \begin{bmatrix} 0 \\ 1 \\ 0 \end{bmatrix} \quad \mathbf{f}_3 = \begin{bmatrix} 1 \\ 0 \\ 0 \end{bmatrix}.$$

Bibliography

[1] H. Anton. *Elementary Linear Algebra.* New York: John Wiley & Sons, 1981. third edition.

[2] W. Boehm and H. Prautzsch. *Geometric Concepts for Geometric Design.* Wellesley, MA: AK Peters Ltd., 1992.

[3] M. de Berg, M. van Kreveld, M. Overmars, and O. Schwarzkopf. *Computational Geometry Algorithms and Applications.* Berlin: Springer–Verlag, 1997.

[4] M. Escher and J. Locher. *The Infinite World of M. C. Escher.* New York: Abradale Press / Harry N. Abrams, Inc., 1971.

[5] G. Farin. *Curves and Surfaces for Computer Aided Geometric Design.* Boston: Academic Press, 1996. fourth edition.

[6] I. Faux and M. Pratt. *Computational Geometry for Design and Manufacture.* West Sussex, England: Ellis Horwood Ltd., 1979.

[7] J. Foley and A. Van Dam. *Fundamentals of Interactive Computer Graphics.* Reading, MA: Addison-Wesley Publishing Company, Inc., 1982.

[8] R. Goldman. Triangles. In A. Glassner, editor, *Graphics Gems, Volume 1*, pages 20–23. Academic Press, 1990.

[9] M. Goossens, F. Mittelbach, and A. Samarin. *The LaTeX Companion.* Reading, MA: Addison-Wesley Publishing Company, Inc., 1994.

[10] D. Hearn and M. Baker. *Computer Graphics.* Englewood Cliffs, NJ: Prentice-Hall, 1986.

[11] F. Hill. *Computer Graphics.* New York: Macmillan Publishing Company, 1990.

[12] J. Hoschek and D. Lasser. *Grundlagen der Geometrischen Datenverarbeitung.* Stuttgart: B.G. Teubner, 1989. English translation: *Fundamentals of Computer Aided Geometric Design*, Wellesley, MA: A K Peters, 1993.

[13] Adobe Systems Inc. *PostScript Language Reference Manual.* Reading, MA: Addison-Wesley Publishing Company, Inc., 1985.

[14] Adobe Systems Inc. *PostScript Language Tutorial and Cookbook.* Reading, MA: Addison-Wesley Publishing Company, Inc., 1985.

[15] L. Johnson and R. Riess. *Numerical Analysis.* Reading, MA: Addison-Wesley Publishing Company, Inc., 1982. second edition.

[16] E. Kästner. *Erich Kästner erzählt: Die Schildbürger.* Hamburg: Cecilie Dressler Verlag, 1995.

[17] L. Lamport. *LaTeX User's Guide and Reference Manual.* Reading, MA: Addison-Wesley Publishing Company, Inc., 1994.

[18] J. Rokne. The area of a simple polygon. In A. Glassner, editor, *Graphics Gems, Volume I,* pages 5–6. Boston, MA: Academic Press, 1991.

Index